中国营造学社

"十二五"国家重点图书出版规划项目

中国古代建筑测绘大系·宗教建筑

华岳庙与道教宫观

清华大学建筑学院 编写

王贵祥 主编

李路珂 刘畅 廖慧农 王贵祥 编著

中国建筑工业出版社

Traditional Chinese Architecture Surveying and
Mapping Series:
Religious Architecture

MOUNT HUA'S YUEMIAO AND TAOIST TEMPLES

Compiled by School of Architecture, Tsinghua University
Edited by WANG Guixiang, LIU Chang, LIAO Huinong, LI Luke

China Architecture & Building Press

Contents

目 录

Introduction

Since its inception in 1946, the School of Architecture at Tsinghua University has been committed to surveying and mapping traditional Chinese buildings, following the practice of the Society for the Study of Chinese Architecture (*Zhongguo Yingzao Xueshe*) that LIANG Sicheng, a driving force in the Society and founder of Tsinghua's architecture department (known as the School of Architecture since 1988), and his assistant MO Zongjiang brought with them to Tsinghua. Between 1930 and 1945, with members of the Society, LIANG visited over two thousand Chinese sites located in more than two hundred counties and fifteen provinces, and discovered, identified and mapped over two hundred groups of traditional buildings, including the famous Tang-period east hall (dating to 857) of Foguang Monastery at Mount Wutai—which was not an easy task because of the harsh working conditions in the secluded and relatively inaccessible villages in the countryside. In that same spirit, despite the difficult political circumstances from the 1950s through the 1970s, the School of Architecture conducted a systematic survey of historical buildings in the New Summer Palace (Yiheyuan). At the beginning of the Cultural Revolution in the late 1970s, all members of the faculty focusing on the history of architecture went to Hebei province under the leadership of MO Zongjiang to measure and draw the main hall of Geyuan Monastery in Laiyuan, an important Liao-period relic hidden in the remote mountains. This was followed by in-depth research and analysis. At the same time, those professors that specialized in Chinese architectural history (MO Zongjiang, XU Bo'an, LOU Qingxi, ZHANG Jingxian, and GUO Daiheng) led a group of graduate students to Zhengding in Hebei province, where they conducted component analysis and research of Moni Hall at Longxing Monastery, a Northern-Song timber-frame structure that had partially collapsed but was then in the process of being rebuilt. They also investigated nearby

导　言

因为前辈学者梁思成及其助手莫宗江两位先生从中国营造学社继承的传统，清华大学建筑学院自创立以来，一直十分注重古代建筑实例的实地考察与测绘。尽管在 20 世纪 50 至 70 年代受到各种因素的影响与冲击，那时的清华大学建筑系，还坚持了对颐和园内一批古代建筑实例的系统测绘。改革开放刚刚开始的 1970 年代末，清华大学建筑历史方向的全体教师，就在莫宗江先生带领下，共同远赴偏僻的河北山区，考察测绘了创建于辽代的涞源阁院寺大殿，并对这座辽代木构建筑进行了系统研究。同是在那一时期，建筑历史教研室的莫宗江、徐伯安、楼庆西、张静娴、郭黛姮等教师，带领研究生赴河北正定，除了对正在落架重修的北宋木构大殿隆兴寺摩尼殿的大木构件进行现场分析研究外，还对正定及周边的古建筑进行了系统考察与调研。这种由老先生带队，

historical buildings in and around Zhengding. This practice of teamwork—senior researchers, instructors, and (graduate) students participating in the investigation and mapping of traditional Chinese architecture side by side—became an academic tradition at the School of Architecture of Tsinghua University.

Since the 1980s, fieldwork has been a crucial part of undergraduate education at the School, and focus and quality of teaching has constantly improved over the past decades. In the 1990s, professors like CHEN Zhihua and LOU Qingxi carried out surveying and mapping in advance of (re)construction or land development on sites all across China that were endangered. Since the turn of the twenty-first century, the two-fold approach—attaching equal importance to practice (fieldwork) and theory (teaching)—was widened and deepened. Sites were deliberately chosen to maximize educational outcome, resulting in a broader geographical scope and spectrum of building types. In addition to expanding on the idea of vernacular architecture, special attention was paid to local (government-sponsored) construction of palaces, tombs, and temples built in the official style (*guangshi*) or on a large scale (*dashi*), and to modern architecture dating to the period between 1840 and 1949. Students and staff have accumulated a lot of experience and created high-quality drawings through this fieldwork.

In retrospect, we have completed surveys of several hundred monuments and sites built in the official dynastic styles of the Song(Jin), Yuan, Ming and Qing all across the country. Fieldwork was always combined with teaching. Among the architecture surveyed are the (single- and multi-story) buildings in front and on the sides of the Hall of Supreme Harmony in the Forbidden City in Beijing; the architecture at Changling, the mausoleum of emperor Jiaqing located at the Western Qing tombs in Yi county, Hebei province; the monasteries on Mount Wutai, Shanxi province, including Xiantongsi, Tayuansi, Luobingsi, Pusading, Nanshansi (Youguosi), and Longquansi; Zhongyue Temple, Songyang Academy, and Shaolin Monastery in Dengfeng, Henan province; Xiyue Temple, Yuquan Court, and the Taoist architecture on the peaks of Mount Hua in Weinan, Shaanxi province; Chongan Monastery, Nanjixiang Monastery, Jade Emperor Temple (Yuhuangmiao) in Shizhang, and the temples of the Two Transcendents (Erxianmiao) in Xiaohuiling and Nanshentou, all situated in Lingchuan county of Shanxi province; and the upper and lower Guangsheng monasteries and the Water God's Temple in Hongdong, Shanxi province. In recent years, we have developed a specialized interest in the study of religious architecture of Shanxi province and investigated almost a dozen privately- or government-sponsored Song and Jin sites

教师与研究生集体参与，对古代建筑进行深入考察与测绘研究的做法，在清华大学形成了一个良好的学术传统。

1980年代以来，清华大学建筑学院始终在本科教学环节中，坚持讲授古代建筑测绘这门经典课程。这一传统在21世纪初的这十几年中始终延续。如果说，20世纪90年代由陈志华、楼庆西等教授带领的测绘教学，将相当的注意力放在了分布于全国多个省、市、自治区大量传统乡土村落建筑的抢救性测绘上，进入21世纪以来，清华大学建筑学院开展的这种结合本科教学的古建筑测绘教学与实践，覆盖的地域范围与建筑类型范围更为宽广：除了进一步拓展乡土建筑的测绘以及近代建筑的测绘之外，在对各地留存的历代官式或大式建筑，如宫殿、陵寝、寺庙等建筑的测绘上，也积累了大量测绘经验、图纸及丰富的调研资料。以古代官式建筑测绘为例，结合本科教学，我们先后完成了北京故宫太和殿前及两侧门殿、楼阁与朝房建筑，河北易县清西陵昌陵完整建筑群，山西五台山显通寺、塔院寺、罗睺寺、菩萨顶、南山寺（佑国寺）、龙泉寺等多座整组寺院建筑群，河南登封中岳庙、嵩阳书院、少林寺古建筑群，陕西渭南华山西岳庙、玉泉院及华山山顶各道观古建筑群，山西陵川崇安寺、南吉祥寺、小会岭二仙庙、南神头二仙庙、石掌玉皇庙，以及山西洪洞广胜上寺、广胜下寺、水神庙等数百座古建筑实例的测绘，其时代的范围覆盖了宋（金）、元、明、清等历代木构建筑遗存实例。近几年，我们又将测绘的重点放在了高平、晋城等晋中及晋东南地区，

located in central Shanxi (Jinzhong) and southeastern Shanxi (Jindongnan), specifically in Gaoping and Jincheng counties. This includes the Youxian, Chongming, and Kaihua monasteries and the Two Transcendents Temple in Xilimen. Additionally, supported by the State Administration of Cultural Relics, the head of the Architecture History and Historic Preservation Research Institute at the School of Architecture, Liu Chang, led a group of students to map and draw the main hall of Zhenguo Monastery in Pingyao, a rare example from the Five Dynasties period. The survey results have been published. Tsinghua fieldwork in Shanxi has become an annual event that is jointly organized almost every summer by the faculty of the School of Architecture, including professors engaged in research on non-Chinese architecture, in cooperation with their graduate students.

It is worth mentioning that since 2007, the School has worked in collaboration with the well-known company China Resources Snow Breweries Ltd., which supports the transmission and dissemination of knowledge on traditional Chinese architecture and provides funds for the School's research and field investigation activities. Drawing on the support from industry allowed us greater initiative and flexibility, and we were thus able to carry out research on and survey often overlooked but no less important Song-Jin monuments in central and southeastern Shanxi.

Our years-long fieldwork has not only enabled us to teach students subject knowledge about scale, material, form, and decoration of traditional Chinese architecture as well as a sense of appreciation for the old, but has also provided us with plenty of data for monument preservation practice and research. China Architecture and Building Press spared no effort in compiling and publishing the results of the fieldwork in 2012. Publication has also been supported by the National Publishing Fund. This highlights not only the importance of our contribution to architectural education at the national level but also shows its significance for the transmission, development, and revival of traditional Chinese architectural culture both at home and abroad. In order to expand the reach of this work to an international audience, *the Traditional Chinese Architecture Surveying and Mapping Series* is being published bilingually. Based on the past ten years of fieldwork, we have now compiled five volumes, namely *Mount Wutai's Buddhist Architecture* (Traditional architecture on Mount Wutai, Shanxi), *Architecture Complex of Songshan* (Traditional architecture in Dengfeng, Henan), *Mount Hua's Yuemiao and Taoist Temples* (Traditional architecture on Mount Hua, Shaanxi), *Architecture Complex of Hongtong* (Traditional architecture in Hongtong, Shanxi), and *Architecture Complex*

对包括高平游仙寺、崇明寺、开化寺、西李门二仙庙等在内的十余座宋金建筑群，进行了全面而系统的测绘。这一期间，在国家文物局的支持下，建筑历史与文物保护研究所刘畅老师还带领研究生对五代时期创建的平遥镇国寺大殿等建筑进行了精细测绘，并出版了测绘研究成果。此外，清华大学建筑学院的测绘工作，几乎每年都是由全体建筑历史教师共同合作，并带领研究生们共同完成的。从事外国建筑史教学的老师，也不例外。

特别值得一提的是，自 2007 年以来，清华大学建筑学院与国家知名企业华润雪花啤酒（中国）有限公司建立了良好的合作关系。该集团不仅支持中国古建筑知识的传承与普及工作，也对清华大学建筑学院中国古代建筑研究及古建筑测绘工作给予了直接的支持，使得我们的古建筑测绘工作变得更为主动和更具选择性。一大批珍贵的山西晋中及晋东南地区宋金时代建筑实例的测绘与研究，就是在这样一个前提下得以顺利开展与完成的。

坚持数十年的古建筑测绘工作，不仅在培养学生对传统中国建筑的尺度、材料、造型与细部装饰的认知与感觉上起到了直接的影响，而且也为各地文物建筑保护与研究工作，提供了相当充分的资料支持。

2012 年，中国建筑工业出版社花大气力组织了汇集全国重点院校建筑系古建筑测绘成果的中国古代建筑测绘大系的编辑出版工作。这一工作也获得了国家出版基金的支持。这不仅是对高校建筑教育成果的一份支持，也是对中国传统建筑文化传承、发展与复兴的一份支持。正是在这样一个背景与前提下，我们对近十余年来考察测绘的古代建筑案例加以整理，分别编汇了包括《五台山佛教建筑》《嵩山建筑群》《华山岳庙与道观》《洪洞建筑群》《高平建筑群》5 册古建筑测

of Gaoping (Traditional architecture in Gaoping, Shanxi). The architectural drawings presented in these books are carefully selected and screened by Tsinghua professors. They only show a part of our comprehensive surveying and mapping work, but still cover a whole spectrum of geographic regions and time periods. Thus, they contain information of high academic value that may serve as a reference for future study and for the protection of cultural heritage. It is hoped that our work will help to promote interest in and improve understanding of traditional Chinese architecture, not only among Tsinghua students (through hands-on experiences in the fieldwork) but also among architectural historians and professionals engaged in monument preservation at home and abroad.

As a final thought, let me shortly address the workflow. The drawings presented here are based on survey and working sketches drawn up on site during several years of fieldwork conducted by Tsinghua professors together with graduate and undergraduate students. Back home, the measured drawings were redrawn over months of diligent work by graduate students with computer-aided software to achieve dimensionally accurate and visually appealing results, a project that was completed under the supervision of LIU Chang, head of Tsinghua's Architecture History Institute, and the Tsinghua professors LIAO Huinong and WANG Nan, as well as TANG Henglu and his colleagues from the WANG Guixiang Studio. We would like to take this opportunity to thank the professors, students and colleagues who participated in the fieldwork and its revision.

Our final thanks go to LI Jing, assistant researcher at the Architecture History Institute here at Tsinghua. Next to participating in surveying and mapping, she organized the development of the book and moreover, made this book possible in the first place.

WANG Guixiang, LIU Chang, LIAO Huinong
Architecture History and Historic Preservation Research Institute, School of Architecture,
Tsinghua University
December 5, 2017

Translated by Alexandra Harrer

004

绘图集，作为这套『中国古建筑测绘大系』的部分成果。尽管这只是我们多年测绘成果的一部分，但也是清华建筑历史学科教师们仔细筛选、认真校对、充分整理之后的较具典型性与参考性的成果。

这些成果对高校建筑系学生们学习古建筑，建筑历史学者研究古建筑，以及文物保护工作者从事文物古建筑的保护与修缮，能够起到积极的推动作用与重要的参考价值。

最后要提到的一点是，除了参与测绘的教师、研究生与本科生多年历尽辛苦的测量与绘图工作之外，此次清华大学建筑学院承担的这5册测绘图集，也经由建筑历史与文物保护研究所刘畅、廖慧农、王南和他们的研究生，以及王贵祥工作室团队的唐恒鲁等同仁们在既有测绘图纸基础上，经过数月认真仔细的线条分层、图面调整、数据校对、图面完善等缜密修复工作，在这里也要向参加测绘图整理的老师、同学和同事们表示感谢。

还应该特别提到的是建筑学院建筑历史与文物保护研究所的助理研究员李菁博士，她不仅参加了多次测绘，还为这套书最后的编辑与出版做了大量相关工作。这里一并表示感谢。

清华大学建筑学院 建筑历史与文物保护研究所

王贵祥、刘畅、廖慧农

2017年12月5日

Preface

Huashan or Mount Hua ("Flower Mountain"), also called Taihuashan ("Great Flower Mountain"), is located south of the city of Huayin in Shaanxi province at the eastern end of the Qin Mountains, a natural boundary between North and South China that overlooks the Yellow River Valley. The noted geographer LI Daoyuan described the mountain in his *Commentary on the Water Classic (Shuijingzhu)* as "seen from afar it resembles a flower", a line that captures the visual form of Huashan and may have served as the origin of the mountain's present name.

The earliest mention of Huashan in historical documents can be traced to as early as the Warring States period. *The Book of Documents (Shangshu*, chapter *Yugong)* mentions Dunwushan as the historical name for Huashan. *The Classic of Mountains and Seas (Shanhaijing,* section *Xishanjing)* describes its steepness and inaccessibility, edged with granite rock face and rugged cliffs, "it rises sheer on all sides/it has a sharpened and square shape, five thousand ren high and ten li wide, birds and beasts lived there". Ancient Chinese believed that Huashan was a dwelling-place for gods, and the spiritual potency gave rise to fear and respect. *The Records of the Grand Historian (Shiji, book on Fengshan sacrifices)* explains that the Yellow Emperor went on excursions to Huashan, one of the famous mountains he visited regularly to contact gods and spirits. *The Complete Works of Han Feizi (Hanfeizi,* section *Waichushuo)* records that in the Warring States period, king Zhao of Qin "ordered a group of workmen to use scaling ladders to climb Huashan… [and construct a giant chess board on the summit] in an attempt to play chess there with a god", which is the earliest record of ancient people getting up the cliffs. Today, the Chess Pavilion located on the East Peak's southern summit is till known as Botai (Broad Terrace), which is related to this legend.

Huashan is the West Mountain of the Five Great Mountains and central to the traditional Chinese culture of wuyue (the Five Peaks or more broadly, the Five Great Mountains). China has a long history of imperial worship and pilgrimage to renowned natural landmarks. Imperial sacrifice prospered in the Tang and Song, and became fully realized

序
言

华山又称『太华山』，位于陕西省华阴市南部，南接秦岭，北瞰黄河，山势陡峭，山形峻秀，郦道元《水经注》述其『远而望之若花（华）状』，华山之得名或缘于此。华山在我国文献中的记载可以追溯到战国时期《尚书·禹贡》所载之『惇物山』。《山海经·西山经》称之为『太华之山』，『削成而四方，其高五千仞，其广十里，鸟兽莫居』，以险峻闻名，在远古人们的心目中即为神灵栖居之所。《史记·封禅书》就记载了远古时期的黄帝，曾以华山为常游的五座名山之一，在此『与神会』。《韩非子·外储说》记载战国时期秦昭王曾『令工施钩梯而上华山……尝与天神博于此』，这是关于古人攀登华山的最早记录，今华山东峰附近的下棋亭又称『博台』，便与此传说有关。

during the Ming and Qing. Although the objects of worship were personifications of natural powers and phenomena in the form of a mountain—all mountains were thought to contain gods and spirits—the actual places of worship were government-sponsored temples built on the foothills in a style that matched palace construction in the capital that matched in both area and rank (with a main hall just one rank lower than that of the imperial palace). An entry in the Qin-Han encyclopedia Erya on the character for mountain (shishan) identifies Huashan as the "Peak of the West", which might be the earliest record of the exact location of one of the Five Great Mountains.

Huashan is a famous Taoist mountain. It was an important place for those seeking immortality, as famous Taoist masters would meditate in the rural seclusion of the mountain. After the Han-Wei-Six-Dynasties period, Huashan was associated with one of the Ten Greater Grotto-heavens (Shi dadongtian) and one of the Thirty-six Lesser Grotto-heavens (Sanshiliu xiaodongtian). During the Northern Wei dynasty, Li Daoyuan's Commentary on the Water Classic (Shuijingzhu) records a mountain trail to climb up the steep cliffs, indicating that no official ascent route had been built yet ("[when climbing and for stability of] ascent or descent, one must climb along a fixed rope, pulling oneself forward"). The text lists several religious buildings that already existed at that time but fails to describe them in more detail—a lower temple, central shrine, south shrine, upper palace-temple, and Huyue Monastery. Claiming descent from the Taoist sage Laozi, Tang emperors granted official status to Taoism, challenging the dominant roles of Confucianism and Buddhism. As a result, Taoism on Mount Hua flourished during this period. The White Cloud Temple was erected as well as the first buildings of Xiangu (Celestial Women) Temple. A dangerous mountain path with steps and railings in the form of stone pillars and iron chains was set up, winding from the valley village to North Peak, known as the Ancient Huashan Trail.

In the Five Dynasties period, the Taoist sage CHEN Tuan moved from Mount Wudang to Mount Hua. After practicing at Yuntai (Cloud Terrace) Temple for nearly forty years, he is said to have transformed – i.e. attained immortality – in a stone chamber in Zhangchao Valley below Lotus Flower Summit, the only summit of the West Peak in 989 (the second year in the Duangong reign period of the Song dynasty). Merging Confucian and Taoist thought ("advocating the Taiji Charts and Yellow River Map as Taoist doctrines relevant to all teachings"), CHEN Tuan had a profound influence upon the idealist philosophy of Song Neo-Confucianism. Emperor Shizong of the Later Zhou and the emperors Taizu and Taizong of Song had great respect for him, which strengthened the position of Huashan

华山为我国五岳之西岳，是我国「五岳」文化的重要组成部分。我国自周代已有帝王祭祀名山大川的传统，至唐宋时期达于鼎盛，明清时期臻于完备。这里的「五岳」，便是我国古代山川祭祀最主要的对象，而山麓的岳庙即是山川祭祀的主要场所，为历朝政府出资修建，平面格局趋近于宫室，其正殿的建筑规制仅次于皇宫一级。秦汉时期的《尔雅·释山》中已列出「华山为西岳」，是「五岳」中最早明确的地点之一。

华山是我国的道教名山。汉魏六朝之后「十大洞天」之西玄洞天、「三十六小洞天」之「总仙洞天」皆位于华山，是道士云游隐居之胜地；从郦道元的《水经注》记载可知，华山在北魏时期并无人造的山路，「升降皆须扳绳挽葛而行」，然而在山中已有下庙、中祠、南祠、胡越寺、上宫神庙等建筑。唐代以老子为鼻祖，奉道教为三教之首，华山道教也因此兴盛，在这一时期修建了自云观和仙姑观等早期宫观，逐渐在山之北坡沿溪谷而上，开凿了一条险道，即后来所称之「自古华山一条路」。

五代末期，道士陈抟从武当山移居华山云台观，在华山修炼近四十年，于宋端拱二年（989年）「化形」于华山莲花峰下张超谷中的石室。陈抟融会儒道之学，「倡太极河洛诸教，作道学纲宗」，对宋代理学影响甚大，曾为后周世宗、宋太祖、宋太宗所敬重，故使华山在道教史上的地位更加突出。金元时期，北方全真道兴起，华山遂成为全真道场，据传王重阳及弟子谭处端、邱处机都曾在华山活动。至元十三年（1276年），全真派华山玉泉院、云台观、希夷洞等，皆为与陈抟有关的遗迹。

in Taoist history. Yuntai (Cloud Terrace) Temple, Yuquan (Jade Spring) Court, and Xiyi (Vacuous Quietness) Cave are cultural monuments related to the philosophical tenets espoused by CHEN Tuan that still exist today. During the rise of the Quanzhen School in North China during the Jin-Yuan period, Huashan came under the control of the Quanzhen branch of Taoism. According to legend, WANG Chongyang and his disciples TAN Chuduan and QIU Chuji lived a secluded life on Huashan. In 1276, the third year in the Zhiyuan reign period of the Yuan dynasty, HE Zhizhen came to Huashan and practiced at Quanzhen Temple built to the west of Jade Spring Court. His disciples carved more than forty caves out of the rock, among them Changkong (Endless Void) Walkway and Dachaoyuan (First Dawn) Cave. After the Ming, the government and the local villagers built new temples and a new ascent route. The cultural landscape as we know it today gradually took shape with buildings blending into their natural environment. In 1982, Huashan was announced as a National Priority Protected Site in the first batch of designated sights.

Because of its natural dignity and power, Huashan became a favorite subject matter of northern Chinese landscape painters. The early-Song master FAN Kuan appreciated the beauty of the landscape on Mount Hua, propagating the tradition of monumental landscape painting in the northern style based on experiences gained through his stays at renowned mountains in Shaanxi province ["living between the deep valleys and forests of Zhongnan and Taihua mountains, amid scenes of changing beauty with dark clouds and mist clearing after rain, (without words in the way) in my silent encounter with the gods"]. His masterpiece *Travelers among Mountains and Streams (Xishan xinglü tu)* perfectly captures the charming scenery of Huashan in this respect. The early-Ming painter WANG Lyu climbed Huashan poor health at the age of fifty-two. In an inscription on his *Paintings of Huashan (Huashan tuce)*, WANG Lyu explained that he most valued what he learnt from his heart, what his heart learnt from his eyes, and what his eyes learnt from Huashan. Through his work, painting became a means of recording things in a way that foreshadowed the individualism of the succeeding Ming-Qing period.

In July 2008, the School of Architecture at Tsinghua University assisted in the preparation of the nomination proposal for inclusion of Huashan in the World Heritage List. Professor WANG Guixiang led a team of teachers and students to Huashan. Among them were the Tsinghua professors LIU Chang, LIAO Huinong, JIA Jun, HE Congrong, and LI Luke, the postdoctoral researcher Marianna Shevchenko, and the graduate students YANG Bo, AO Shiheng, BAI Zhaoxun, XIN Huiyuan, DUAN Zhijun, ZHAO Xiaomei, YUAN Lin,

道士贺志真入华山修道，在玉泉院西筑全真观居住，又率徒在华山开凿岩洞 40 余处，其中最有名的是华山绝壁上的长空栈道和大朝元洞。明代以后，官府与民间在山上广建宫观洞府，开辟登山道路，逐渐形成令之人工与自然相融的景观格局。1982 年，华山被国务院批准列入首批国家级风景名胜区之一。

华山又以其浑厚雄强，成为中国北派山水美学与艺术的源泉。宋初山水画家范宽即长年「卜居于终南太华岩隈林麓之间，而览其云烟惨淡风月阴雾难状之景，默与神遇」，他以陕西华山及终南山为蓝本，「对景造意，不取繁饰，写山真骨，自为一家」，遂成刚劲雄浑之北派山水宗师，范宽传世的《溪山行旅图》，即有华山风景的神韵。明初画家王履在 52 岁时抱病登华山，曾慨叹道：「余自少喜画山，模拟四五家，常以不得逼真为恨。及登华山，见奇秀天出，非模拟者可模拟。于是摒去旧习，以意匠就天出」，最终画出传世不朽的《华山图册》，王履的名言「吾师心，心师目，目师华山」，亦成明清中国山水绘画试图摆脱宗派和程式之束缚的宣言。

2008 年 7 月，为配合编制《华山申报世界自然文化遗产文本及保护管理规划》，清华大学建筑学院由王贵祥教授带队，组织师生前往华山进行古建筑测绘。其中，教师还有刘畅、廖慧农、贾珺、贺从容、李路珂；博士后有玛丽安娜；研究生有杨博、敖仕恒、白昭薰、辛惠园、段智君、赵晓梅、袁琳、郑亮、徐桐、黄文镐、陈迟。承担主要测绘任务的学生为 2005 级本科生 68 人。此次

ZHENG Liang, XU Tong, HUANG Wengao and CHEN Chi. Sixty-eight undergraduate students (Class of 2005) carried out the surveying and mapping of the historical buildings. The field work focused on architecture that was key to understanding the cultural traditions of the Five Great Mountains, Taoism, and landscape painting, and resulted in two hundred and seventy-six architectural drawings, including the Temple to the West Peak (Xiyuemiao) and Jade Spring Court on the northern mountain slope; Cuiyun (Green Cloud) Palace on Huanshan's West Peak, Yunyu (Jade Maiden) Palace on Central Peak, Qunxian (Immortals) Temple on North Peak, and several Taoist caves such as First Dawn Cave. I would like to thank the professors of the department of landscape design (YANG Rui, ZHUANG Youbo, and WU Dongfan) and one of their graduate students, ZHAO Zhicong, for their help and support. I further owe gratitude to party secretary BAI Xinmiao and director WU Jianfeng, staff BIAN Youbing from the Huashan Scenic Area Management Committee and to the comrades in arms who have made this work possible.

The following paragraphs introduce the traditional architecture on Huanshan that we surveyed:

Xiyuemiao or Xiyue Temple to the West Peak (referring to the West Mountain of the Five Great Mountains), also known as Huashan Shrine (ci) or Huayue Temple (miao), is located on the northern slope of the mountain five kilometers east of Huayin. It was the place where the emperors worshipped and offered sacrifices to the gods and spirits of Huashan. The first temple – Jilinggong, or Palace of Gathering Spirits – was constructed under emperor Wu of Han and located in Huangshen Valley to the east of the present site of imperial sacrifices at Xiyue Temple. It moved during the Three-Kingdoms period in 220, the first year in the Huangchu reign period of emperor Wen of Wei; the original location without a view on South Peak was not ideal. The existing architecture covers an area of more than 120.000 m sq. Renovated on more than twenty occasions, most of the remaining structures date to the Ming-Qing period. In 1988, Xiyue Temple was put under state protection, announced as a National Priority Protected Site in the third batch.

Following a traditional urban planning pattern, Xiyue Temple was built in the style of a double city, which gave it its nickname of Small Forbidden City of Shaanxi. The main architecture is situated in the north, facing southward toward South Peak. Because of its long history of imperial worship, the scattered buildings were protected by high walls, with a barbican (wencheng) attached to the front. Access was provided through a double gate—only after entering through the barbican gate, one would pass through

测绘涵盖了华山在五岳文化、道教文化及山水美学方面的主要古建筑遗存，包括华山北麓之西岳庙、玉泉院，主峰区之西峰翠云宫、中峰玉女宫、北峰群仙观，以及道教洞窟大朝元洞、无上洞等，共完成图纸276张。此次测绘工作得到了清华大学景观学系教师杨锐、庄优波、邬东璠和研究生赵智聪等人的支持，以及华山风景名胜区管理委员会白新淼书记及吴剑锋主任、边有兵等同志的帮助，于此致谢。

以下对本卷收入的几组重要的古建筑作一简要介绍：

西岳庙，又称华山祠、华岳庙，位于华山北麓5公里华阴市区，是历代帝王祭祀华山神祇的庙宇，其建制始于汉武帝时期，原在现址以东的黄神谷，称『集灵宫』，但由于其位置不理想、无法望见华山主峰，三国时魏文帝黄初元年（220年）迁至现址，后历经整修扩建20余次。现存建筑占地面积逾12万平方米，多为明清时期之遗构，1988年被公布为第三批国家级重点文物保护单位。

西岳庙的总体格局为宫苑重城式，坐北朝南，面向华山主峰；由于是历代帝王西巡驻跸之地，

the principal temple gate and walk into the front courtyard. A historical map in *Huayue Gazetteer* from 1762, the twenty-seventh year of emperor Qianlong of Qing, illustrates the position of Xiyue Temple on Huashan (Fig. 1). Another map from 1779, the forty-fourth year of Qianlong, depicted on the Xiyue Temple stele erected by imperial order, shows the layout of Xiyue Temple at its zenith (Fig. 2). The extant buildings of the temple are aligned along the central axis, and comprise, from south to north, a glazed-tile screen wall (Ming), Haoling Gate (Ming-Qing), Wu (Meridian) Gate (destroyed in the reign of the Qing emperor Tongzhi; modern reconstruction), Lingxing Gate (Qing), a stone archway in the front courtyard (Ming), Jincheng Gate (Ming-Qing), Golden River Bridge with water pool, Haoling Hall (rebuilt in the Tongzhi reign), an imperial library, stone archway, and Wanshou Pavilion (destroyed in 1932; modern reconstruction). Along the east-west axis are aligned a bell tower, a drum tower, Lingguan Hall, Mingwang Hall, Lüzu Hall, Wanghua Pavilion, and a stele pavilion.

Haoling Hall, the principal building of the temple, is situated on a platform with a protruding front known as *yuetai*; an imperial way (*yudao*) leads to the front of the hall. The illustration on the temple stele tells us that Haoling Hall originally had a gong-shaped plan (like the Chinese character *gong*), because it was connected with a rear hall (*qingong*, "residential palace") through a corridor. Haoling Hall in its current condition is a seven-bay wide, eleven-rafter deep rectangular structure with a single-eave hip-gable roof and surrounding corridor. An inscription on the seven-rafter beam suggests repair in the Tongzhi period under the supervision of ZUO Zongtang, governor of Shaanxi and Gansu provinces. From a photo taken by the French Sinologist Émmanuel-Édouard Chavannes in 1907, we know how Haoling Hall and Wanshou Pavilion actually looked like in the late Qing period (Fig. 3, Fig. 4).

Yuquan (Jade Spring) Court, located on the northern slope of Huashan, was built in the Huangyou reign period of the Northern Song dynasty by JIA Desheng to commemorate his teacher CHEN Tuan. It was named after a pure spring, which, according to legend, entered the courtyard from a "mysterious gorge" below Shansun (Aromatic Plant) Pavilion and connected to the Jade Well at Zhenyue (Mountain Guarding) Palace on the mountaintop. The extant architecture, rebuilt in the Qianlong period, covers an area of 9.000 square meters and is a unique example of Taoist landscape design. Since 1992, Jade Spring Court is registered as a Priority Protected Site at provincial level, and together with the Eastern (Dongdao) Court and Zhenyue Palace as a National Priority Protected Site at state level. The historical layout of Jade Spring Court is depicted on a map in Huayue

因此周围筑有城墙，城正南辟瓮城，设二重城门。关于西岳庙与华山的位置关系，可以参见清乾隆二十七年（1762年）刻印的《华岳志》中的《县境图》（图1）；关于西岳庙盛期的格局，可以参见乾隆四十四年（1779年）所立的《敕建西岳庙图碑》（图2）。西岳庙现存的主要建筑位于中轴线上，自南向北依次是琉璃影壁（明）、灏灵门（明清）、午门（毁于清同治年间，现代重建）、棂星门（清）、前院石牌楼（明）、金城门（明清）、金水桥、泮池、灏灵殿（清同治年间重建）、御书楼、石牌楼、万寿阁（毁于1932年，现代重建）；中轴线两侧还有钟楼、鼓楼、灵官殿、冥王殿、吕祖堂、望华亭及碑亭等建筑。

其中灏灵殿是建筑群中的主要殿堂，坐落于一个『凸』字形的月台上，台前有御道，从乾隆敕建图碑可以看到，殿后原有寝宫与大殿以通廊相连，作『工』字形布局，现存建筑为七间十一檩周围廊单檐歇山式，殿内七架梁上有清同治年间陕甘总督左宗棠主持重修题记。1907年，法国学者沙畹在西岳庙拍摄了灏灵殿和万寿阁的照片，可见这两座主要建筑在晚清时的面貌（图3、图4）。

玉泉院，位于华山北麓，传为北宋皇祐年间道士贾得升为纪念其师陈抟而建，昔日有泉水自山苏亭下『玄峡』流入院内，泉与西岳峰顶镇岳宫玉井潜通，故名。现存建筑占地面积约9000平方米，为清乾隆年间重建，为一处独具特色的道教园林，1992年被列为省级重点文物保护单位，并与山内之东道院、镇岳宫一同列为全国重点道教宫观。

关于玉泉院盛期的格局，可以参见清道光

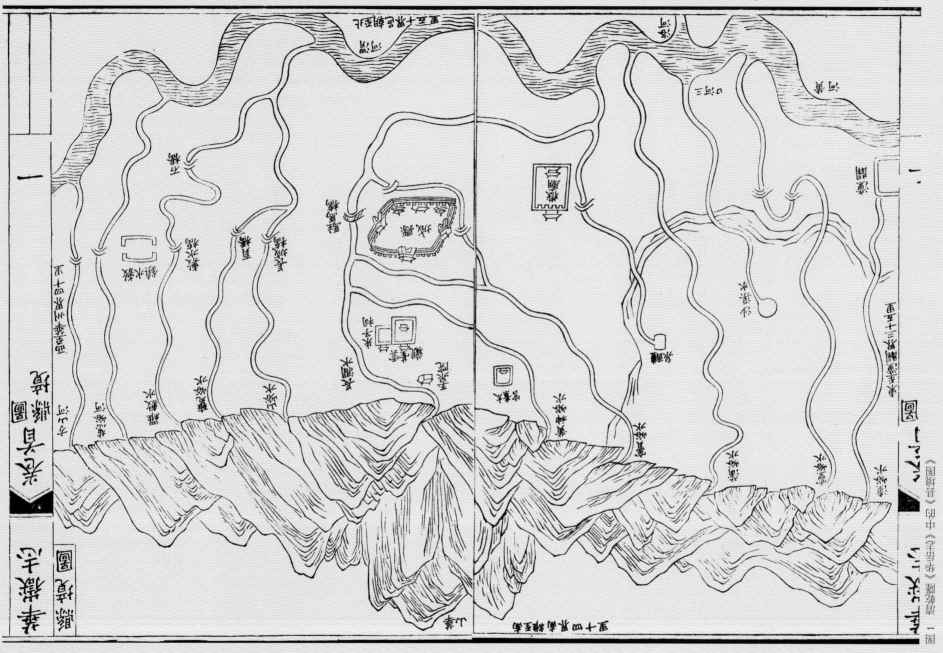

图2　清乾隆敕建西岳庙图碑

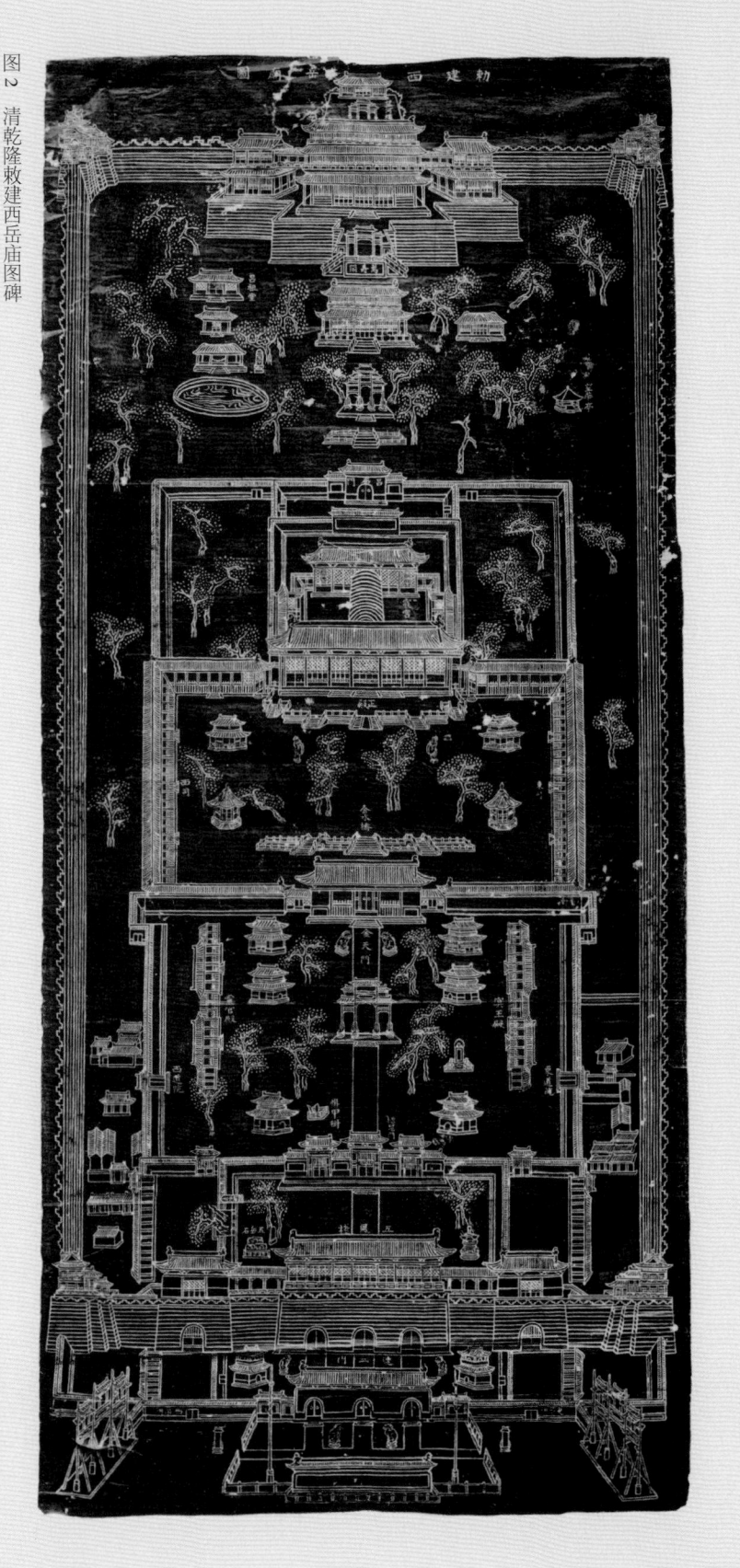

图3　法国学者沙畹拍摄的西岳庙灏灵殿《北中国考古图录》

图4　法国学者沙畹拍摄的西岳庙万寿阁《北中国考古图录》

Fig.2　The Xiyue Temple Stele erected by emperor Qianlong of Qing Dynasty

Fig.3　The Haoling Hall of Xiyue Temple photographed by the French Sinologist Émmanuel-Édouard Chavannes (*Archaeological Catalogue of Northern China*)

Fig.4　The Wanshou Pavilion of Xiyue Temple photographed by the French Sinologist Émmanuel-Édouard Chavannes (*Archaeological Catalogue of Northern China*)

Gazatteer from 1831, the eleventh year of the reign of the Qing emperor Daoguang (Fig. 5).

At Huashan, there are dozens of places of Taoist worship. The main architectural structure of each peak belongs in fact to Taoism, regardless if it is nestled in a valley or sitting on a summit or even cut into the cliff high up in the air. Buildings tend to be small in size. They use construction materials and methods in line with local practices and fit harmoniously into the landscape. The *Mustard Seed Garden Painting Manual (Jieziyuan huazhuan)* describes how architecture can enrich the natural scenery through its profile. Looking back on a thousand-year long history, the architecture at Huashan's main peak area is the perfect example of this idea. The marvelous blend of natural and man-made environment is illustrated on two Qianlong-period maps included in *Huayue Gazetteer* (Fig. 6, Fig. 7). But the architecture of the main peak area was destroyed and rebuilt on several occasions, not least because of the mass tourism in the twenty-first century. Today only a few Taoist buildings (Green Cloud Palace on West Peak, Jade Maiden Palace on Central Peak, and Immortals Temple at the lower end of Laojun's Furrow) have retained their original design, in addition to a handful of Taoist caves, for example, First Dawn Cave on South Peak.

West Peak is also called Lotus Flower Summit and is the most beautiful summit of Huashan, as described in a line by the Tang poet LI Bai ("the rock formation [appears] as a lotus flower above a terrace of clouds, as though, the White Emperor's spirit itself had descended there"). Cuiyun (Green Cloud) Palace, built in the early Qing period, stands on top of West Peak and is partly hidden by Lotus Flower Cave and Ax Cutting Stone. Originally, there was also a Taoist library, but it was destroyed by fire in 1932 (together with the scriptures it housed) and completely rebuilt in the next year. Buildings at Cuiyun Palace are arranged around a courtyard, but due to the gradually rising terrain, the courtyard front appears as a four-storied structure whereas the rear side two-storied. From the northern end of the courtyard, one can climb up to the highest point of West Peak (Zhaixingshi, "Reaching-for-the-Stars Stone"). Here, architecture effectively makes use of the physical terrain. The design is intentionally kept simple and straightforward in order to contrast the natural beauty of the mountain peak. A map in *Huayue Gazetteer* from the Qianlong period depicts the original form of West Peak and Cuiyun Palace (Fig. 8).

Yunyu (Jade Maiden) Palace on Central Peak was built to commemorate Nong Yu, daughter of Duke Mu of Qin, She fell in love with a poor young man who was good at playing the bamboo flute in the Warring States period. They married and lived at Huashan for the rest of their life together. Jade Maiden Palace was first built in the Kangxi period

十一年（1831年）刻印的《华岳志》中之《玉泉院图》（图5）。

除上述宫观之外，华山还有道教宫观洞府数十处，各主峰的点景建筑亦多为道教建筑，有的安于山峪中群山环抱之地，有的嵌于山巅一侧，有的甚至开凿于绝壁，凌空于深渊，宛如神居。这些建筑往往体量不大，材料做法因地制宜，巧妙地利用地形安排空间，又以丰富朴实的轮廓为山水增色，使天然之山水，呈现『可居』『可游』之情趣。正如《芥子园画传》所论山水中屋宇的写照……

『凡山水中之有堂户，犹人之有眉目也……眉目虽佳，亦在安放得宜。眉目不可少……凡房屋画法，必须端详山水之面目所在，天然自有结穴。……山水有人居，则生情；庞杂人居，则纯市井气。』

华山主峰区之古建筑营建历经千余年，渐次形成之景观，成为人工与自然相融的典范，这一景象，可以从清乾隆《华岳志》中的《太华山阴图》与《太华山阳图》中略知概貌（图6、图7）。但主峰区的建筑屡毁屡建，尤其在21世纪为发展大众旅游，改变甚巨，目前仅有西峰之翠云宫、中峰之玉女宫、老君犁沟之群仙观略存其旧貌。另有南峰大朝元洞、金锁关无上洞、北峰日月岩尚存，可为华山道教洞府之代表。

西峰又称『莲花峰』，为华山诸峰最为秀、丽者，李白的诗句『白帝金精运元气，石作莲花云作台』即诵此地。翠云宫嵌于西峰之巅，半隐于莲花洞与斧劈石之后，始建于清初，原有藏经楼、藏道教真经，1932年建筑与经卷毁于火患，现存建筑为1933年重修。建筑为合院形式，前部四层，后部二层，自院落北侧上行即可攀上西峰最高点『摘星石』；建筑巧妙利用地形，造型简洁古朴，对主峰起到烘托和点缀的作用。西峰与翠云宫之原状，可见于清乾隆《华岳志》中的《西峰图》（图8）。

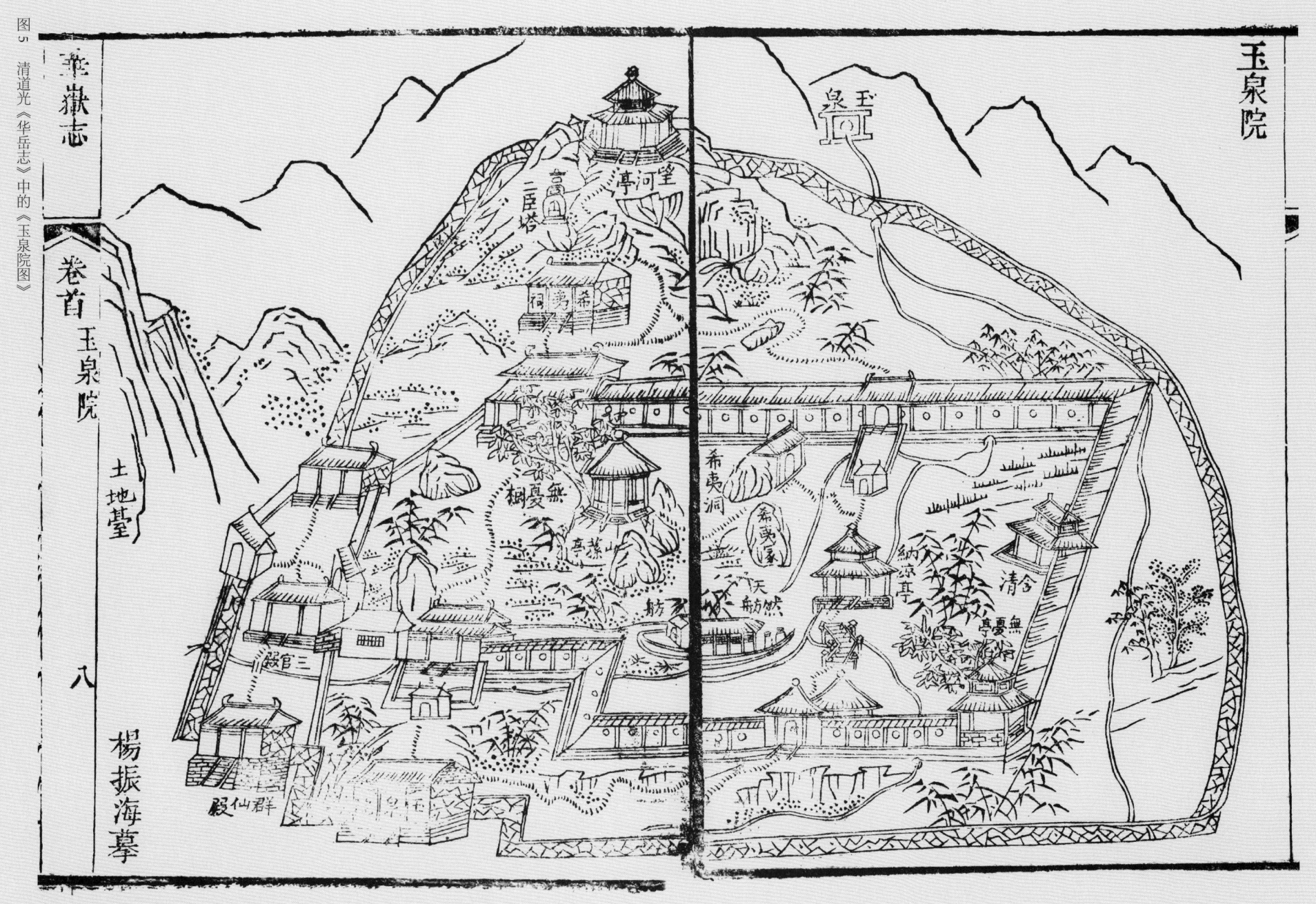

图5 清道光《华岳志》中的《玉泉院图》

Fig.5 "The Map of Yuquan Courtyard" in *Huayue Gazetteer* by emperor Daoguang of Qing Dynasty

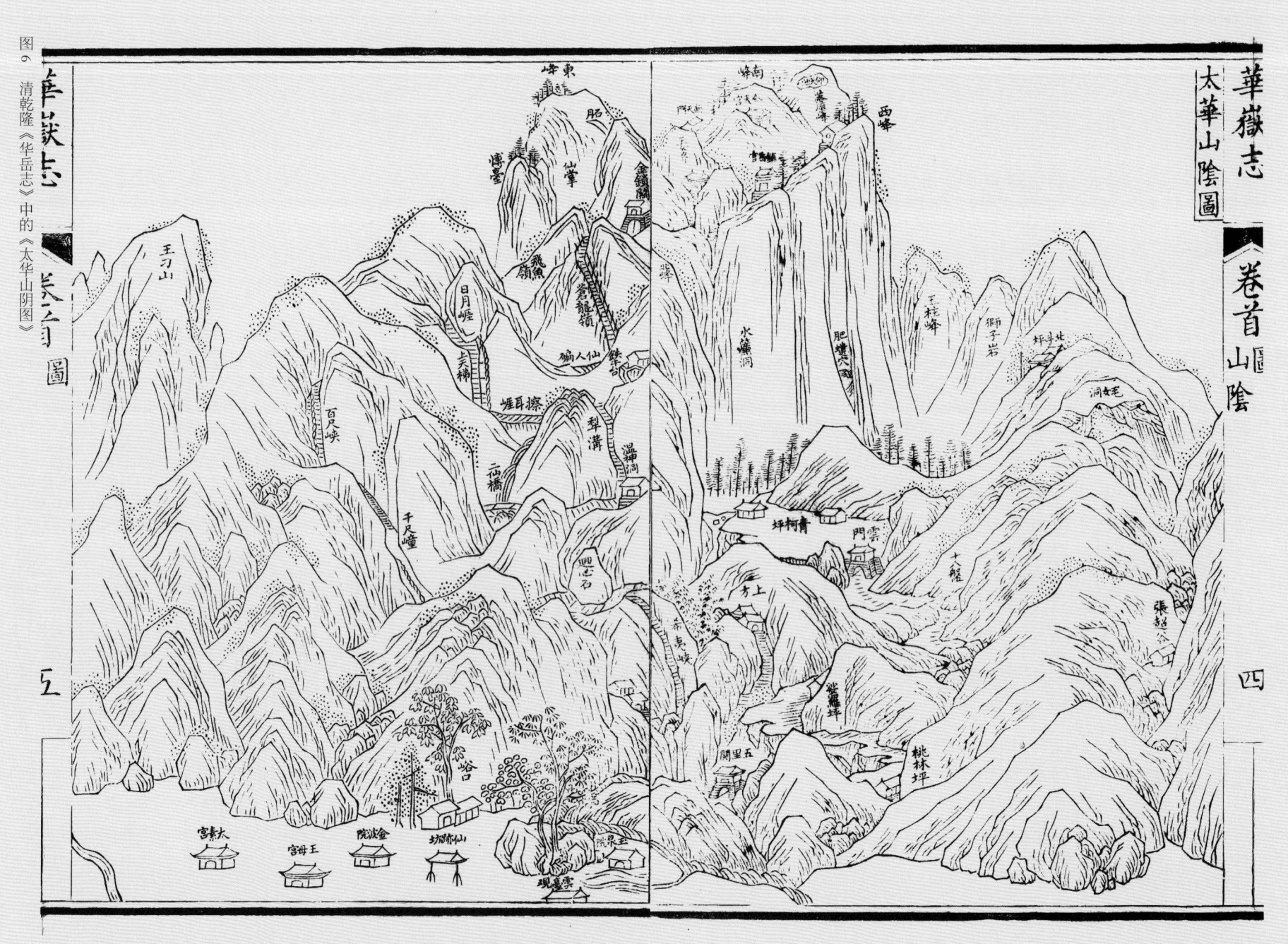

图6 清乾隆《华岳志》中的《太华山阴图》

Fig.6 "The North Elevation of Taihua Mountain" in *Huayue Gazetteer* by emperor Qianlong of Qing Dynasty

图7 清乾隆《华岳志》中的《太华山阳图》

Fig.7 "The South Elevation of Taihua Mountain" in *Huayue Gazetteer* by emperor Qianlong of Qing Dynasty

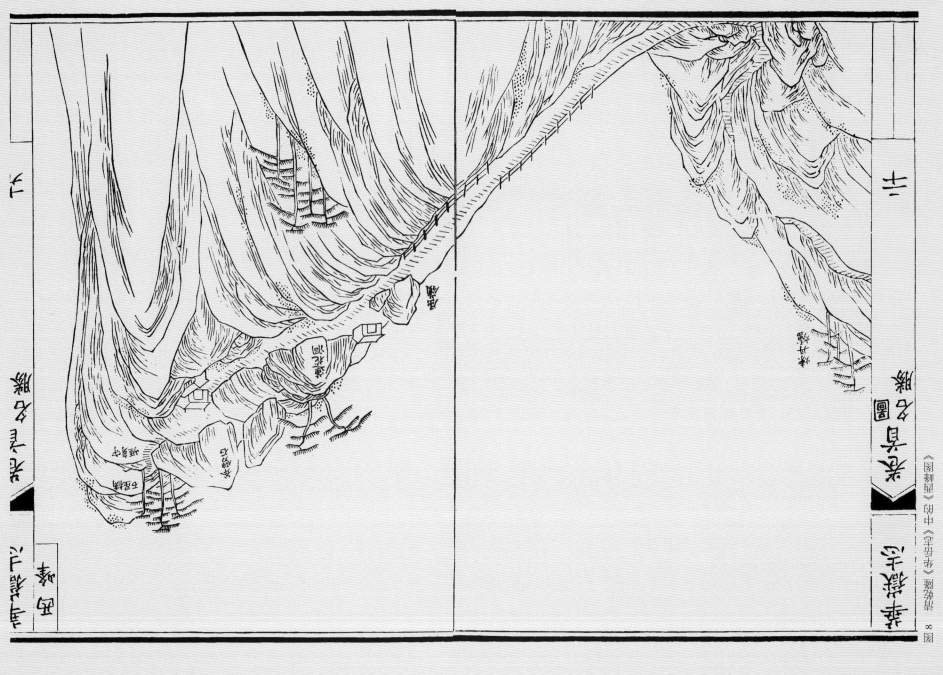

图8 清乾隆《华岳志》中的《西峰图》

of the Qing dynasty but was destroyed during the 'Cultural Revolution'. It originally had a guest room furbished with imperial utensils for the emperor's lodging when traveling, including a bed of state and an embroidered robe. The extant structure was rebuilt in recent years but retained its original form and scale. The single-story front hall stands atop a huge stone shaped like a turtle; the two-story rear hall stands to the north of the turtle-shaped stone. Underneath the stone is Jade Maiden Cave, a place for the spiritual practice of nuns. The buildings of Jade Maiden Palace are comparatively low and compact, yet they have a far-reaching view over the stunningly beautiful scenery around Central Peak. Located half way up the mountain at the lower end of Laojun's Furrow, the Temple of the Immortals (Qunxiansi) also fits harmoniously into the mountain landscape.

First Dawn Cave and Endless Void Walkway of South Peak are striking and more permanent examples—compared to the fragile wooden Taoist temples—that devoted disciples cut out of the massive rock to provide Taoist masters with a place for seeking spiritual perfection. A stele from 1325, the second year in the Taiding reign period of the Yuan dynasty, tells us that First Dawn Cave was hewn out of the middle of a massive cliff for the Quanzhen Taoist master HE Yuanxi, who had first taken residence at Quanzhen Temple situated west of Jade Spring Temple. First Dawn Cave was only to be reached by Endless Void Walkway. Afterwards, Hezu Cave was hollowed out below to commemorate the famous Taoist master. The caves were absolutely quiet and secluded locations (See fig. 7).

I would like to express my thanks to TANG Henglu from WANG Guixiang Studio, in charge of the architecture plans that were modified by his colleagues MAI Linlin, SHAN Menglin, and HU Jingfu according to the original survey and mapping drawings drawn by students.

LI Luke
Architecture History and Historic Perservation Research
Institute, School of Architecture, Tsinghua University

Translated by Alexandra Harrer

中峰玉女宫相传为纪念战国时期秦穆公女『吹箫引凤』的故事而建，始建于清康熙年间，原有客堂，为皇帝驻跸准备了龙床、黄罗伞、蟒袍等物，均毁于『文革』。现存建筑为近年重修，其形式和尺度基本维持了旧貌。前殿建筑憩于一龟形巨石之上，高一层；后堂立于巨石北侧，高二层，巨石下有玉女洞，为道姑修真之处。玉女宫建筑群为原本较为低矮敦实，环抱于群峰之内的中峰塑成丰富优美的轮廓线。

位于老君犁沟下山腰的群仙观，亦是依山就势的成功之作。

南峰大朝元洞与长空栈道，是华山道人在极险峻处开辟修仙处所的典范。据元泰定二年（1325年）井道泉所撰《太华山创建朝元洞之碑》可知，大朝元洞为全真派道士贺元希所创，初时他在玉泉观以西筑全真观，后以该处不足以『振宗风、崇德化』，又『登华山之巅，辟山膺而洞焉。其肇基也，聚葛而悬，踞蘽以凿』，在绝壁上凿出『长空栈道』，开凿大朝元洞，后又在崖下开贺祖洞，皆为极其幽静之所（见图7）。

本册测绘图纸是由王贵祥老师工作室的唐恒鲁负责，由买琳琳、单梦林及胡竞芙根据当年学生的测绘图进行修改和整理完成。在此表示感谢。

李路珂

清华大学建筑学院　建筑历史与文物保护研究所

Figure

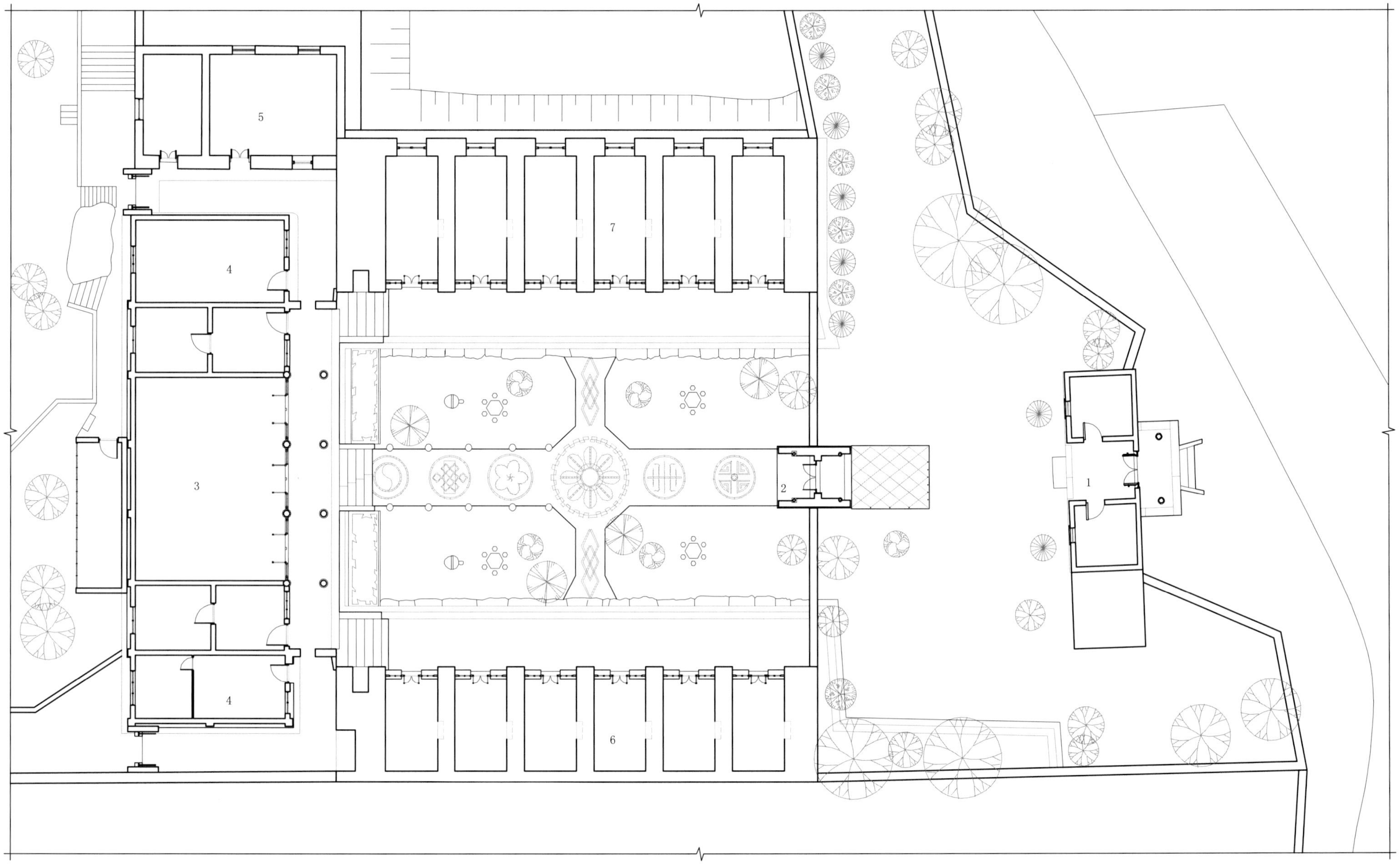

1 大门　2 二门　3 纯阳宝殿　4 耳房　5 附属用房　6 东六洞　7 西六洞

纯阳观总平面图
Site plan of Chunyang Taoist Temple

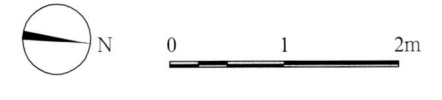

N　0　1　2m

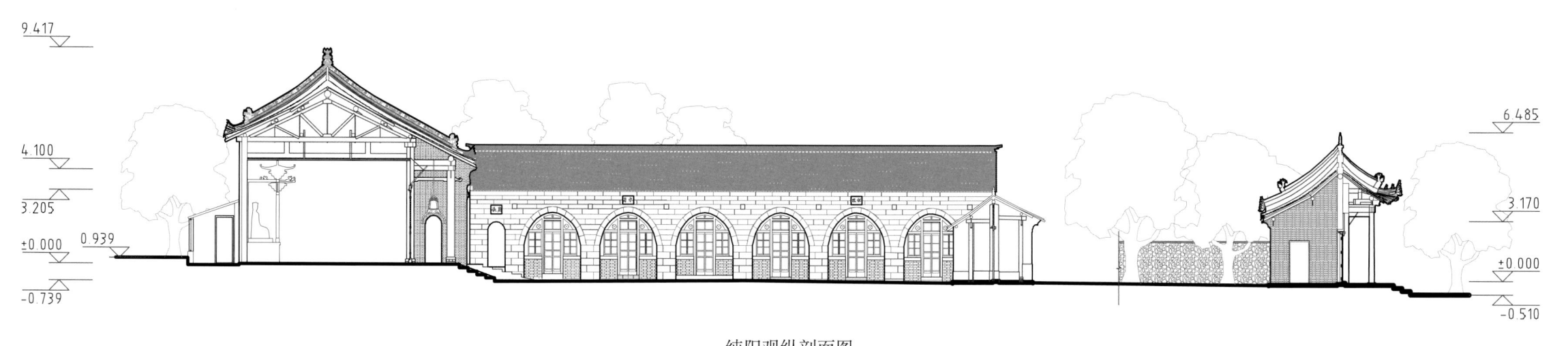

9.417

6.485

4.100

3.205

3.170

±0.000   0.939

±0.000

-0.739

-0.510

纯阳观纵剖面图
Longitudinal section of Chunyang Taoist Temple

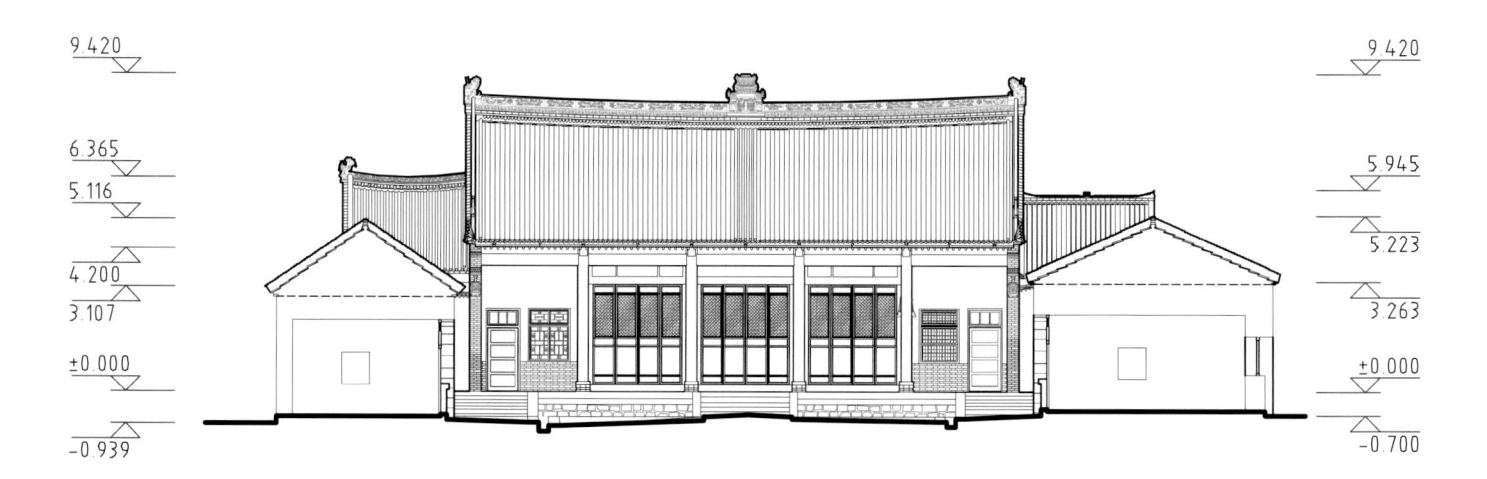

9.420

9.420

6.365

5.945

5.116

5.223

4.200

3.263

3.107

±0.000

±0.000

-0.939

-0.700

纯阳观横剖面图
Cross-section of Chunyang Taoist Temple

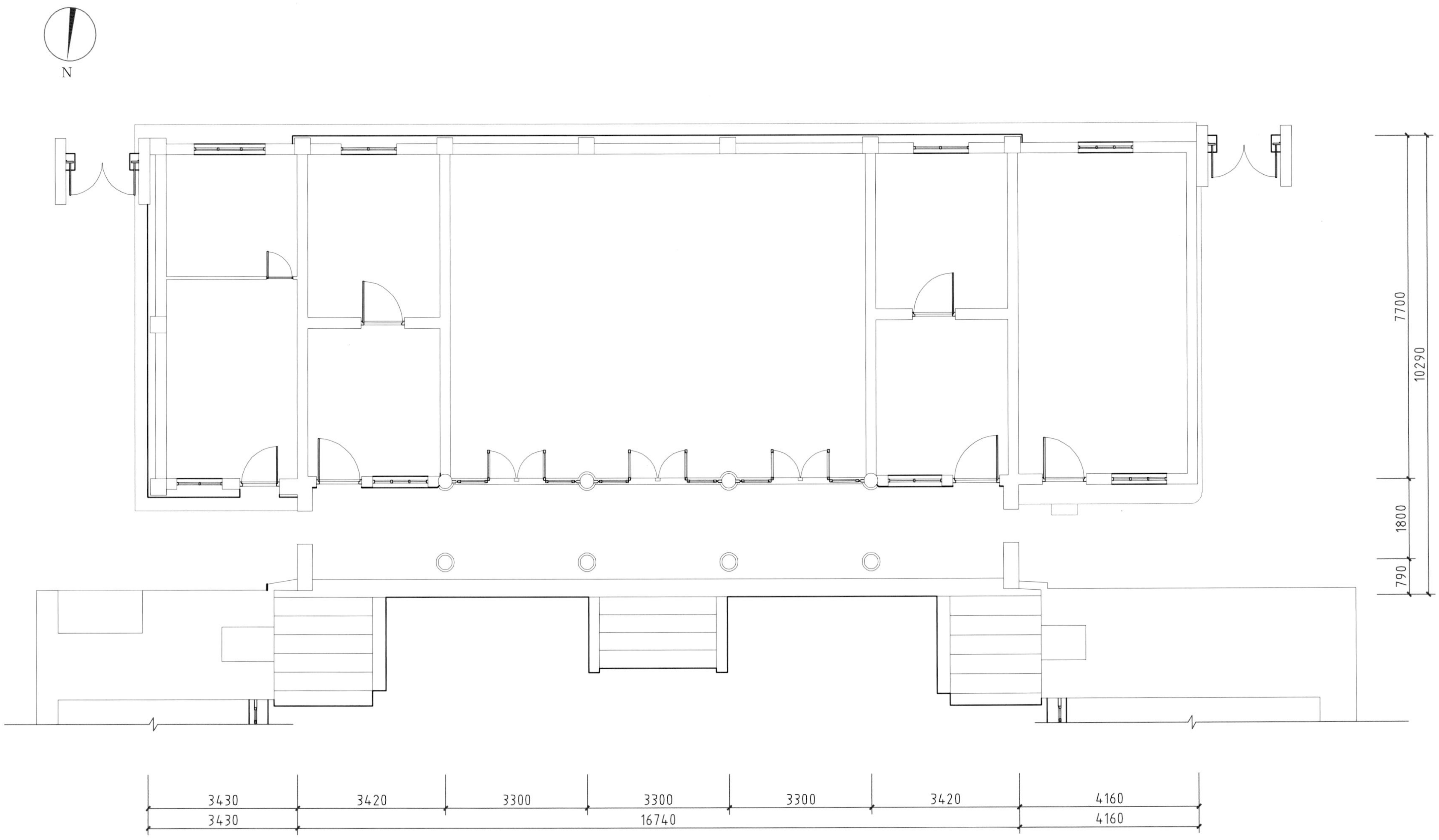

N

7700

10290

1800

790

3430 3420 3300 3300 3300 3420 4160

3430 16740 4160

纯阳观纯阳宝殿及耳房首层平面图
Plan of ground floor of Chunyang Taoist Temple's Chunyang baodian and *erfang*

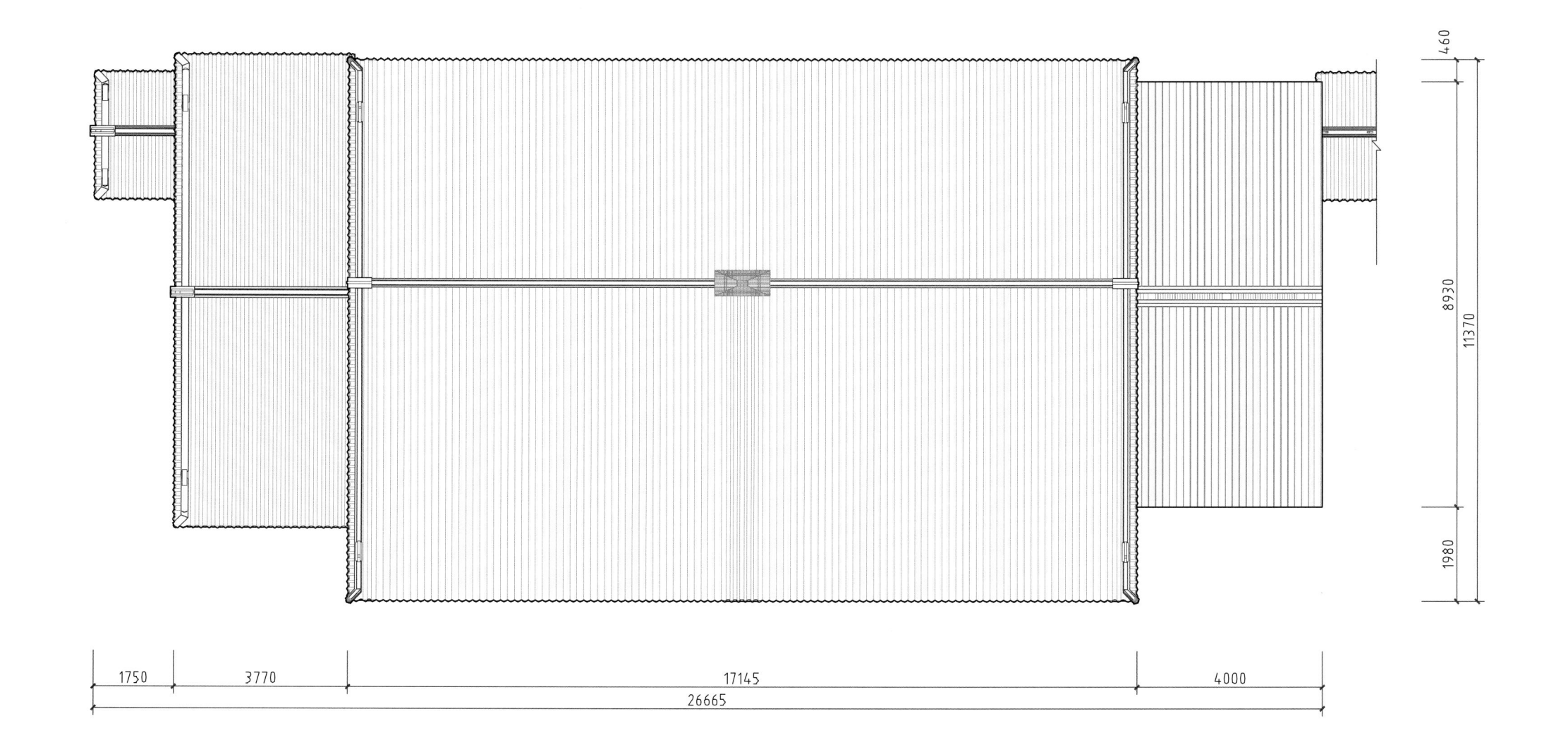

460

8930

11370

1980

1750   3770   17145   4000

26665

纯阳观纯阳宝殿及耳房屋顶平面图
Roof plan of Chunyang Taoist Temple's Chunyang baodian and *erfang*

9.420

6.365

4.200
3.220

2.965

±0.000

−0.740

0　　2　　4m

纯阳观纯阳宝殿及耳房正立面图
Front elevation of Chunyang Taoist Temple's Chunyang baodian and *erfang*

9.420

6.365

4.200
3.220

2.965

±0.000

纯阳观纯阳宝殿及耳房纵剖面图
Longitudinal section of Chunyang Taoist Temple's Chunyang baodian and *erfang*

0　　2　　4m

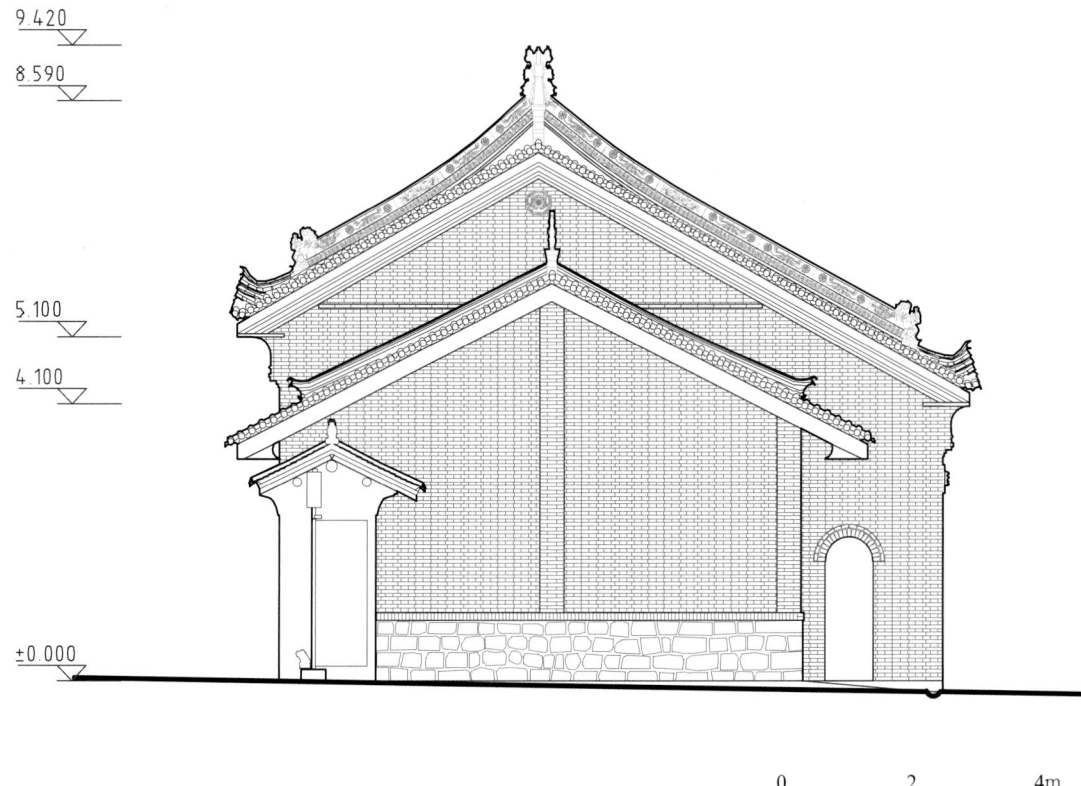

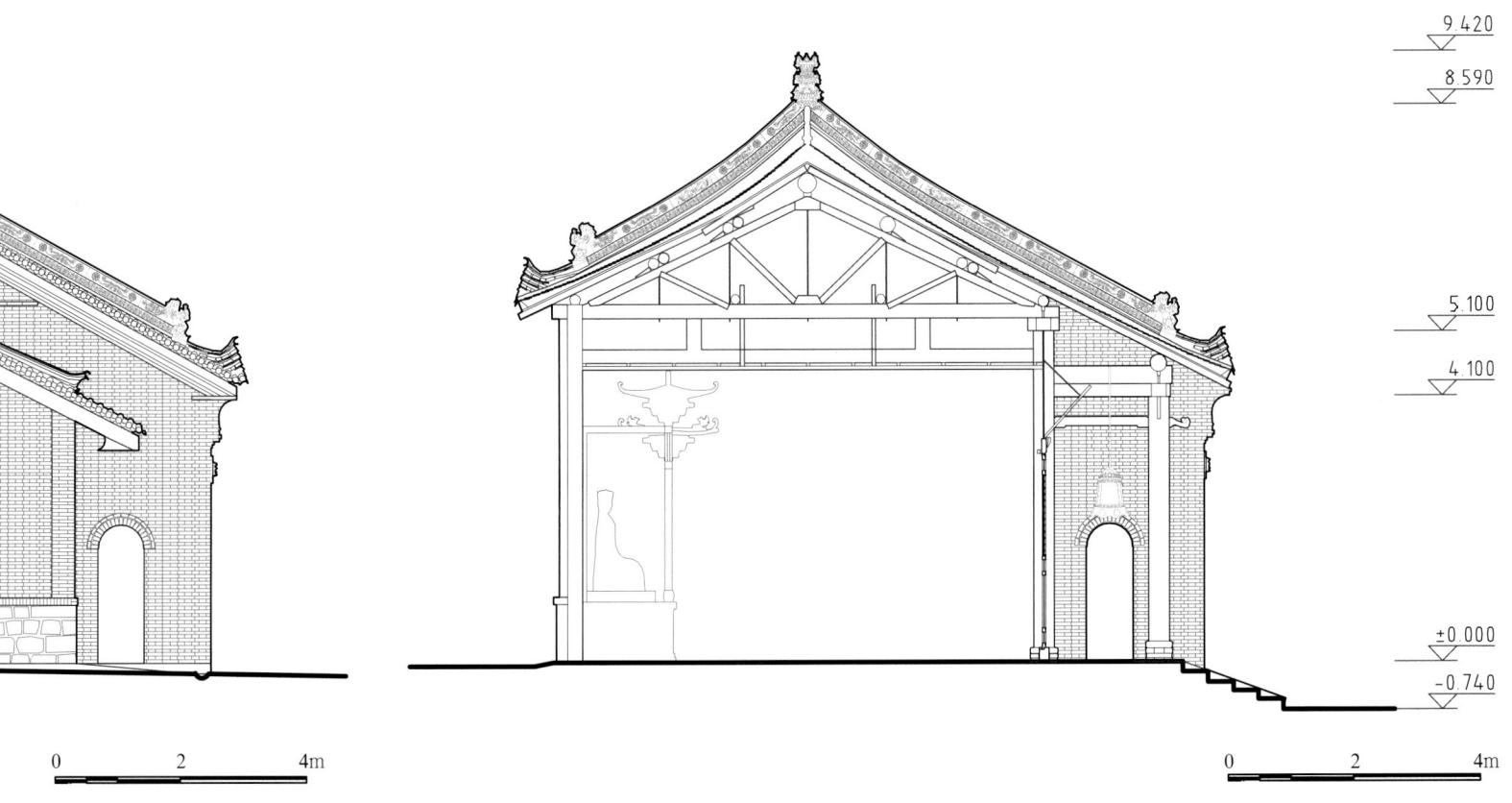

纯阳观纯阳宝殿及耳房侧立面图
Side elevation of Chunyang Taoist Temple's Chunyang baodian and *erfang*

纯阳观纯阳宝殿及耳房横剖面图
Cross-section of Chunyang Taoist Temple's Chunyang baodian and *erfang*

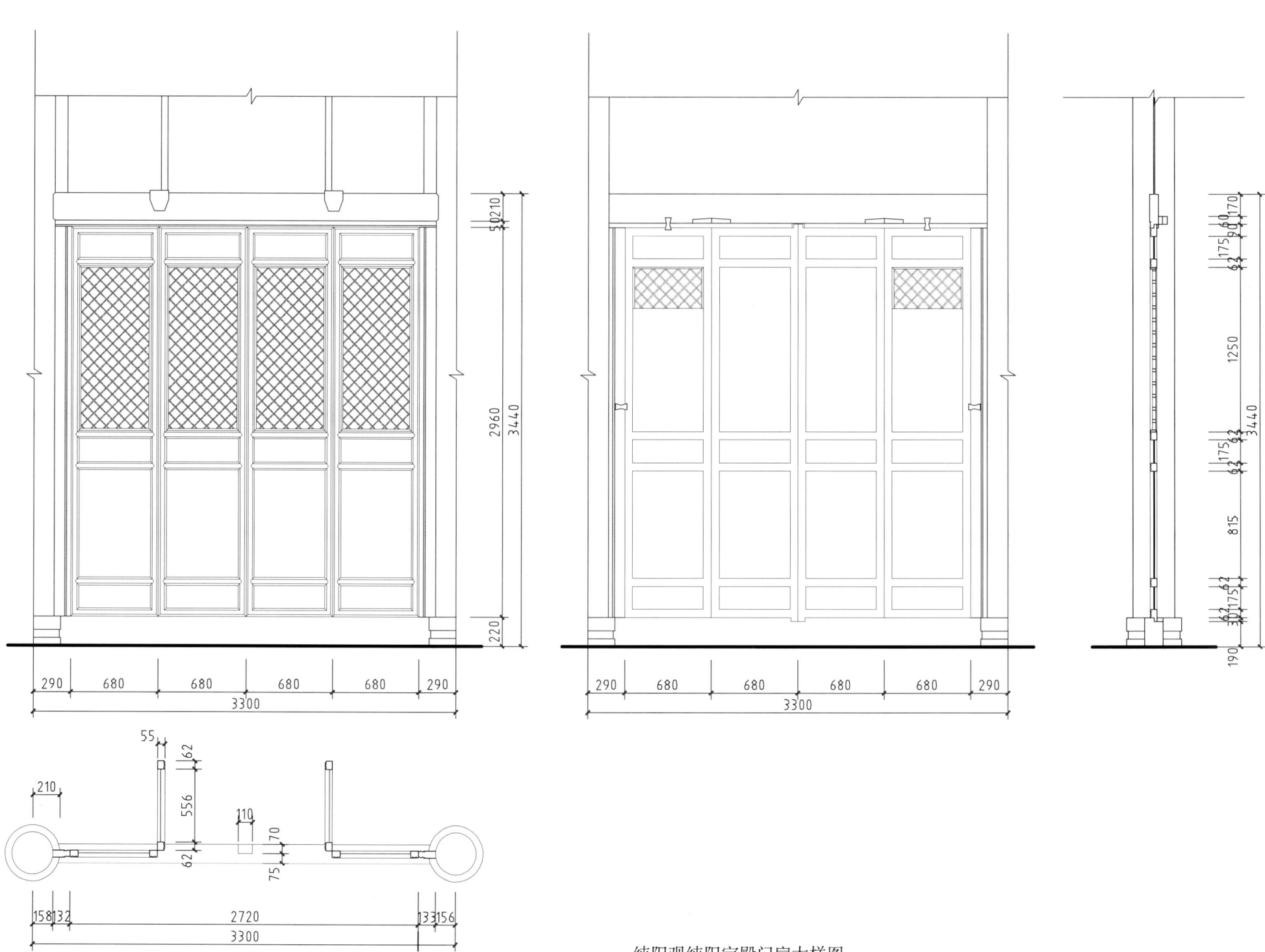

纯阳观纯阳宝殿门扇大样图
Menshan of Chunyang Taoist Temple's Chunyang baodian

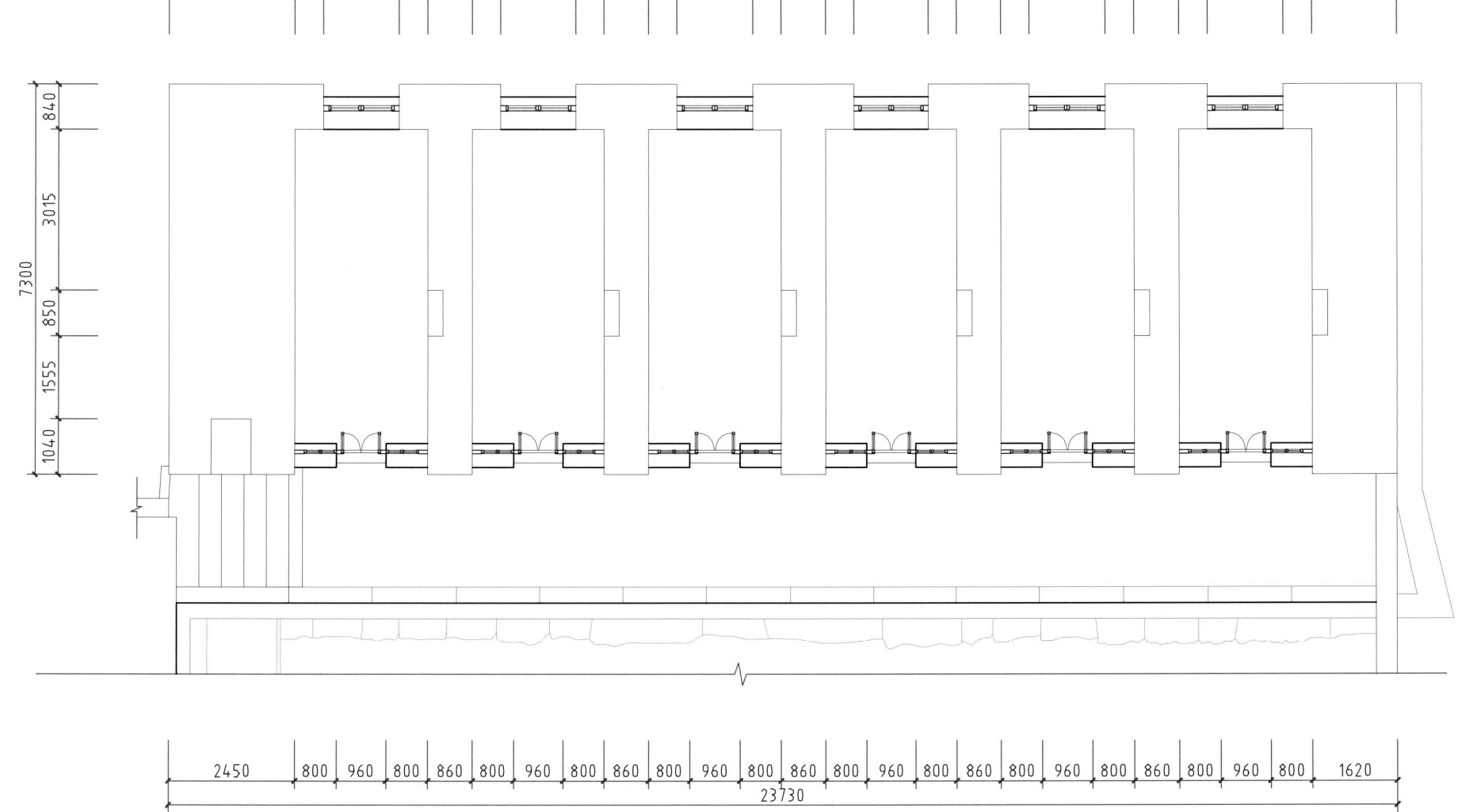

纯阳观西六洞平面图
Plan of Chunyang Taoist Temple's Xiliu Cave

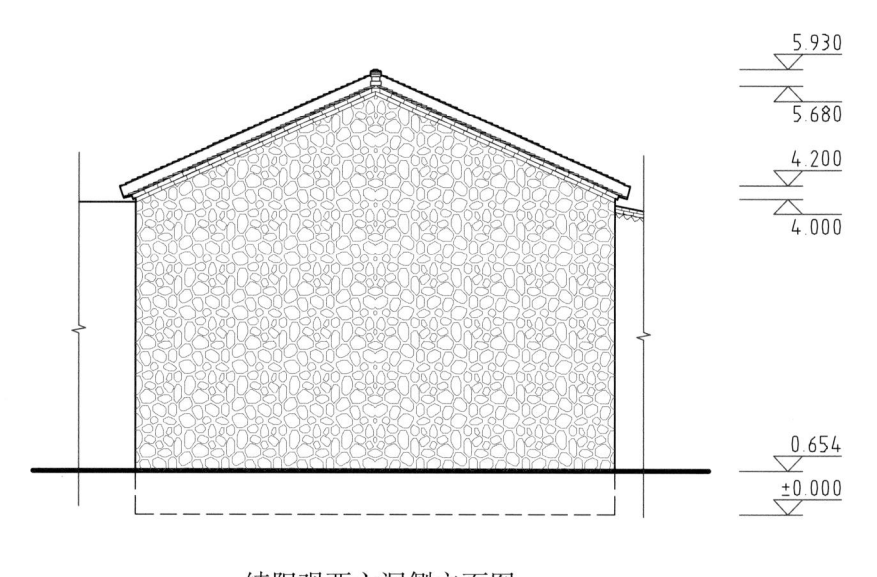

纯阳观西六洞侧立面图
Side elevation of Chunyang Taoist Temple's Xiliu Cave

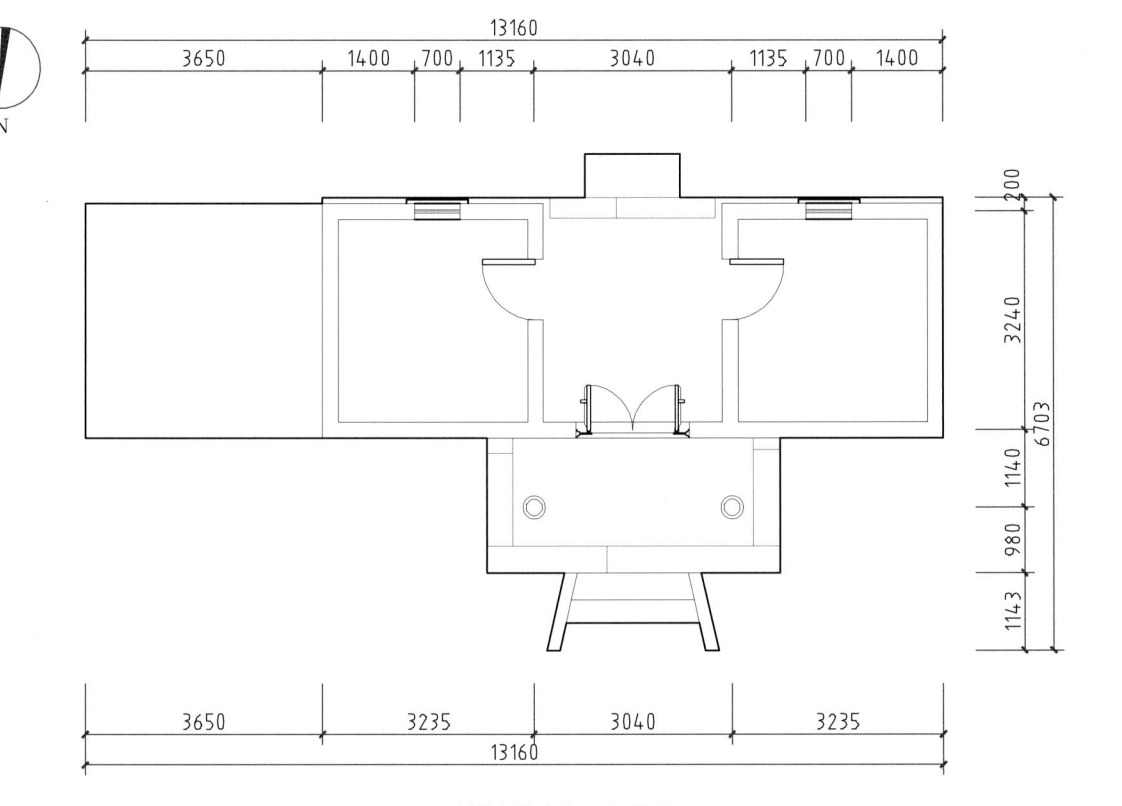

纯阳观山门平面图
Plan of Chunyang Taoist Temple's shanmen

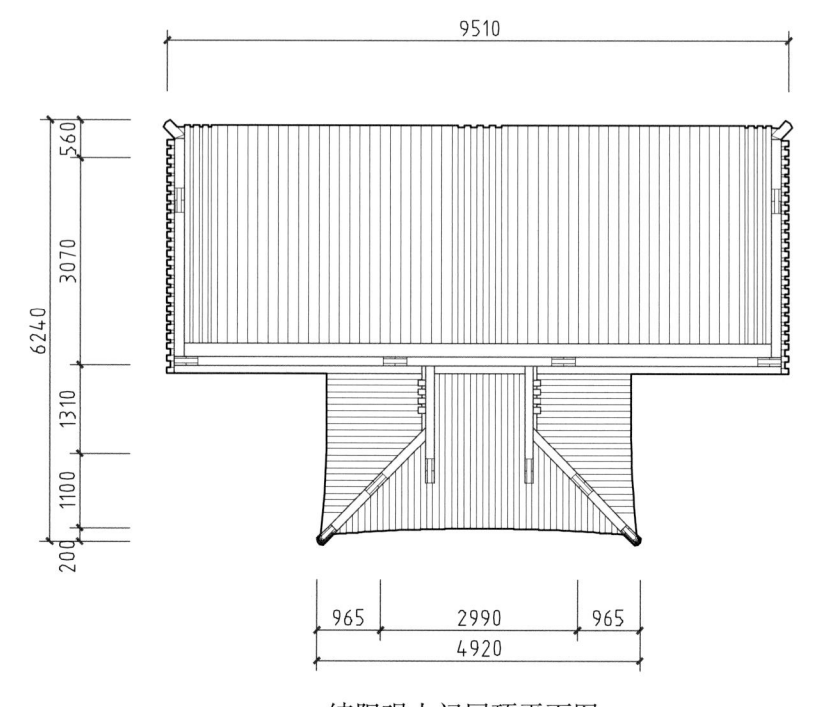

纯阳观山门屋顶平面图
Roof plan of Chunyang Taoist Temple's shanmen

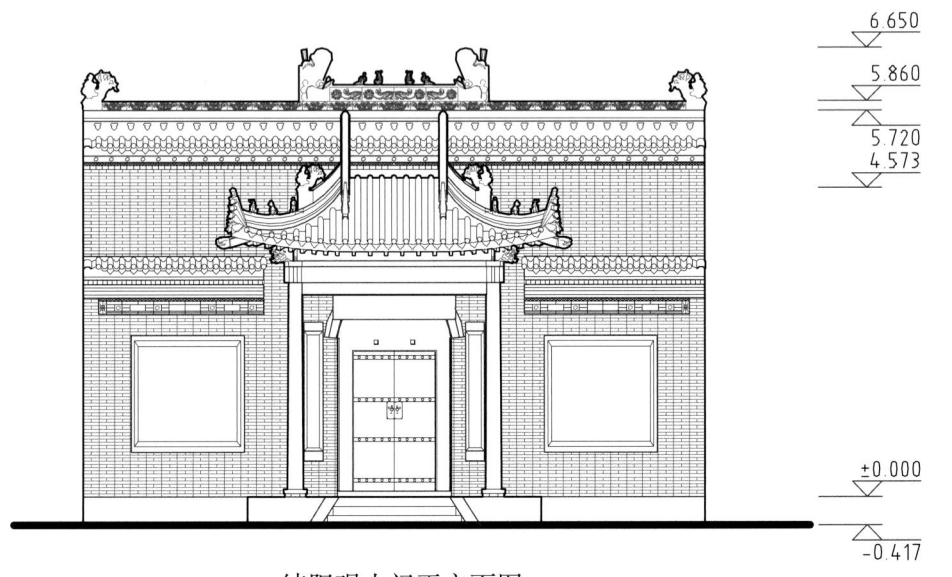

纯阳观山门正立面图
Front elevation of Chunyang Taoist Temple's shanmen

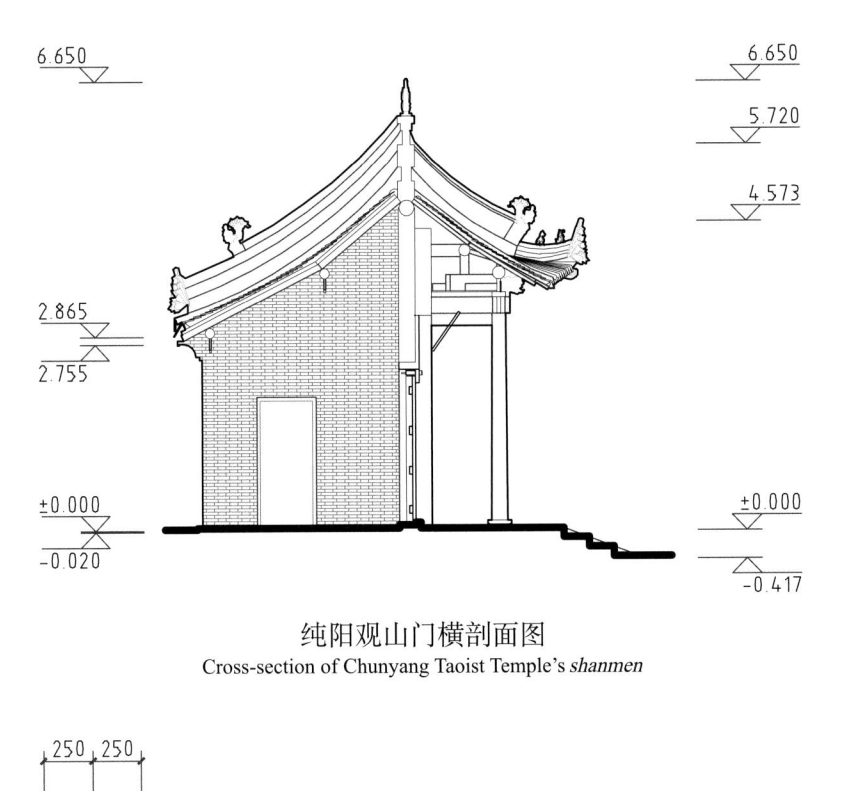

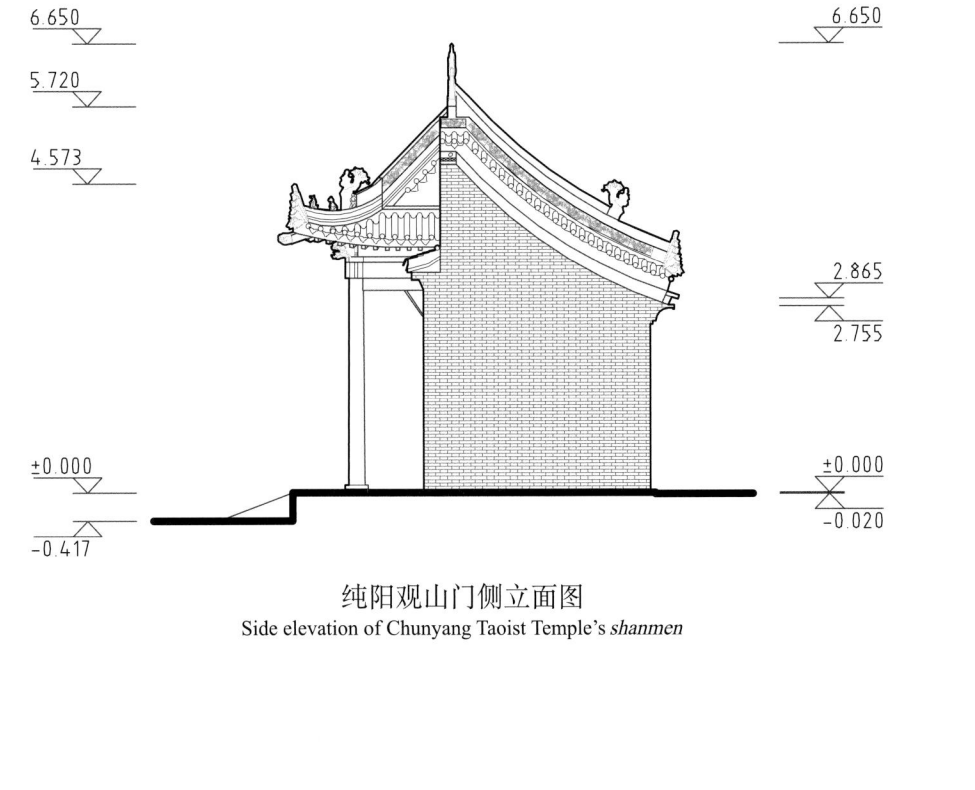

纯阳观山门侧立面图
Side elevation of Chunyang Taoist Temple's *shanmen*

纯阳观山门横剖面图
Cross-section of Chunyang Taoist Temple's *shanmen*

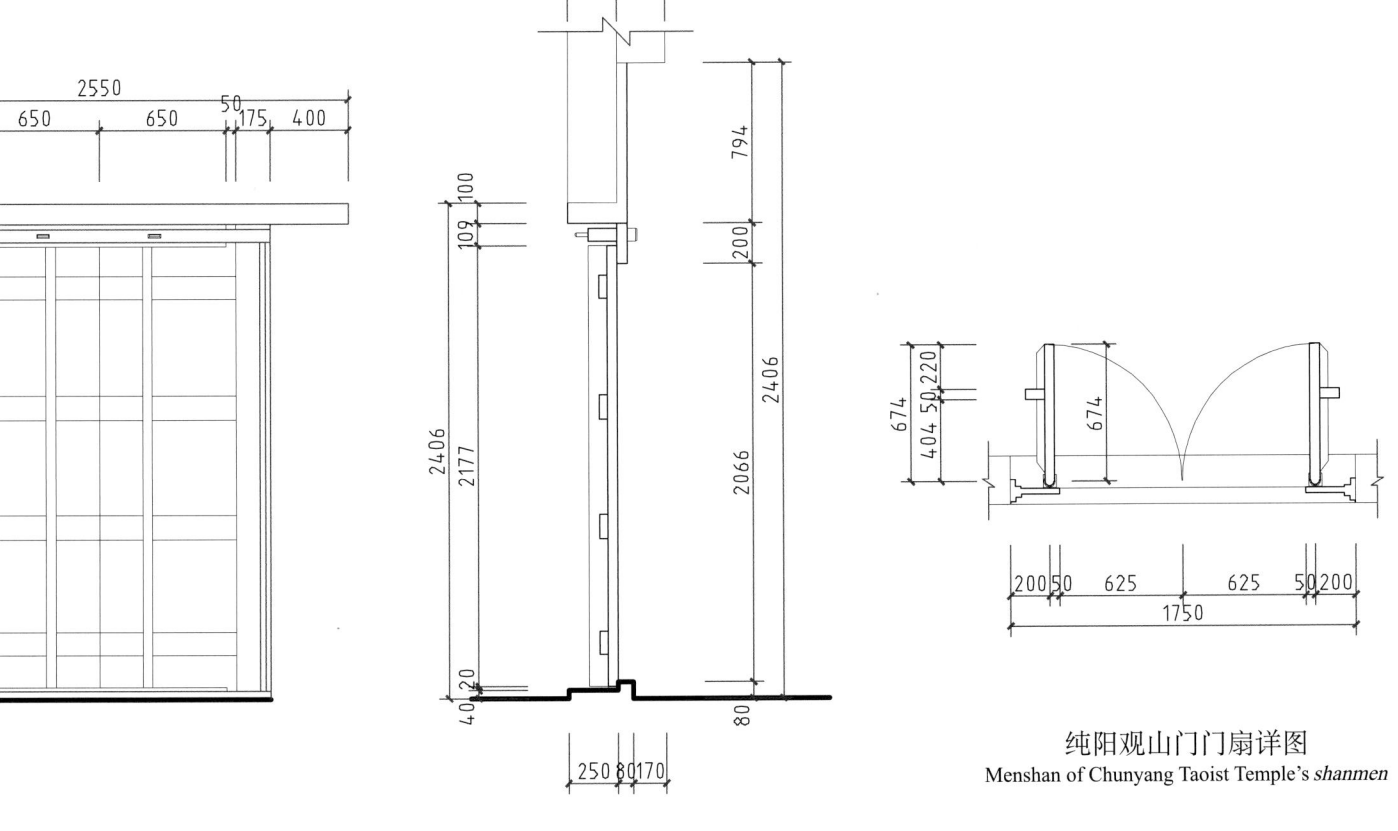

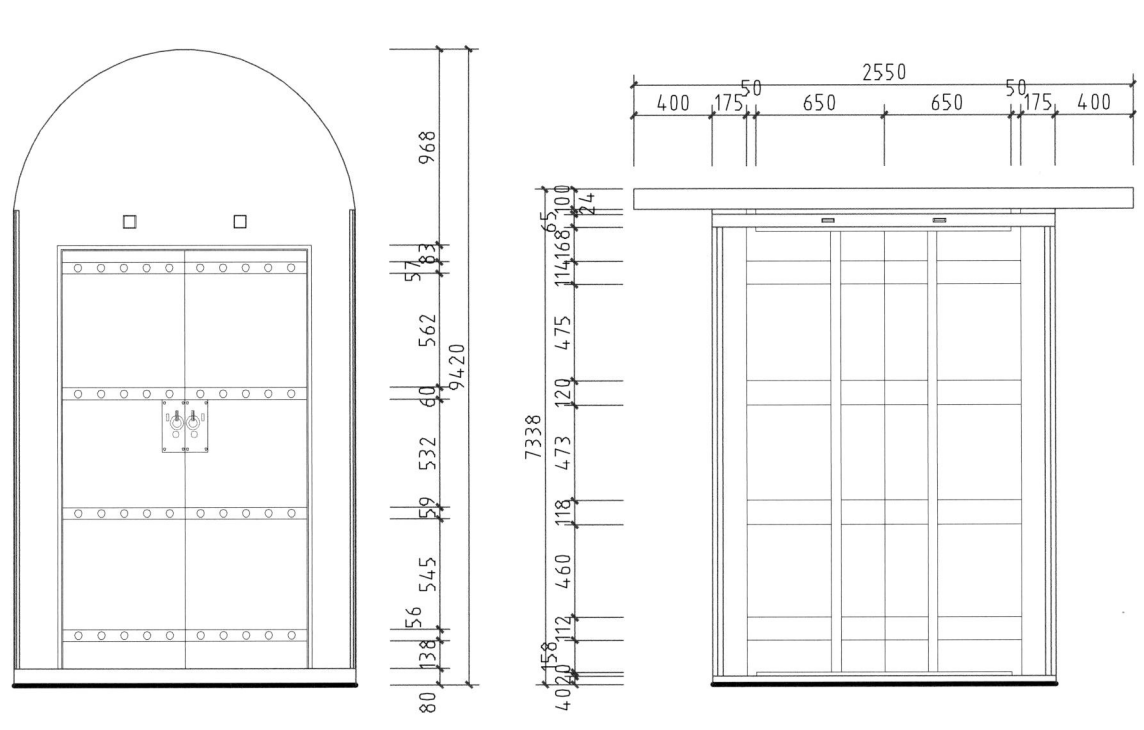

纯阳观山门门扇详图
Menshan of Chunyang Taoist Temple's *shanmen*

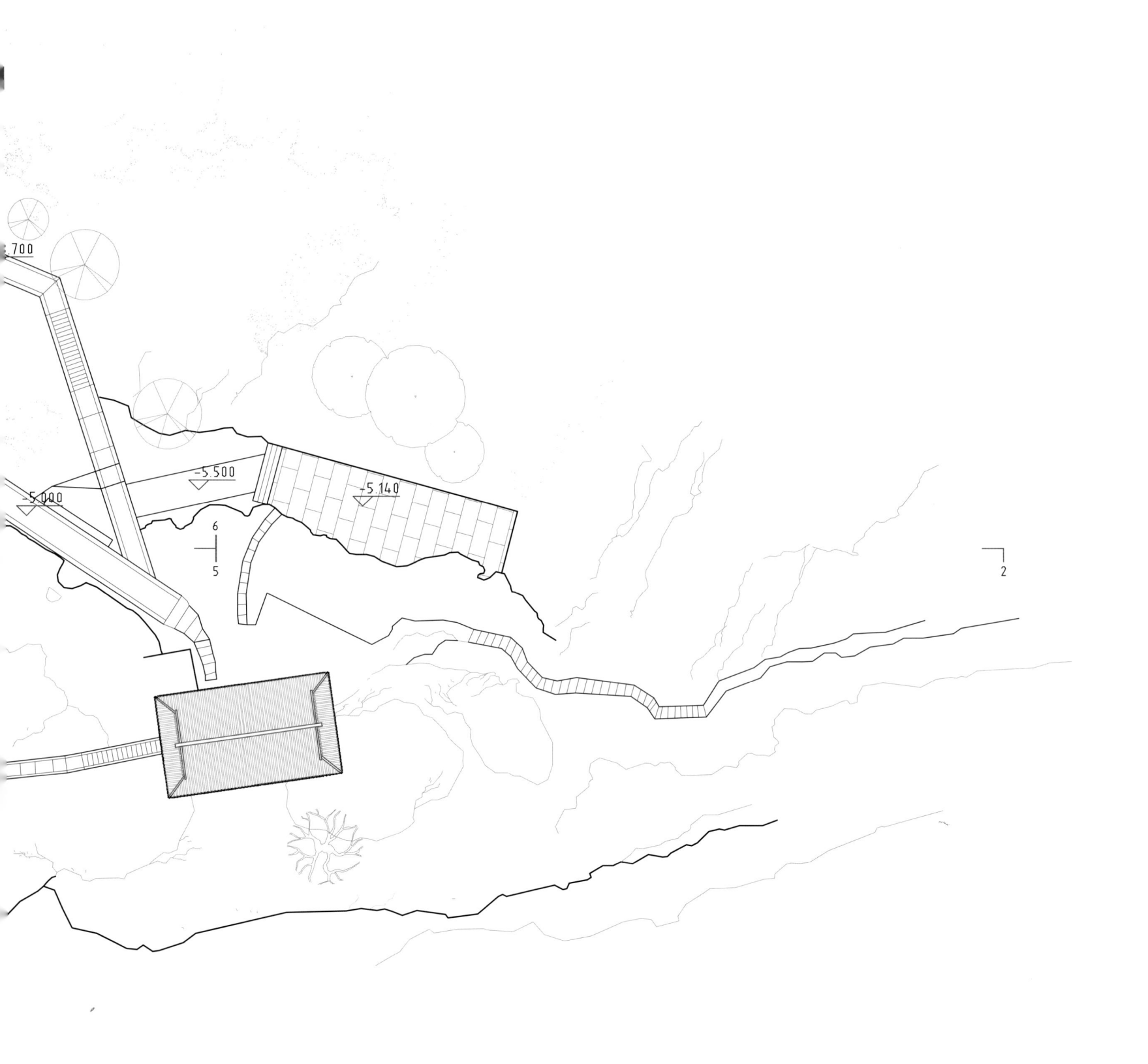

翠云宫总平面图
Site plan of Cuiyun Palace

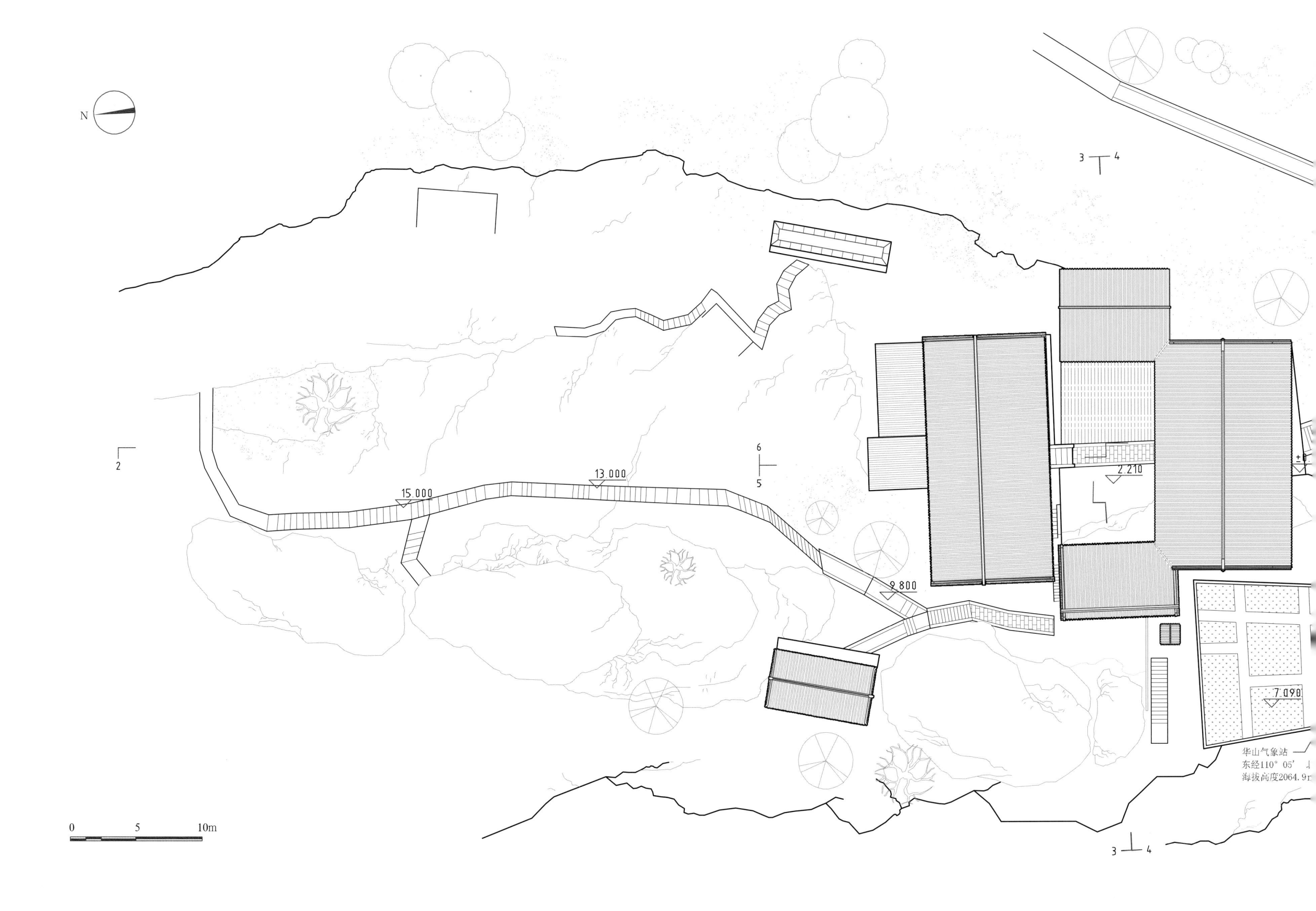

N

3ㄒ4

2

6
5

15 000

13 000

9 800

2 210

7.090

3丄4

3丄4

华山气象站
东经110°05′
海拔高度2064.9m

0   5   10m

翠云宫东侧总立面图
East site elevation of Cuiyun Palace

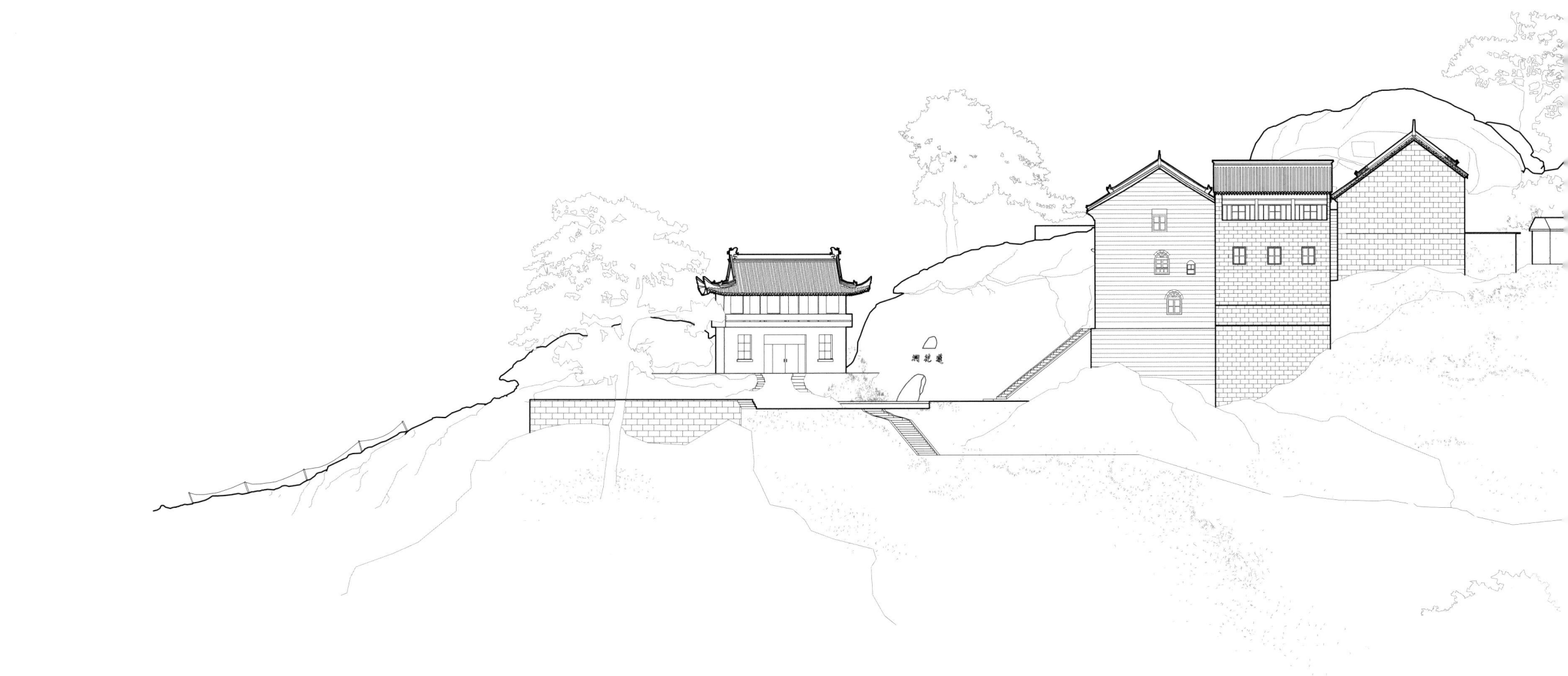

洞乾遁

0   5   10m

翠云宫南侧总立面图

South site elevation of Cuiyun Palace

0

5

10m

中国古建筑测绘大系·宗教建筑——昆明真庆观与太和宫

034

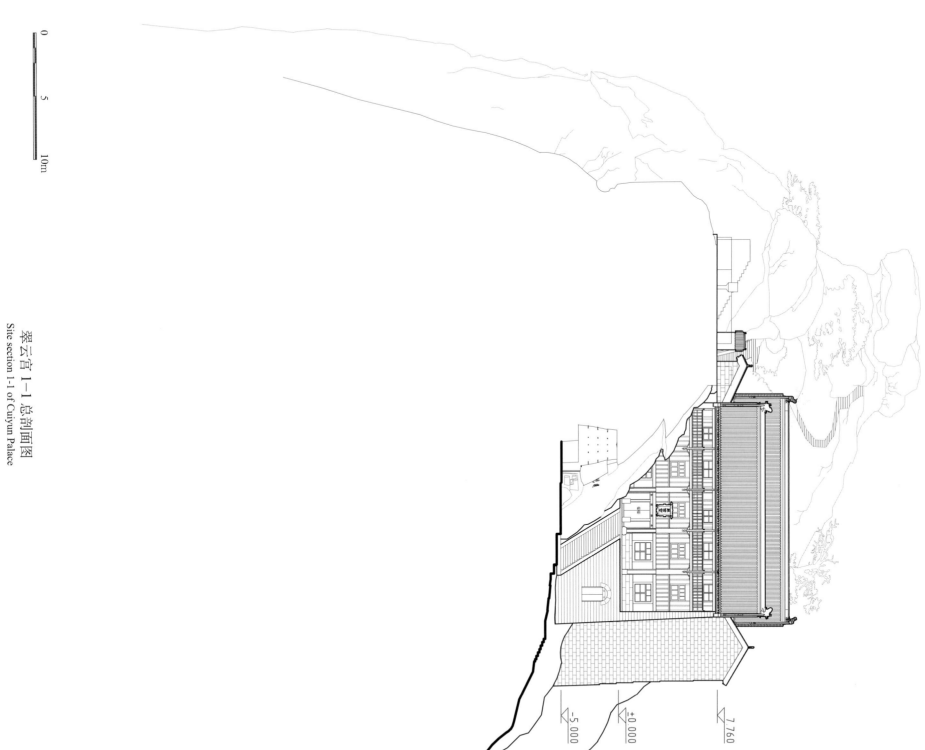

翠云宫 1—1 总剖面图
Site section 1-1 of Cuiyun Palace

0
5
10m

-5.000
±0.000
7.760

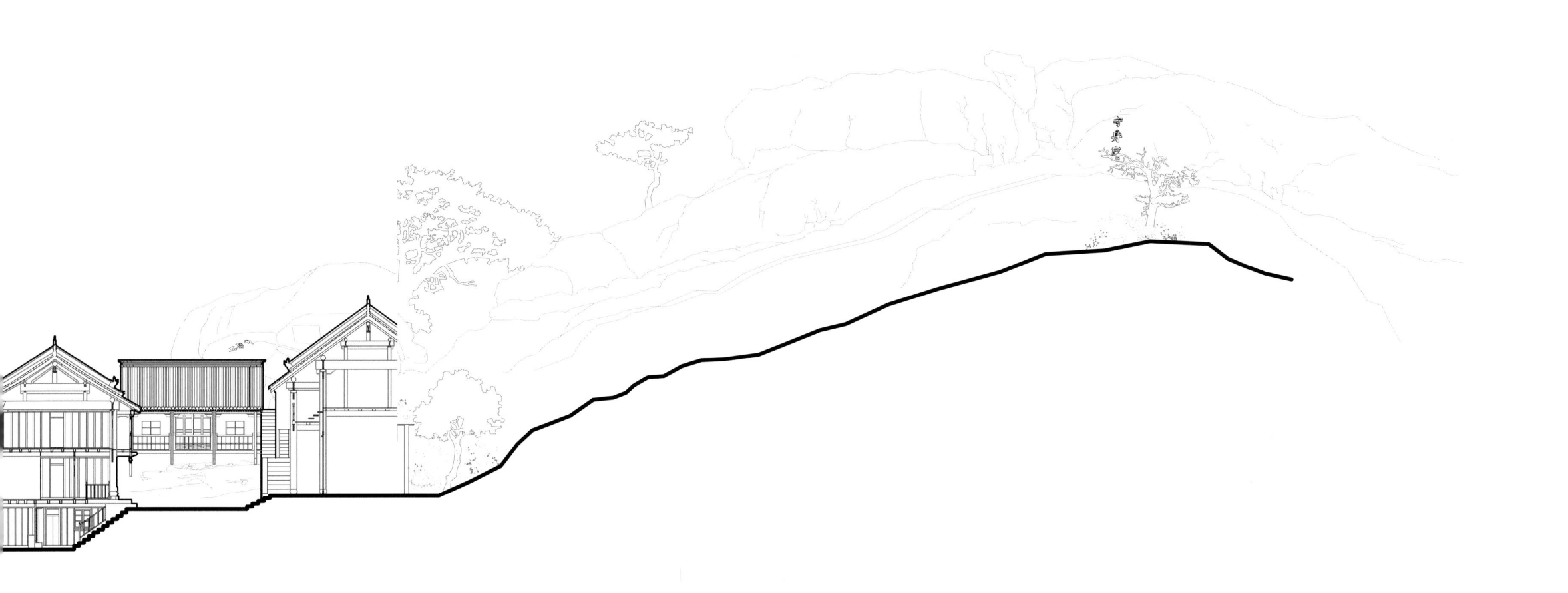

翠云宫 2-2 总剖面图
Site section 2-2 of Cuiyun Palace

27.640

17.770

2.210

±0.000

-5.000

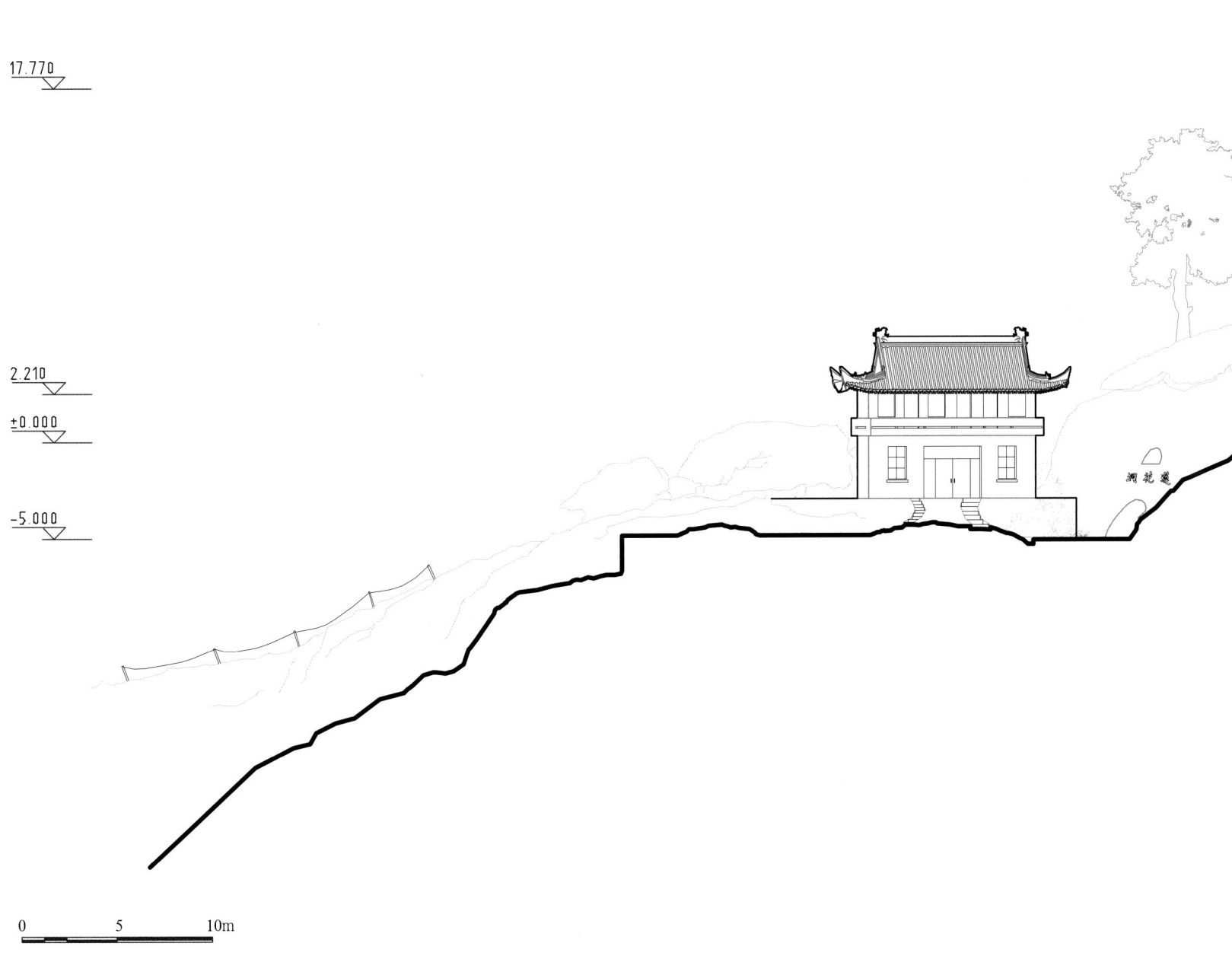

0   5   10m

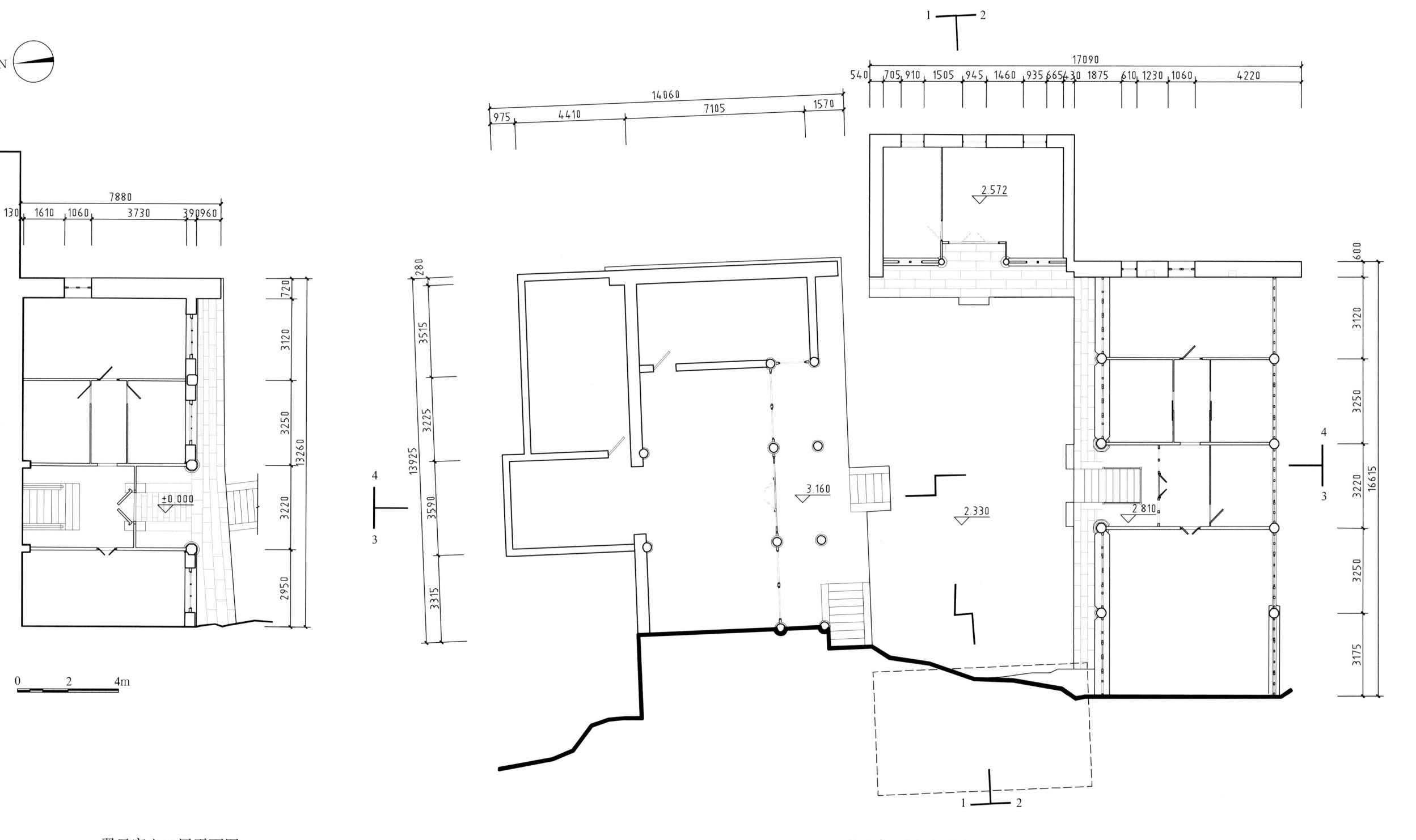

翠云宫入口层平面图
Plan of Cuiyun Palace's entrance level

翠云宫一层平面图
Plan of first floor of Cuiyun Palace

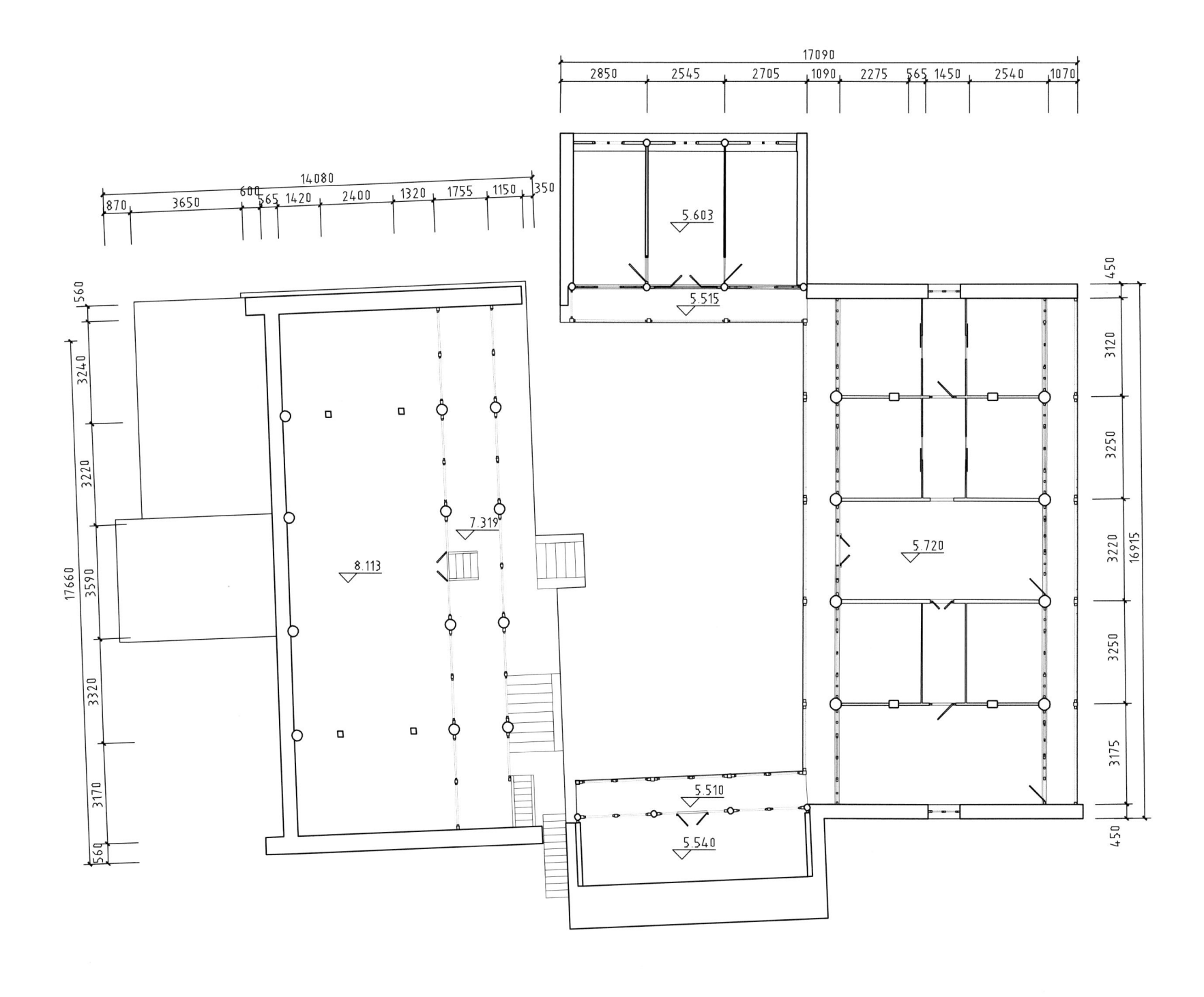

17090

2850　2545　2705　1090　2275　565　1450　2540　1070

450

3120

16915

3250

3220

3250

3175

450

5.603

5.515

5.720

14080

870　3650　600　565　1420　2400　1320　1755　1150　350

560

3240

3220

17660

3590

3320

3170

560

8.113

7.319

5.510

5.540

翠云宫二层平面图
Plan of second floor of Cuiyun Palace

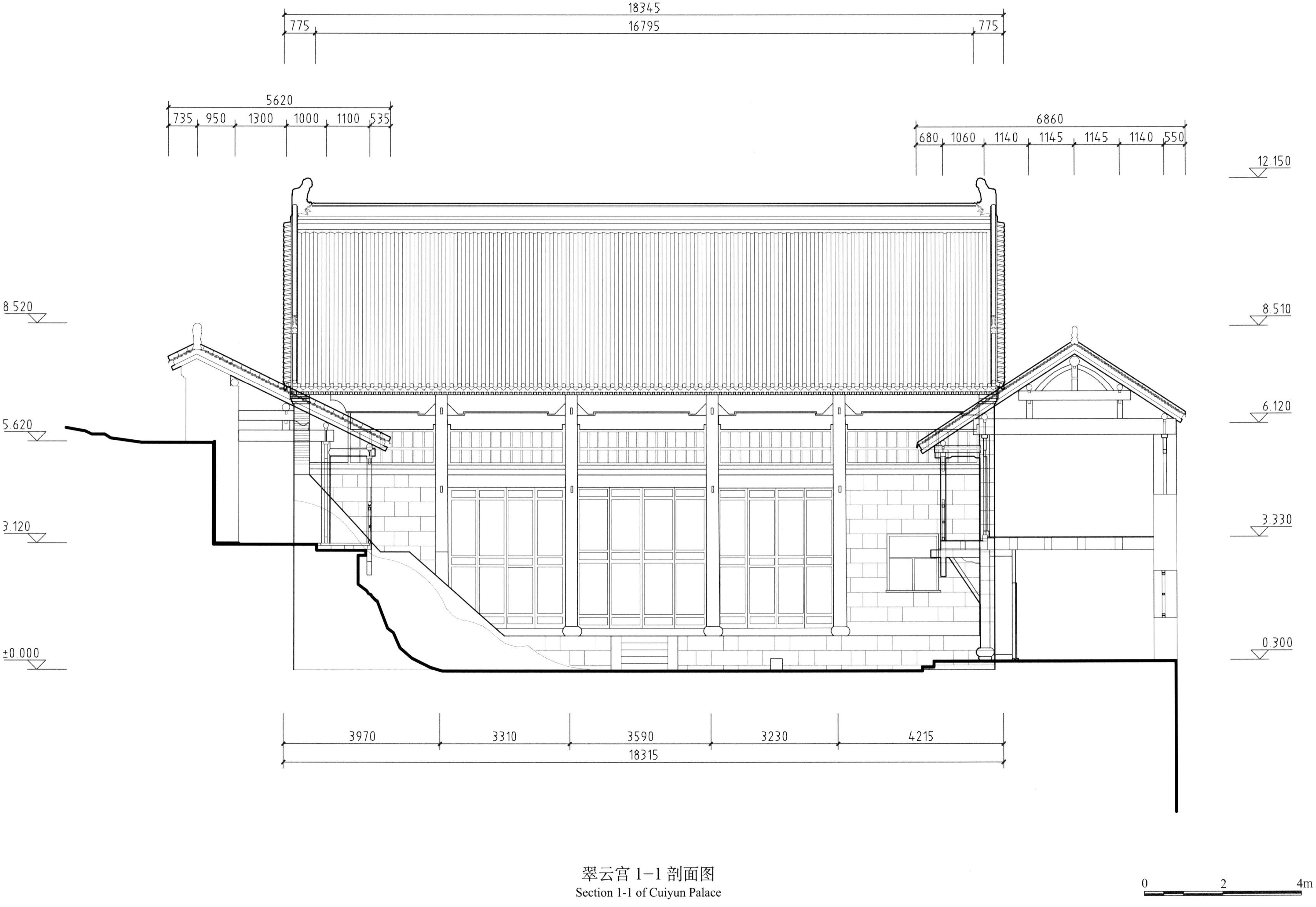

18345
775 16795 775

5620
735 950 1300 1000 1100 535

6860
680 1060 1140 1145 1145 1140 550

12.150

8.520

8.510

6.120

5.620

3.120

3.330

0.300

±0.000

3970 3310 3590 3230 4215
18315

翠云宫 1-1 剖面图
Section 1-1 of Cuiyun Palace

0 2 4m

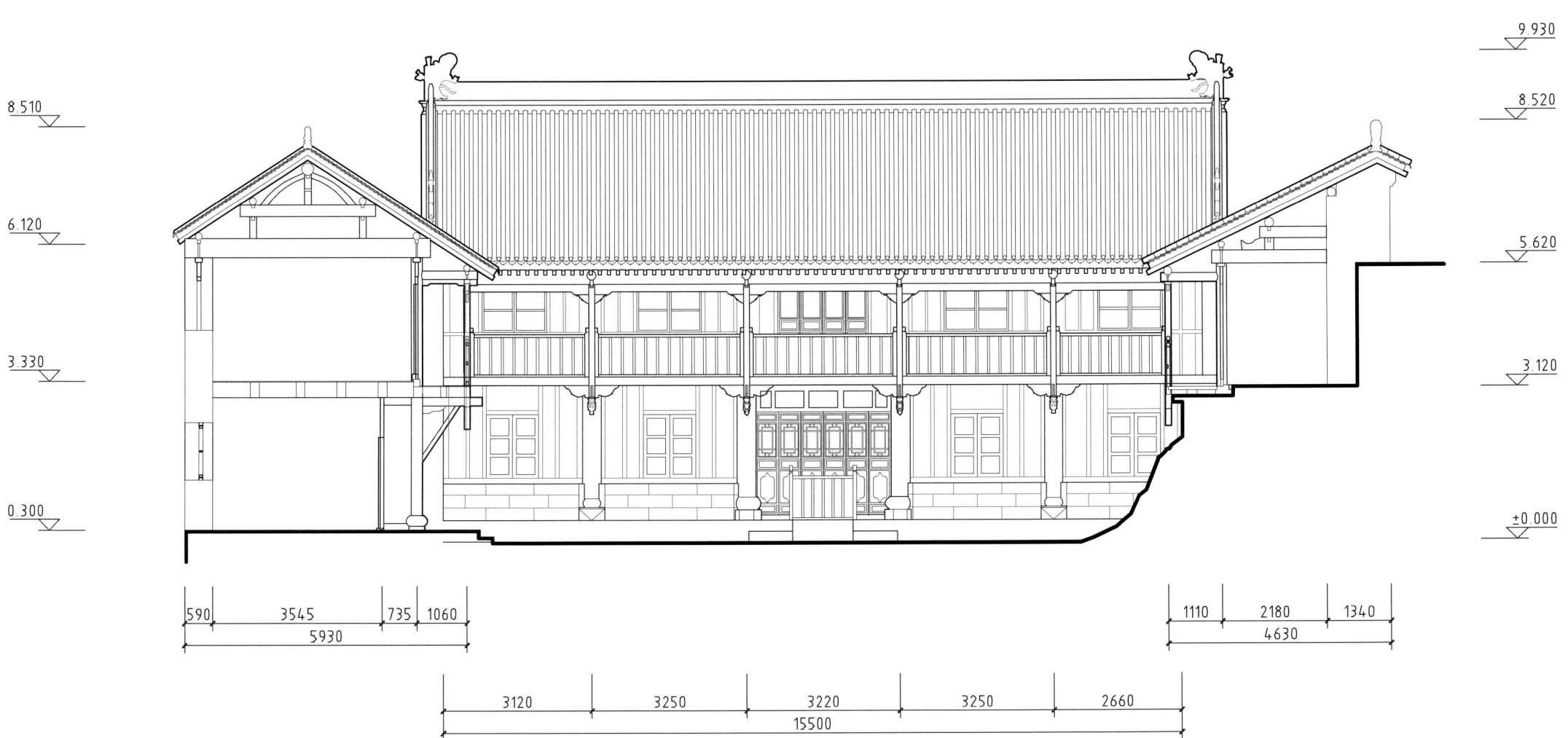

12.150

9.930

8.510

8.520

6.120

5.620

3.330

3.120

0.300

±0.000

590 3545 735 1060
5930

1110 2180 1340
4630

3120 3250 3220 3250 2660
15500

翠云宫 2-2 剖面图
Section 2-2 of Cuiyun Palace

0   2   4m

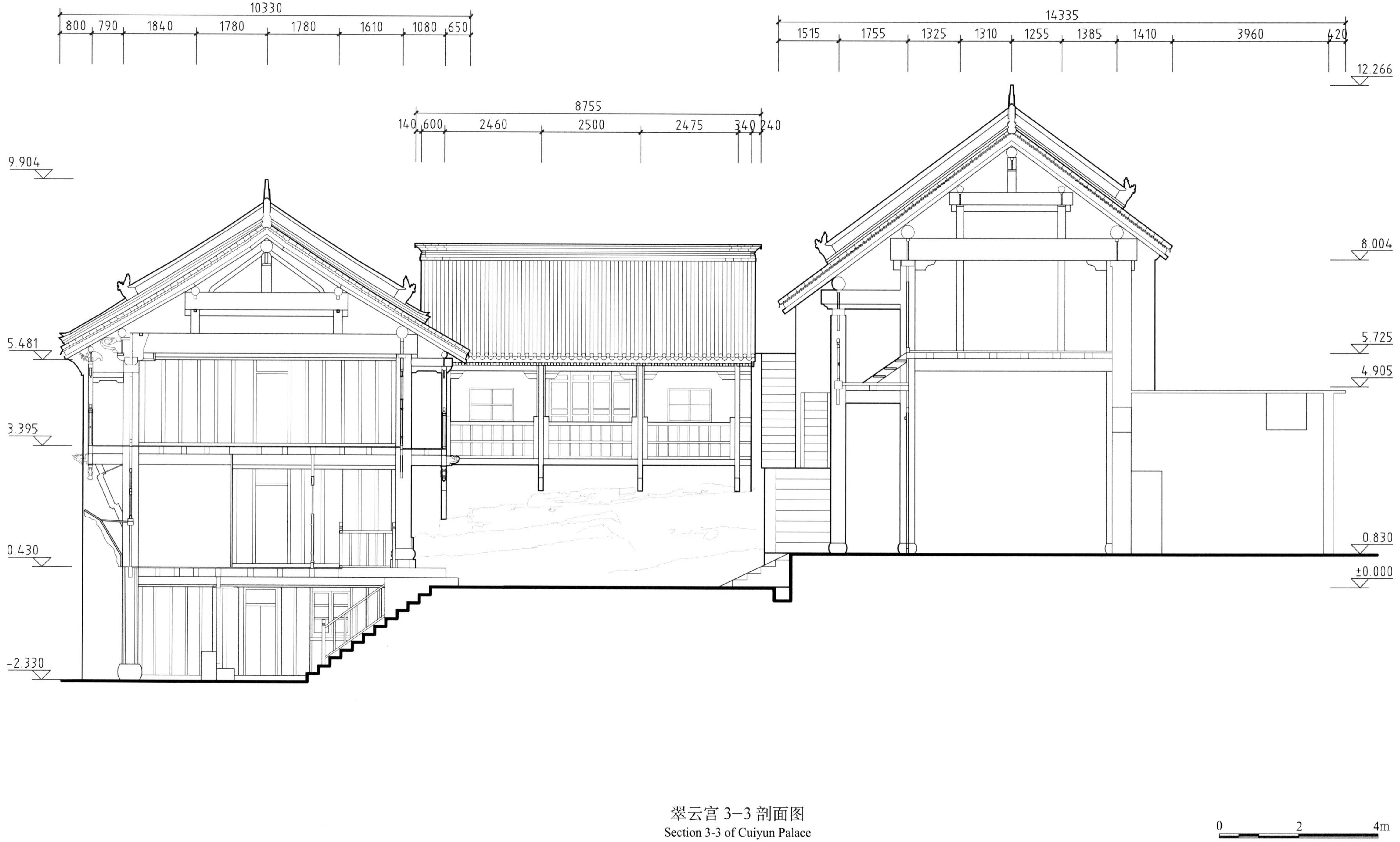

10330

| 800 | 790 | 1840 | 1780 | 1780 | 1610 | 1080 | 650 |

14335

| 1515 | 1755 | 1325 | 1310 | 1255 | 1385 | 1410 | 3960 | 420 |

12.266

8755

| 140 | 600 | 2460 | 2500 | 2475 | 340 | 240 |

9.904

8.004

5.725

5.481

4.905

3.395

0.830

±0.000

0.430

-2.330

翠云宫 3-3 剖面图
Section 3-3 of Cuiyun Palace

0   2   4m

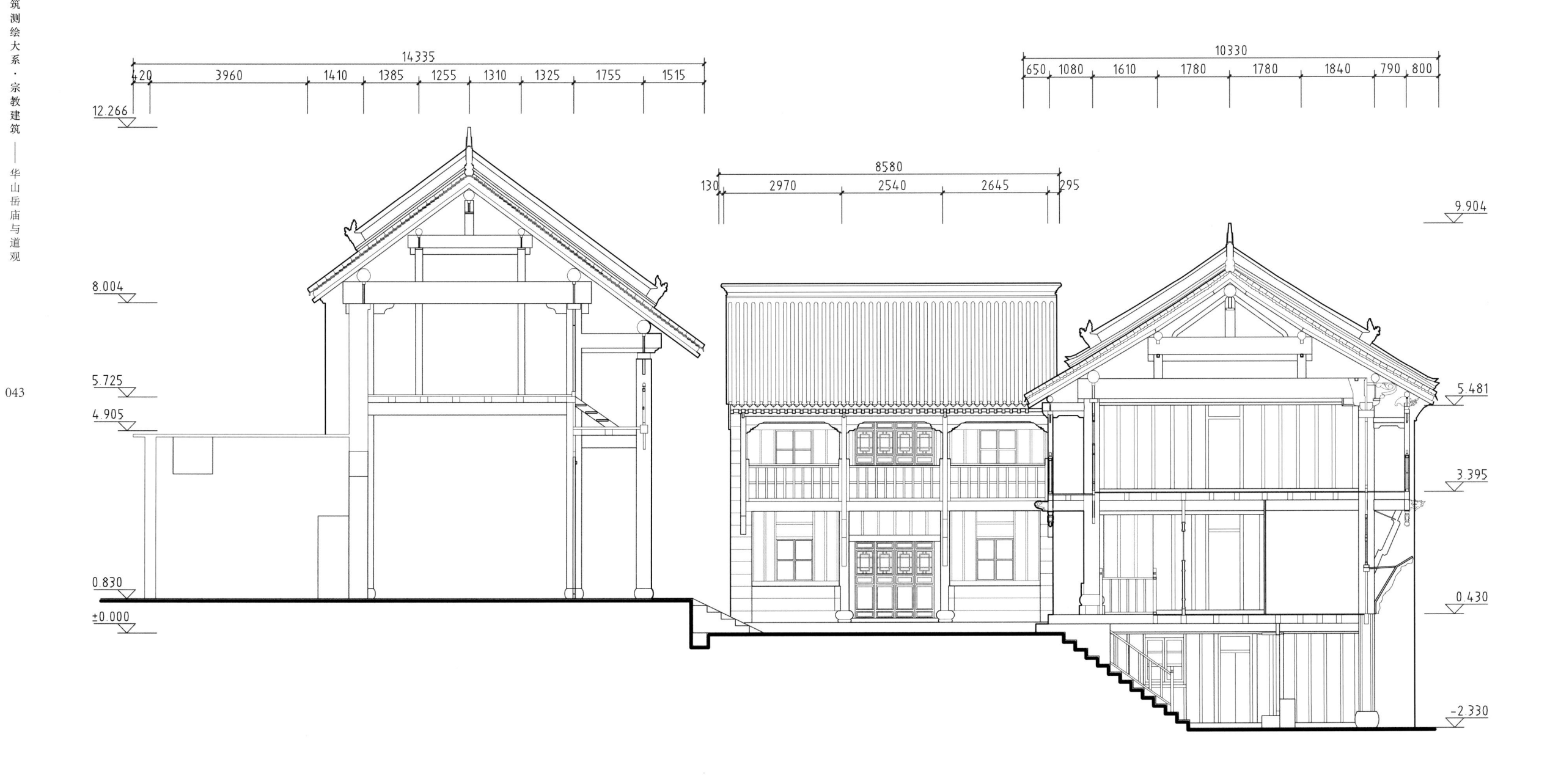

翠云宫 4-4 剖面图
Section 4-4 of Cuiyun Palace

0   2   4m

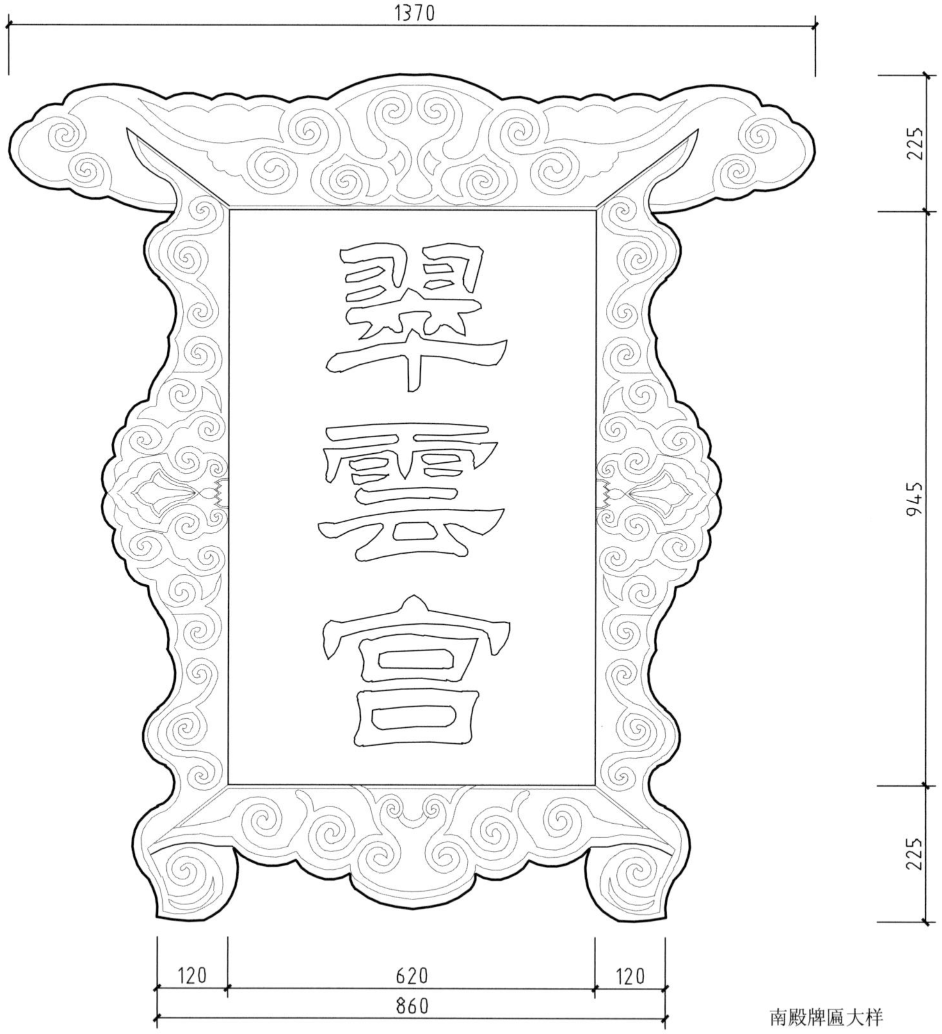

1370

225

945

225

翠雲宮

120 620 120

860

南殿牌匾大样

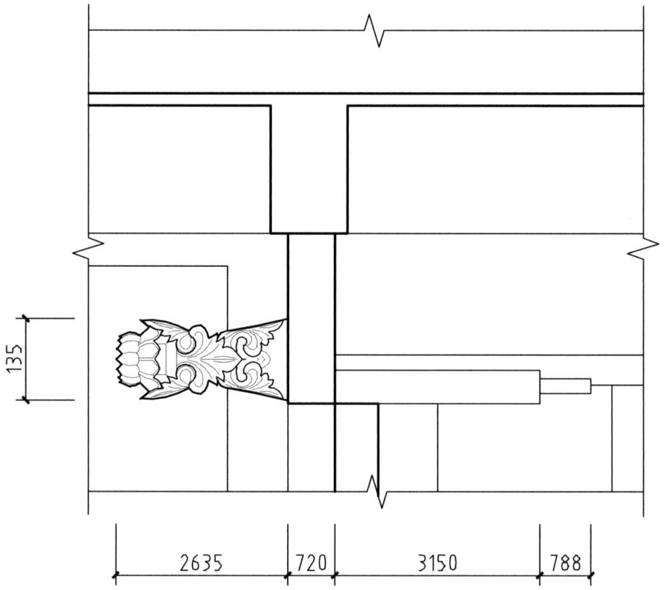

135

2635 720 3150 788

南殿门头门簪大样一

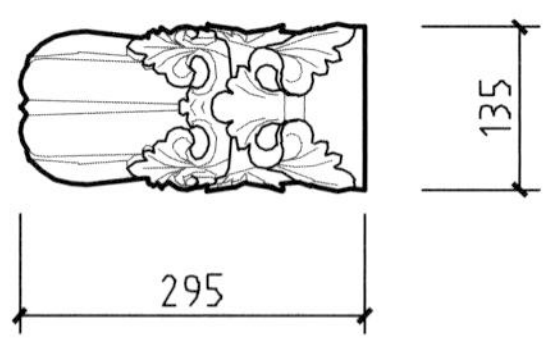

135

295

南殿门头门簪大样二

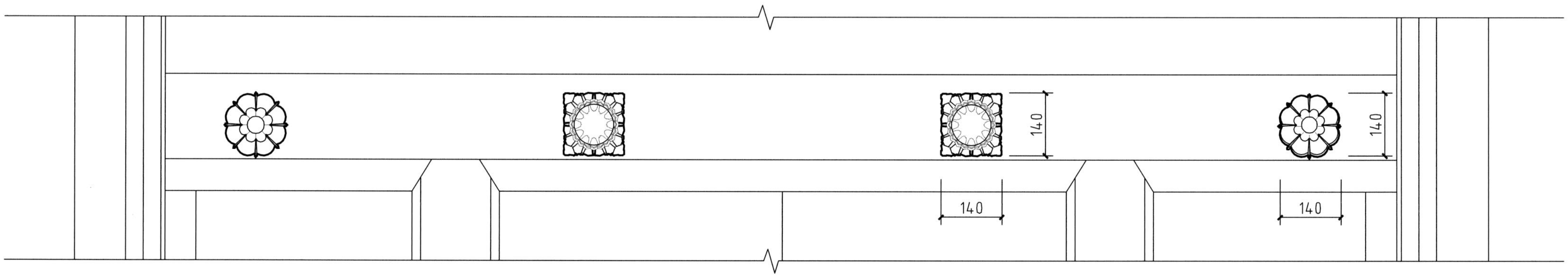

140

140

140

140

南殿门头门簪大样三

翠云宫大样图（一）

Detail of Cuiyun Palace (1)

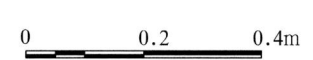

0 0.2 0.4m

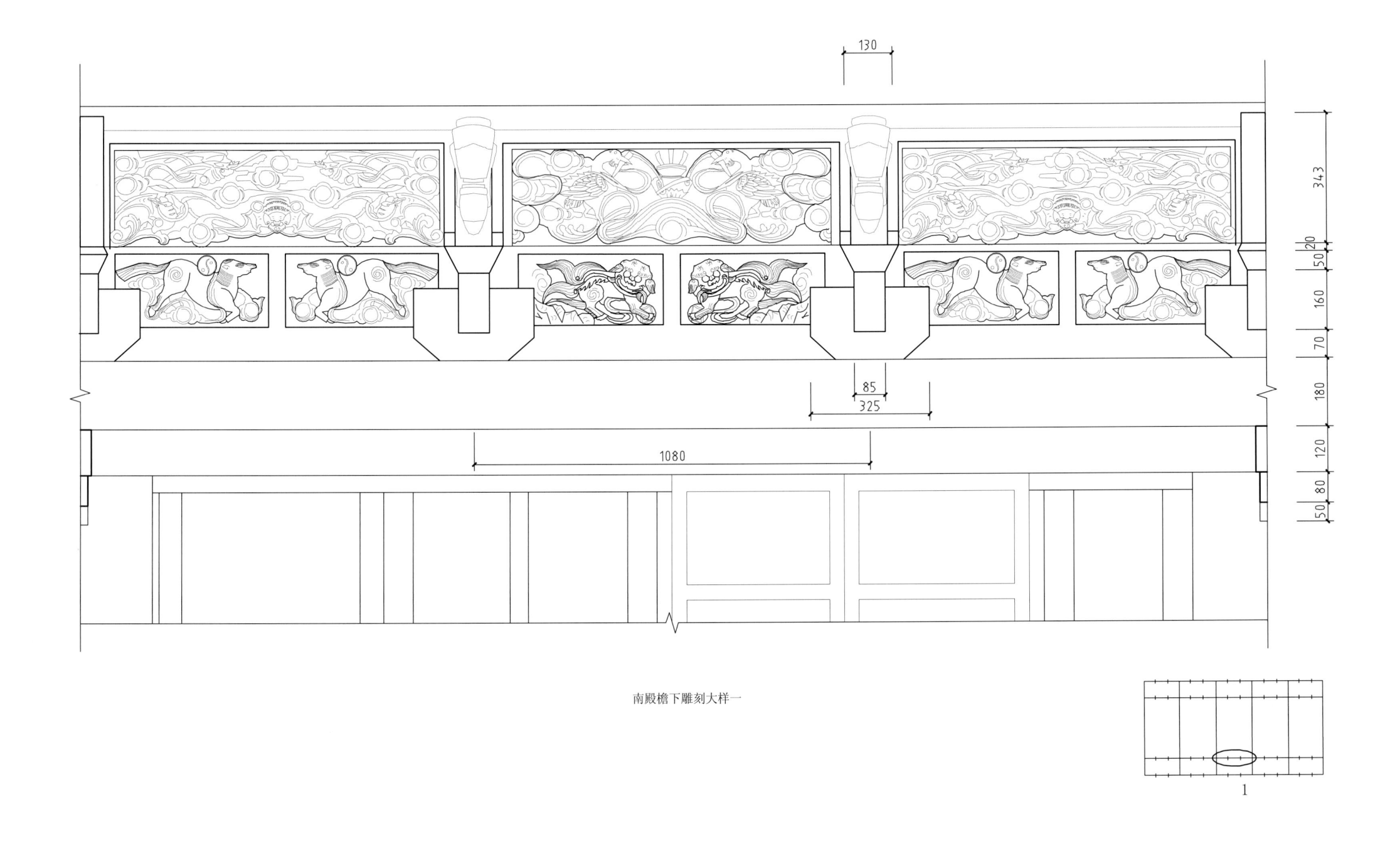

南殿檐下雕刻大样一

翠云宫大样图（二）
Detail of Cuiyun Palace (2)

0   0.2   0.4m

245

250 | 125

南殿檐下雕刻大样三

365

470

南殿檐下雕刻大样四

530

66

80

195

180 | 460

南殿檐下雕刻大样二

220

70

305

70

120

175

360

1099

2200

630

810

2070

30 | 50

240 | 45 | 30 | 290

南殿檐下雕刻大样五

0   0.2   0.4m

翠云宫大样图（三）
Detail of Cuiyun Palace (3)

5 4
2   3

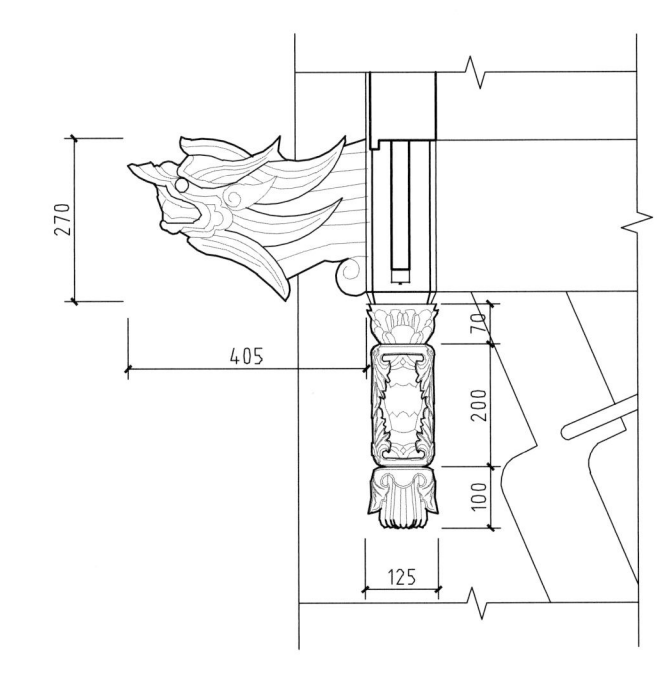

270

405

70

200

100

125

南殿檐下雕刻大样六

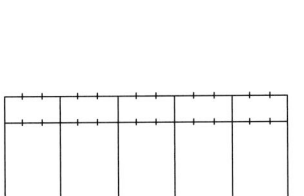
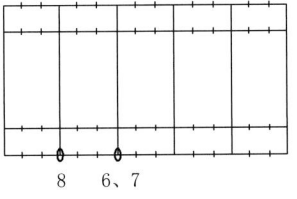

40

200

1080

5496

南殿檐下雕刻大样七

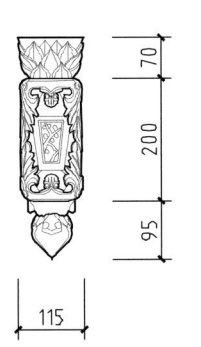

70

200

95

115

南殿檐下雕刻大样八

8　　6、7

翠云宫大样图（四）
Detail of Cuiyun Palace（4）

0　　　0.2　　　0.4m

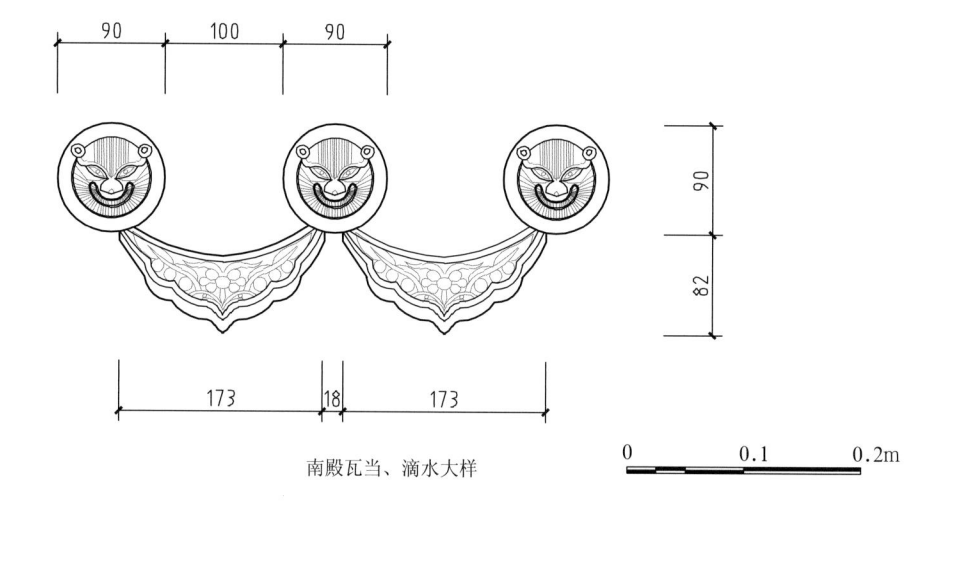

南殿瓦当、滴水大样

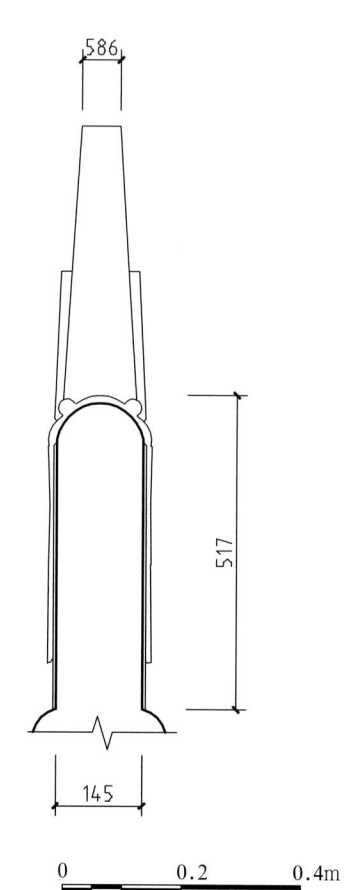

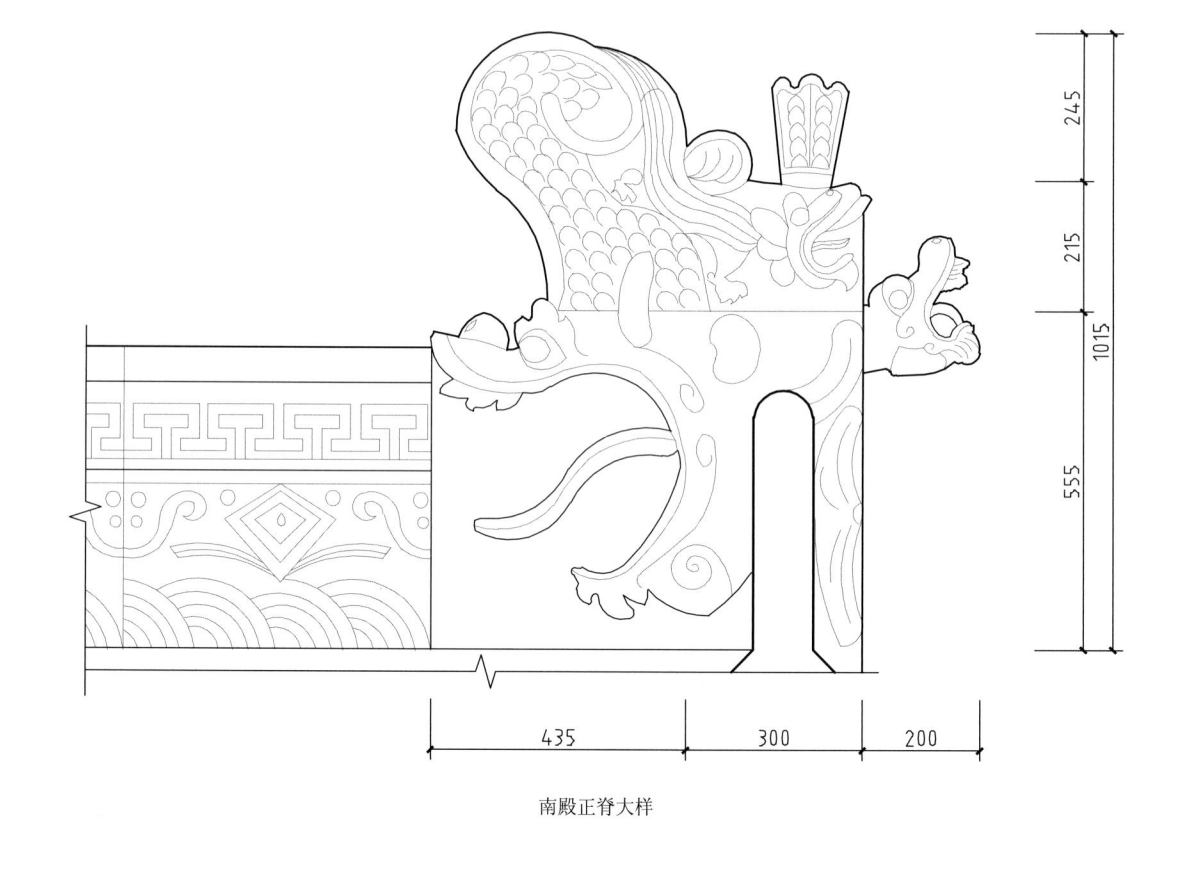

南殿正脊大样

翠云宫大样图（五）
Detail of Cuiyun Palace (5)

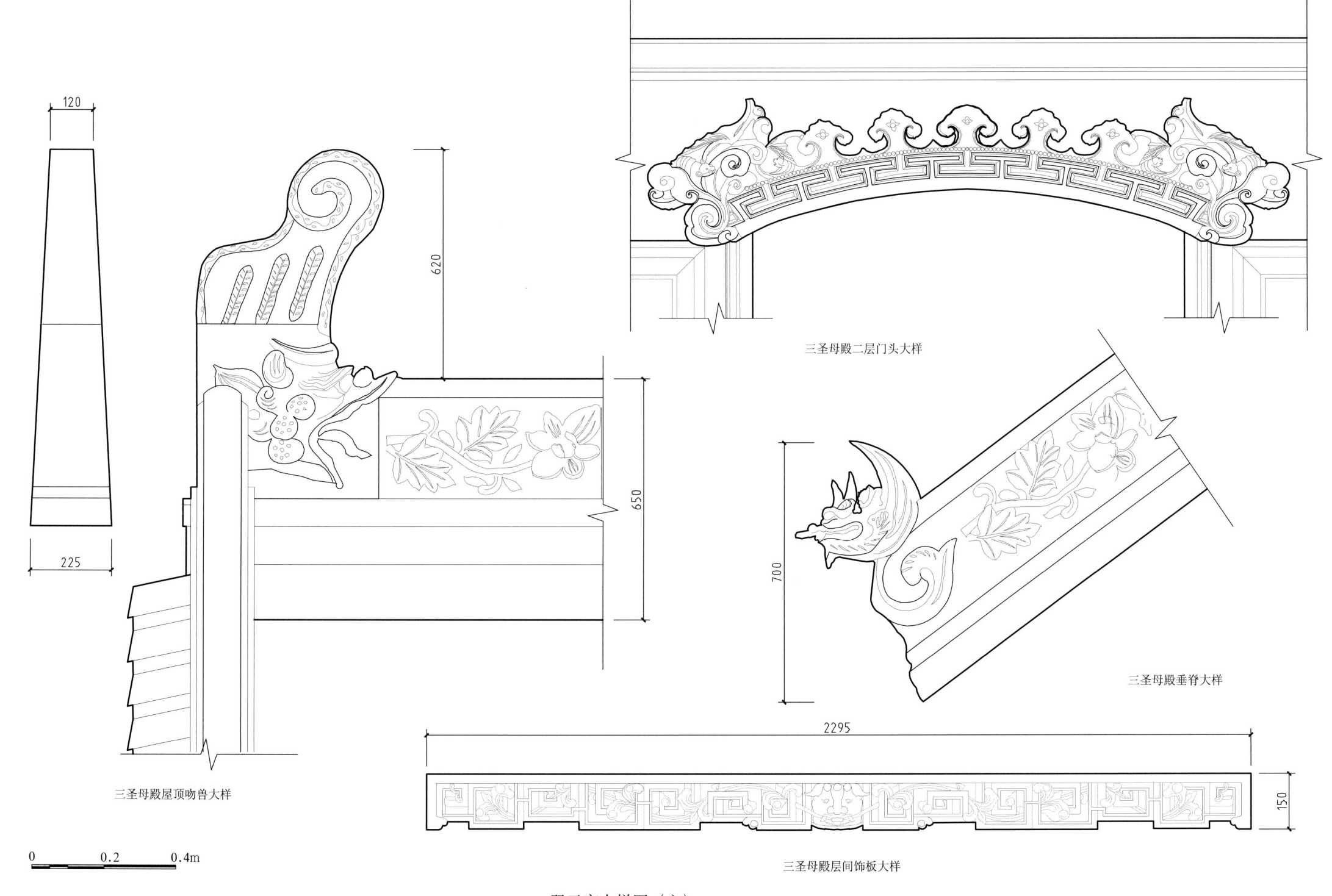

120

225

620

650

700

2295

150

0   0.2   0.4m

三圣母殿二层门头大样

三圣母殿垂脊大样

三圣母殿屋顶吻兽大样

三圣母殿层间饰板大样

翠云宫大样图（六）
Detail of Cuiyun Palace（6）

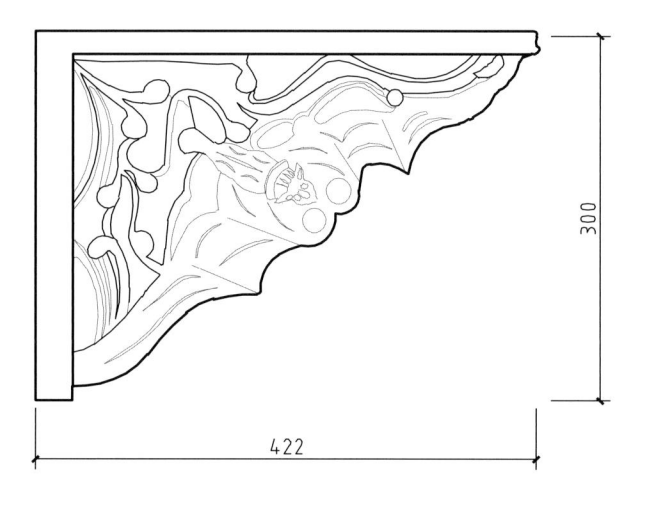

西殿雀替大样一

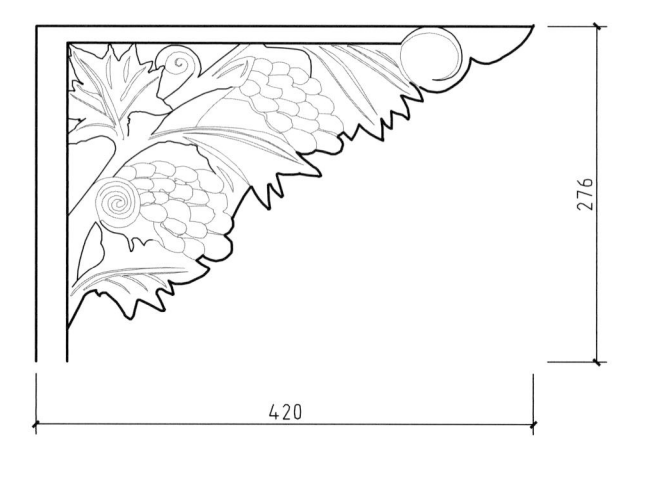

西殿雀替大样二

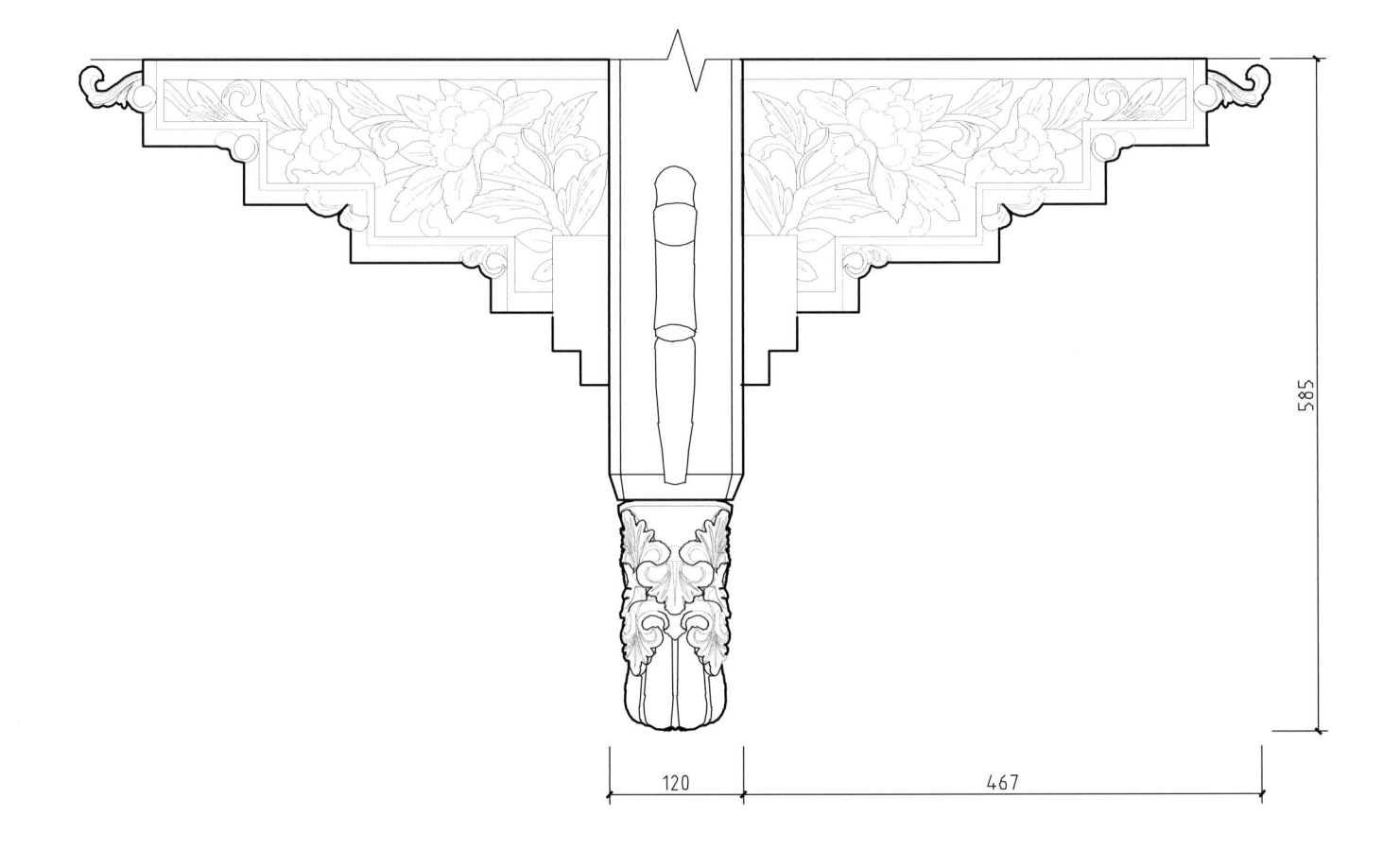

三圣母殿柱头雀替大样

翠云宫大样图（七）
Detail of Cuiyun Palace (7)

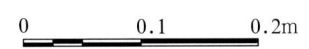

0　　0.1　　0.2m

大朝元洞
First Dawn Cave

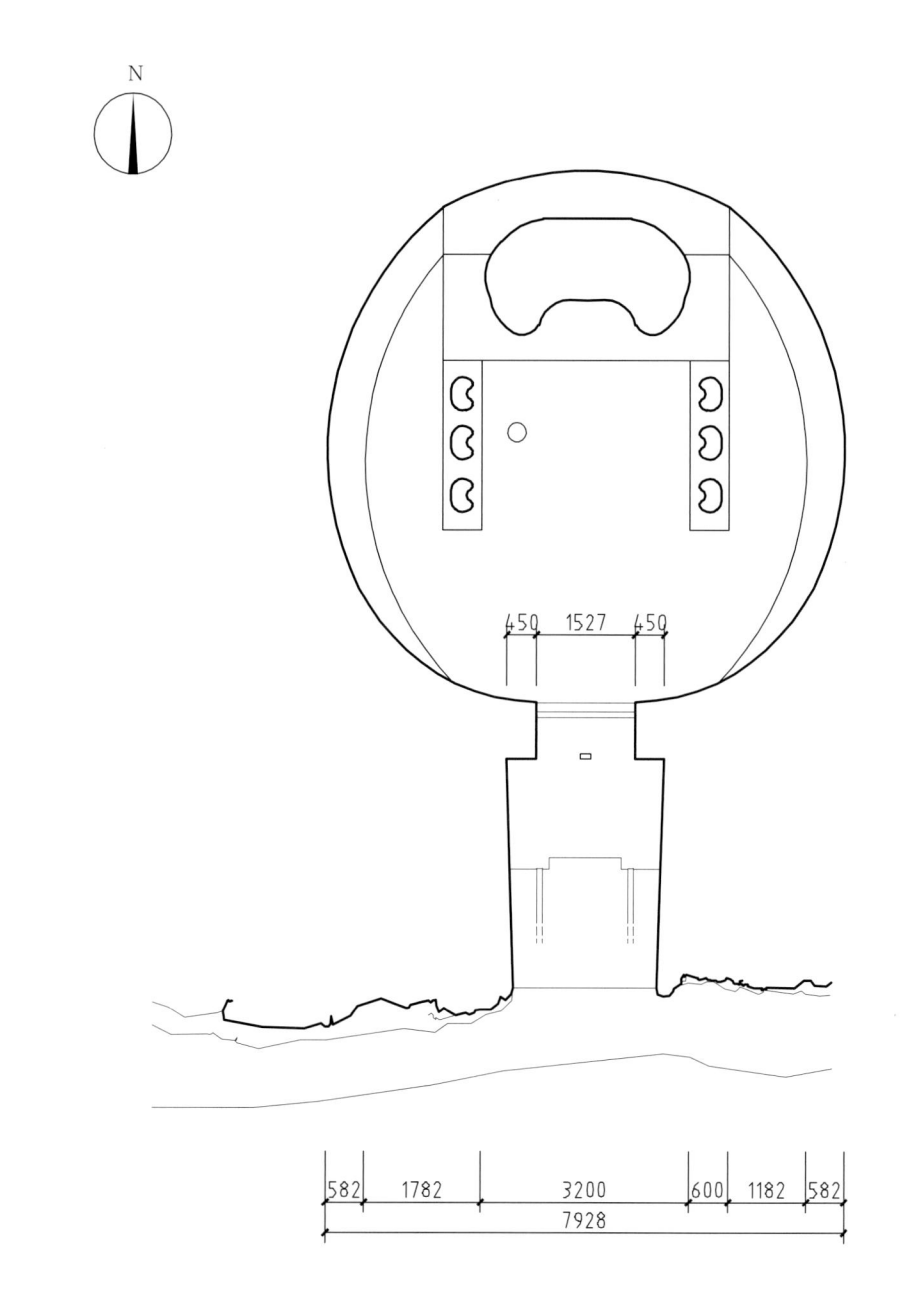

N

450 1527 450

582 1782 3200 600 1182 582
7928

大朝元洞平面图
Plan of First Dawn Cave

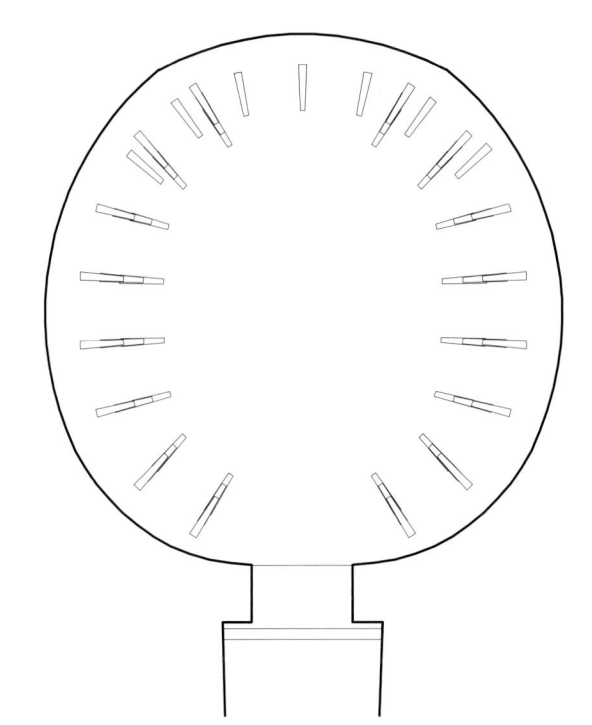

大朝元洞屋顶仰视图
Plan of roof of First Dawn Cave as seen from below

0 2 4m

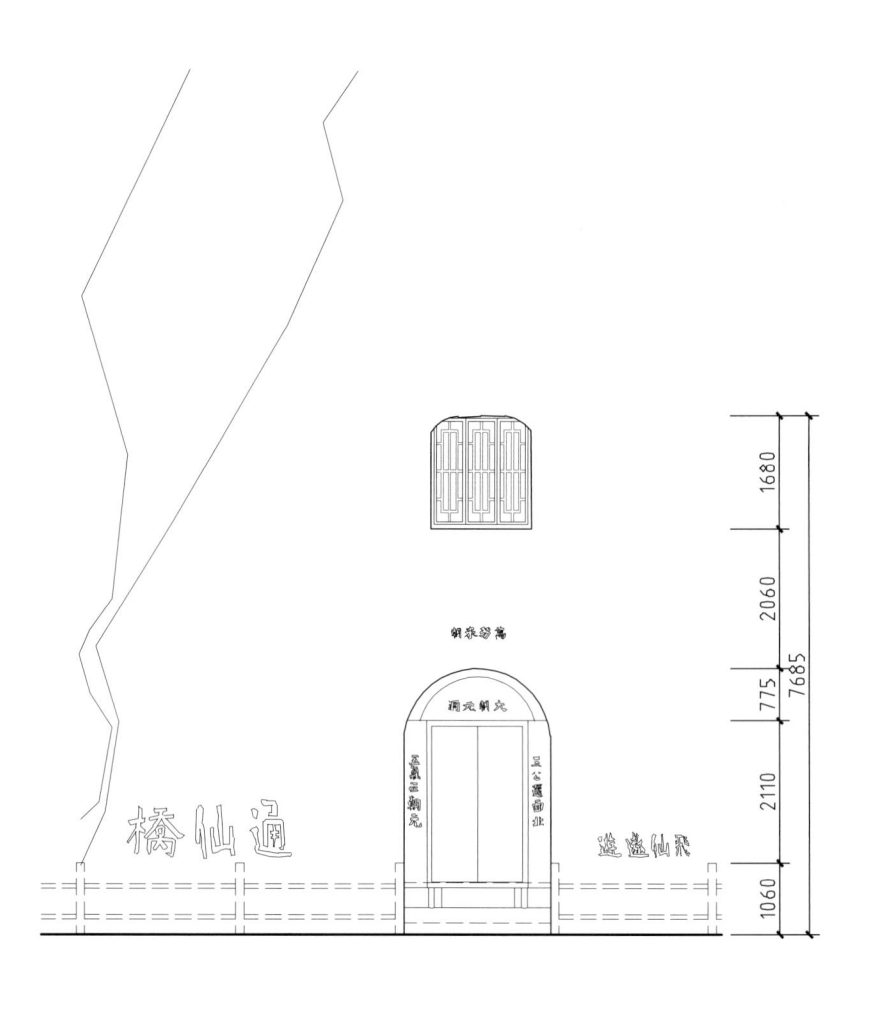

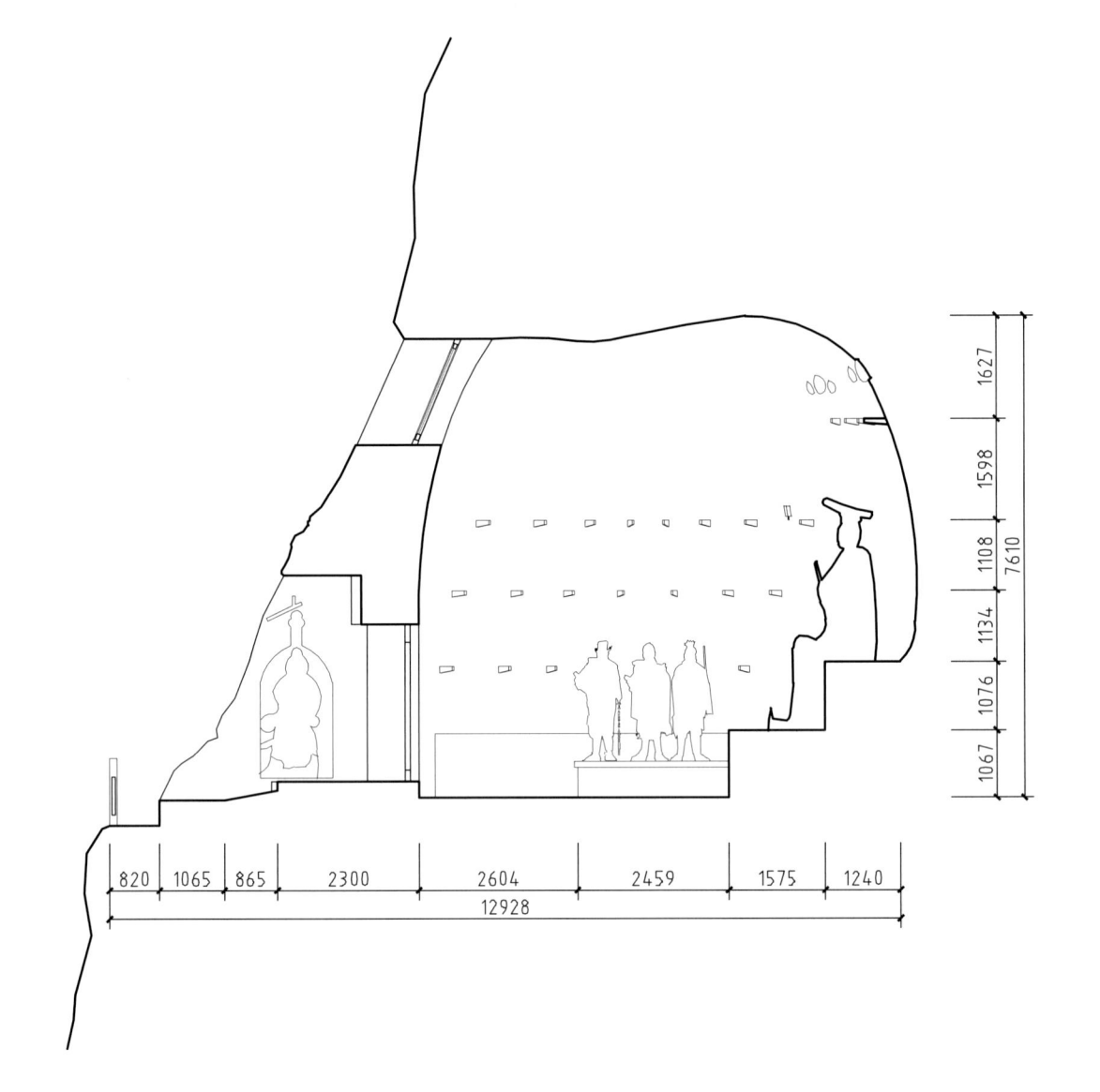

大朝元洞立面图
Elevation of First Dawn Cave

大朝元洞剖面图
Section of First Dawn Cave

0　　2　　4m

大朝元洞总剖面图
Site section of First Dawn Cave

中国古建筑测绘大系·宗教建筑——大理剑川石窟

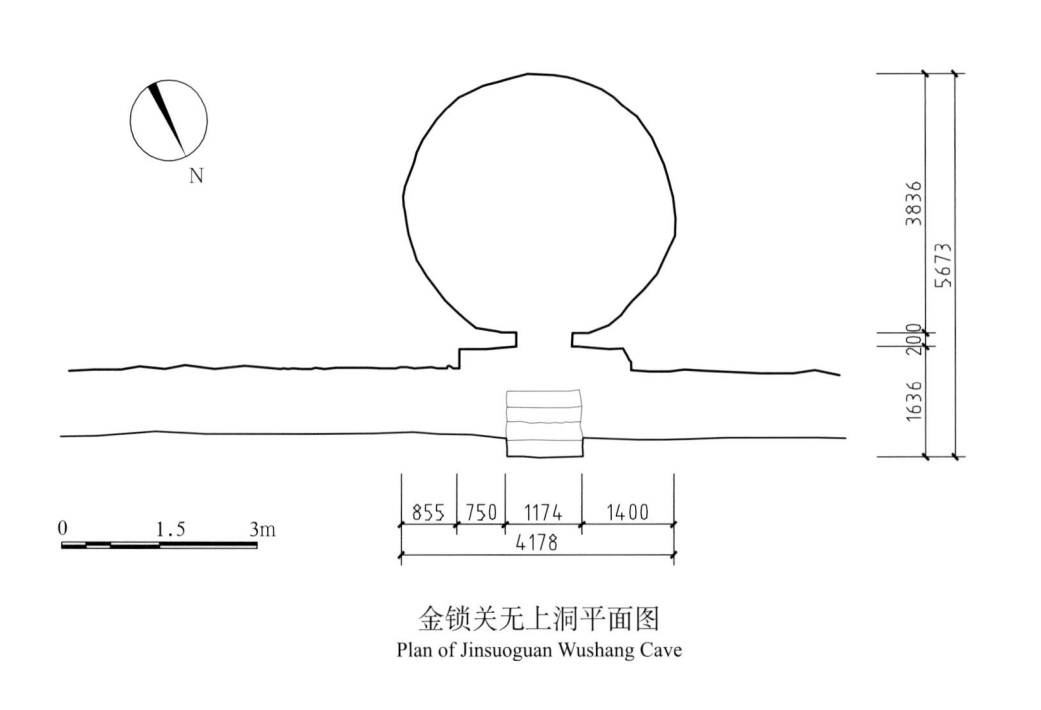

金锁关无上洞平面图
Plan of Jinsuoguan Wushang Cave

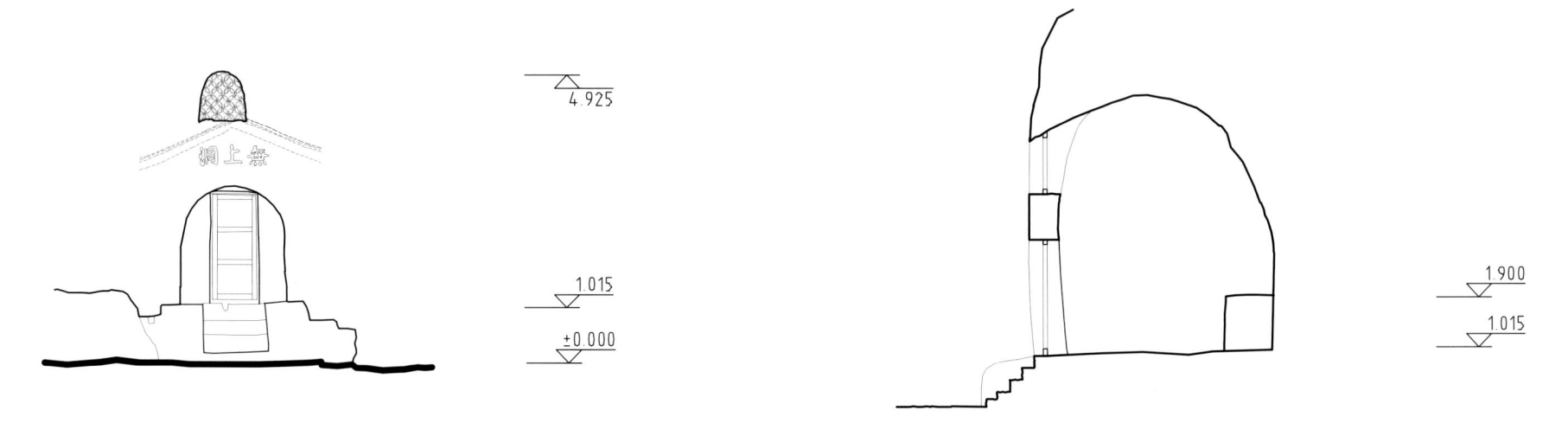

金锁关无上洞立面图
Elevation of Jinsuoguan Wushang Cave

金锁关无上洞剖面图
Section of Jinsuoguan Wushang Cave

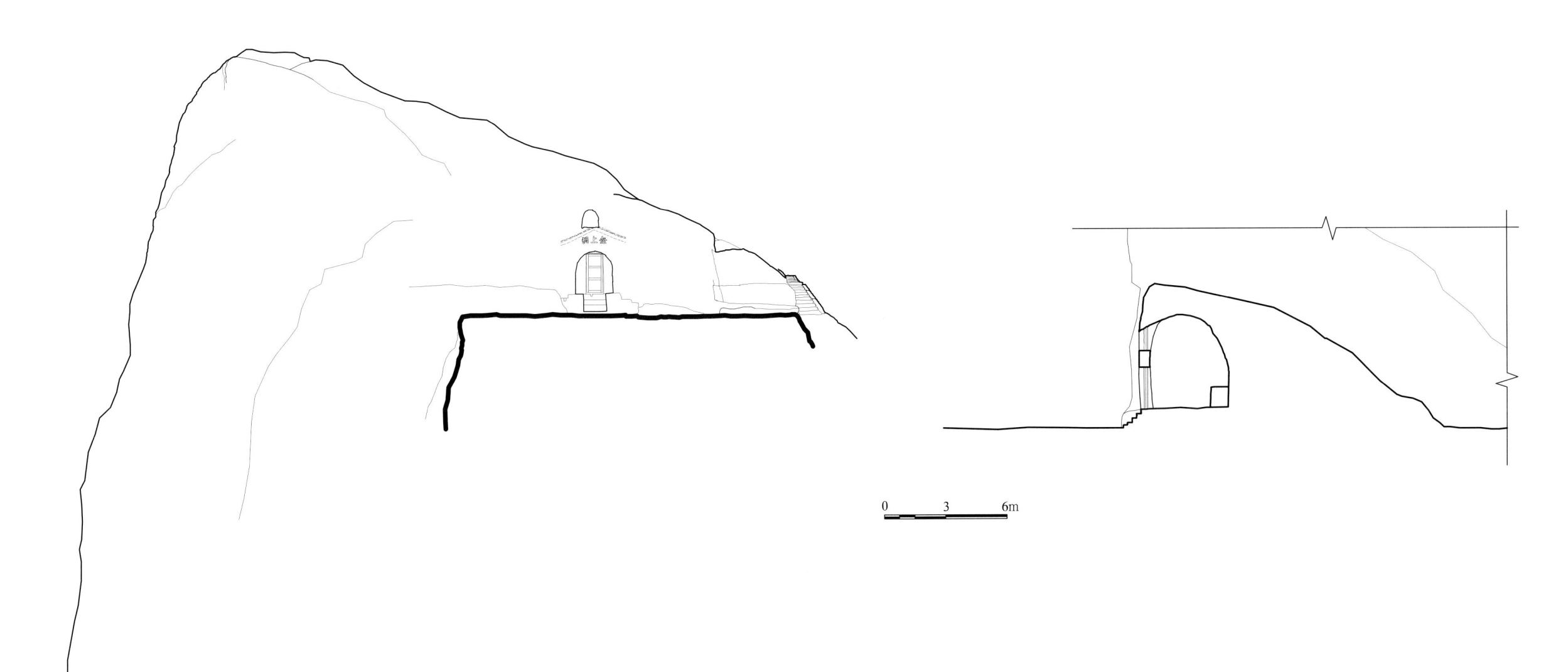

金锁关无上洞总立面图
Site elevation of Jinsuoguan Wushang Cave

金锁关无上洞总剖面图
Site section of Jinsuoguan Wushang Cave

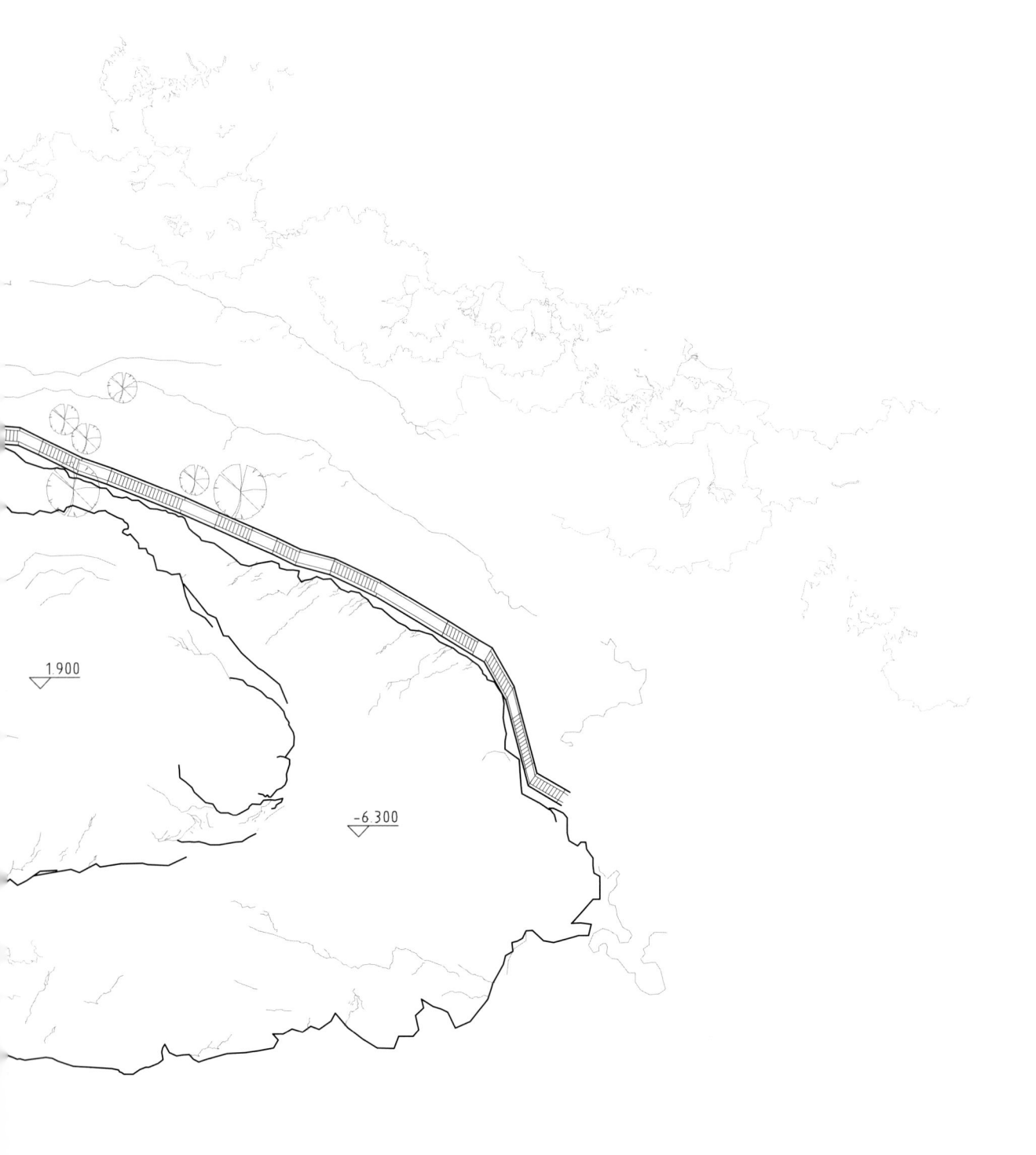

▽ 1 900

▽ -6 300

玉女宫总平面图
Site plan of Yu'nv Palace

N

±0.000

1.800

7.200

6.100

6.700

2.450

2.000

2.600

1

1

2

2

0   6   12m

0   8   16m

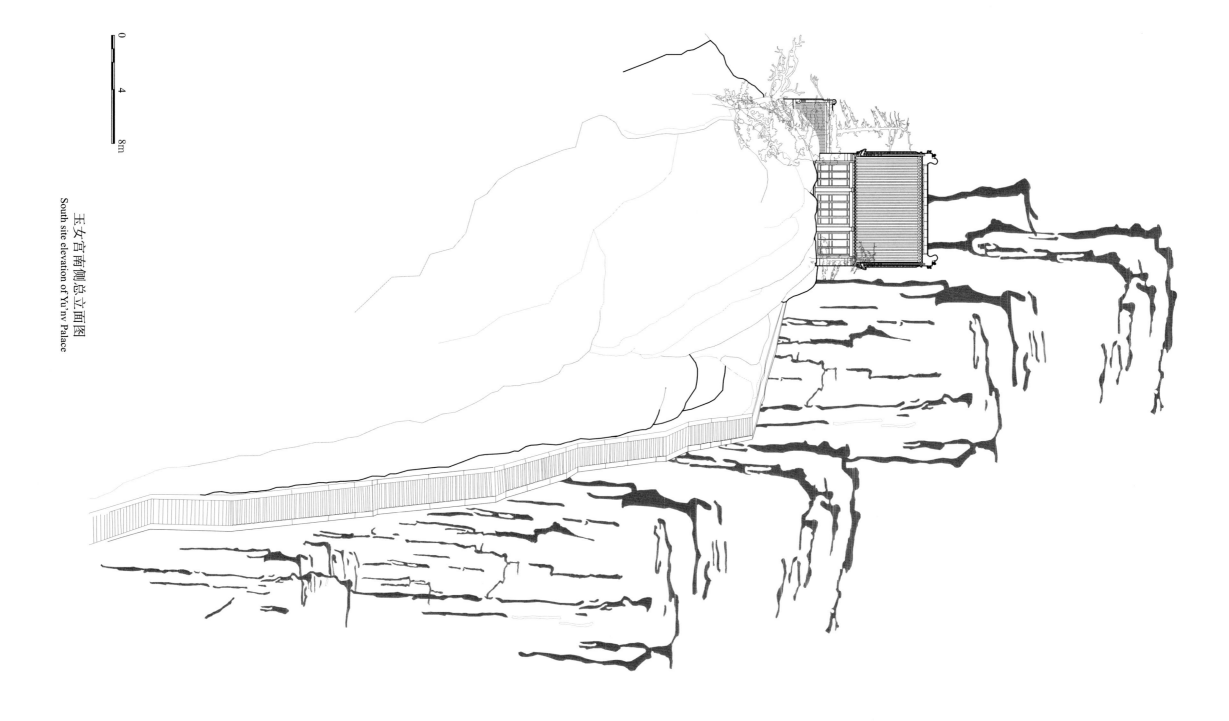

王女官南侧总立面图
South site elevation of Yü'nv Palace

0 4 8m

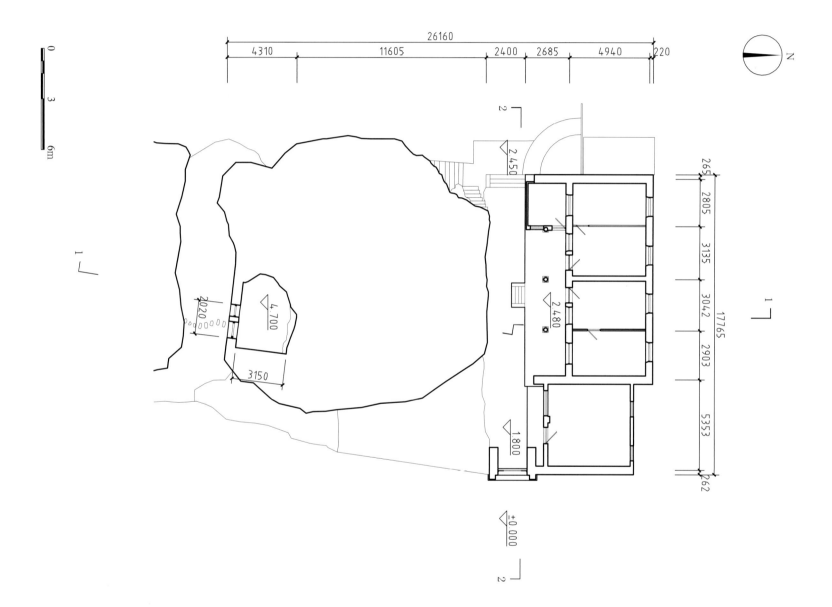

王女宫一层平面图
Plan of first floor of Yü'nv Palace

26160
4310　11605　2400　2685　4940　220

N

17765
265　2805　3135　3042　2903　5353　262

2.450
2.480
4.700
2020
3150
1.800
±0.000

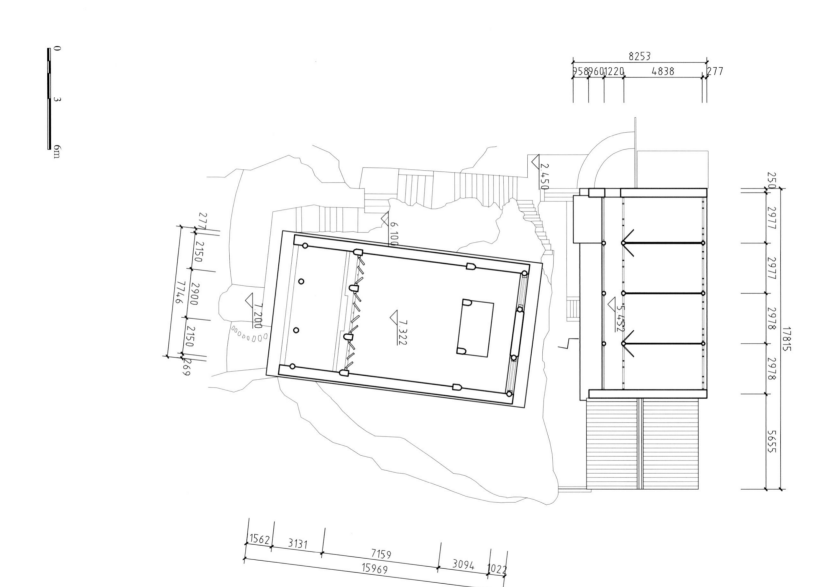

玉女宫二层平面图
Plan of second floor of Yu'nv Palace

中国早期建筑遗产·学术研究丛书——中国早期楼阁与楼阁式塔

8253
958 960 1220 4838 277

277
2150
7746 2900
2150
1269

250
2977
2977
17815
2978
2978
5655

2.450

6.100

7.200
7.322

5.452

0
3
6m

1562 3131 7159 3094 1022
15969

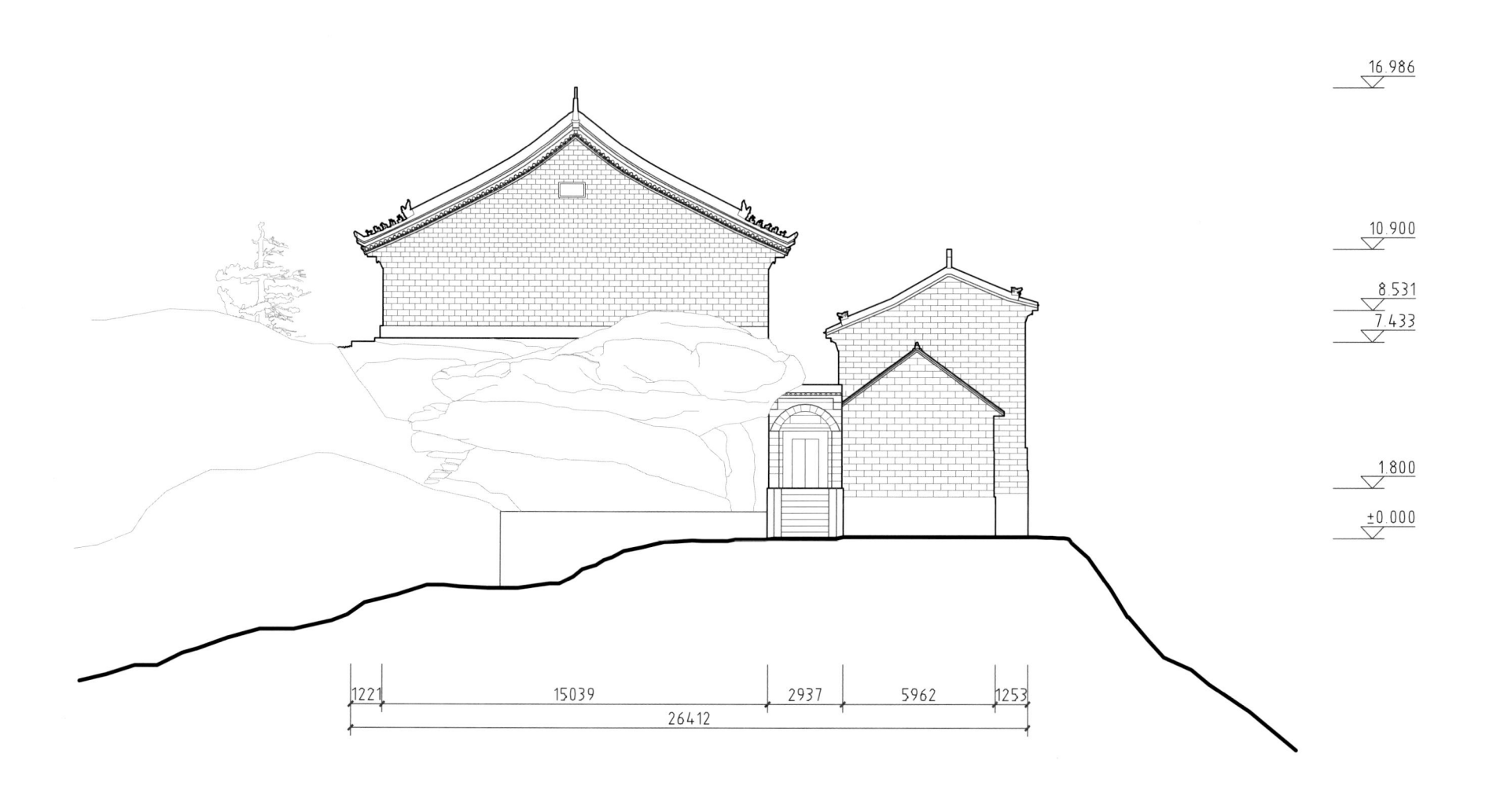

16.986

10.900

8.531
7.433

1.800

±0.000

1221　15039　2937　5962　1253

26412

玉女宫东立面图
East elevation of Yu'nv Palace

0　　3　　6m

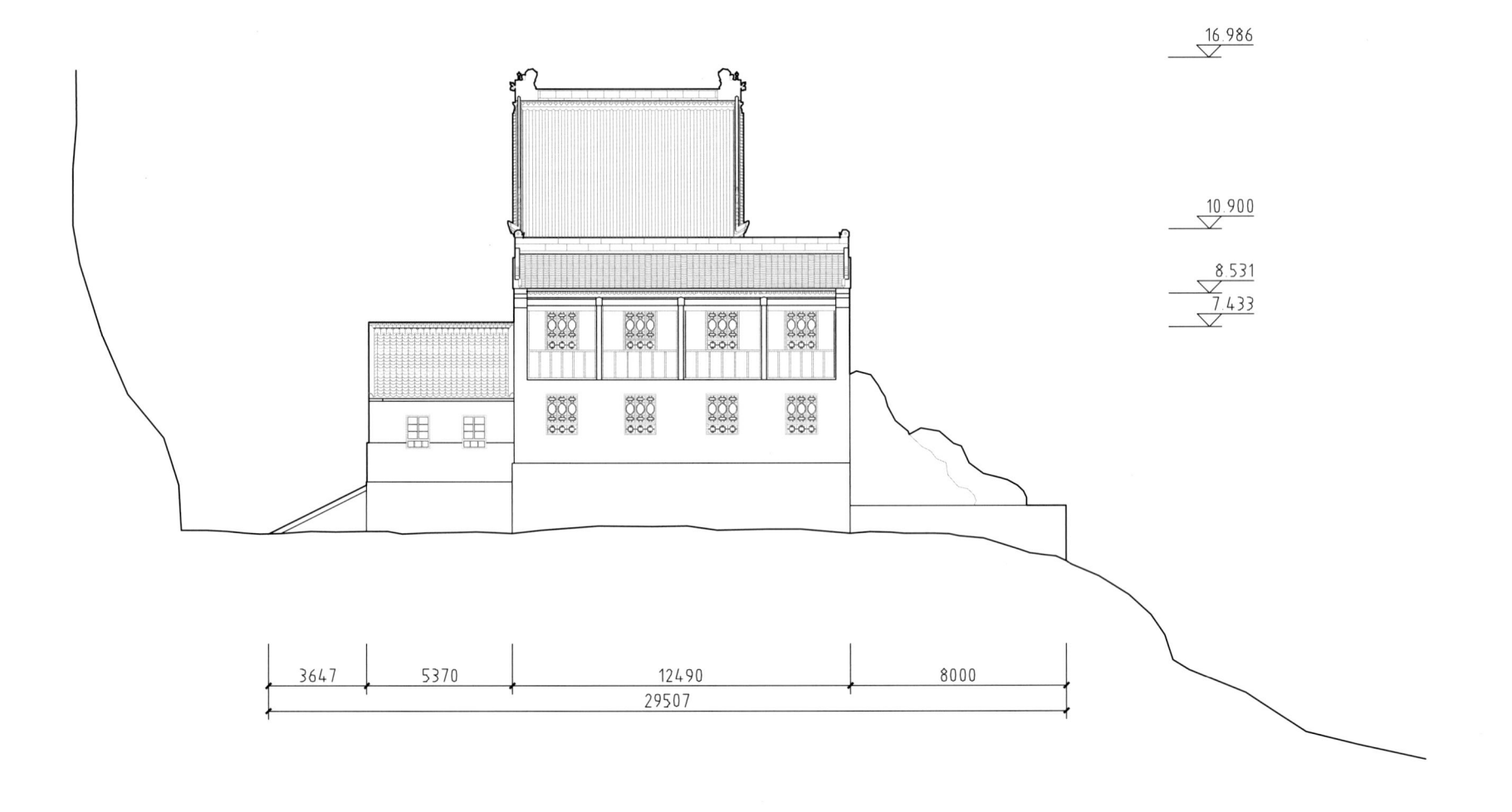

16 986

10 900

8 531
7 433

3647　5370　　　12490　　　8000
29507

玉女宫北立面图
North elevation of Yu'nv Palace

0　　3　　6m

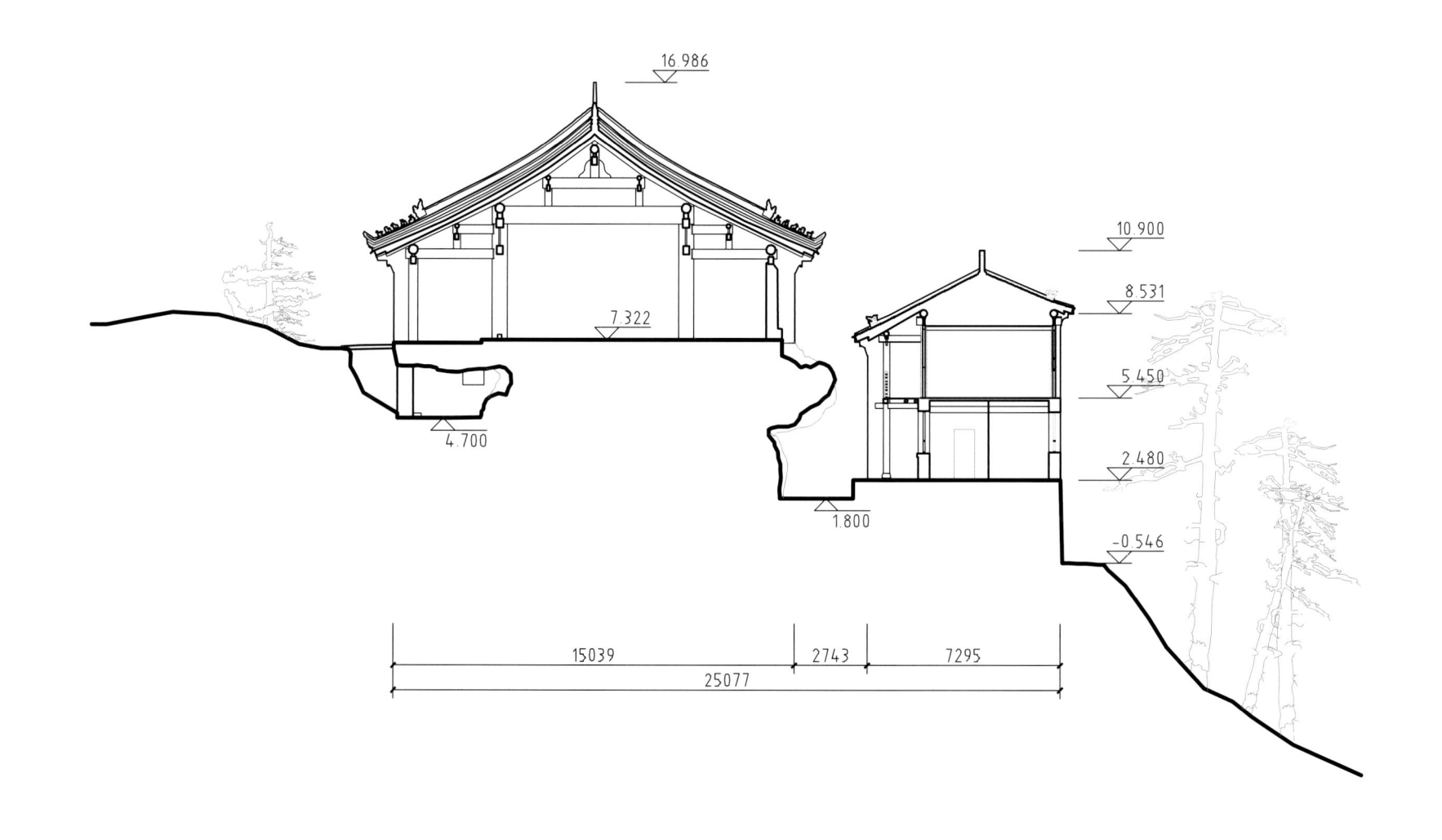

16.986

10.900

8.531

7.322

5.450

4.700

2.480

1.800

-0.546

15039　　2743　　7295

25077

玉女宫 1-1 剖面图

Section 1-1 of Yu'nv Palace

0　　3　　6m

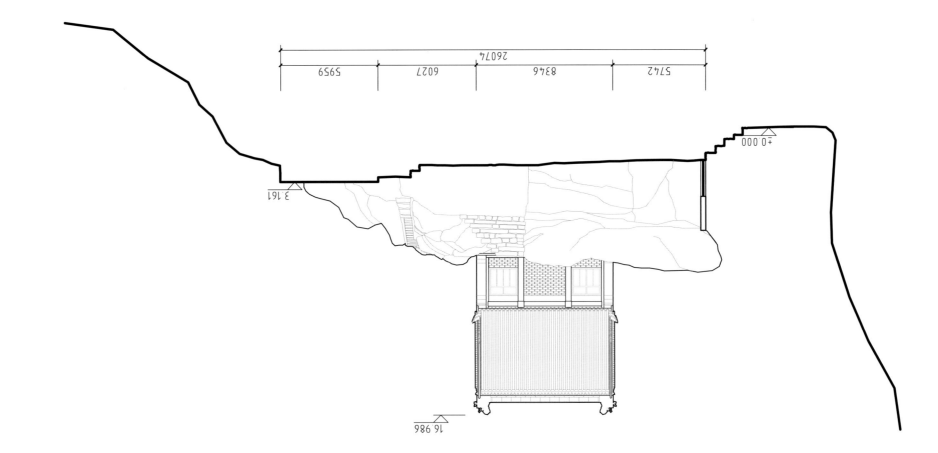

26074

5742　8346　6027　5959

3 161

+0 000

16 986

0　3　6m

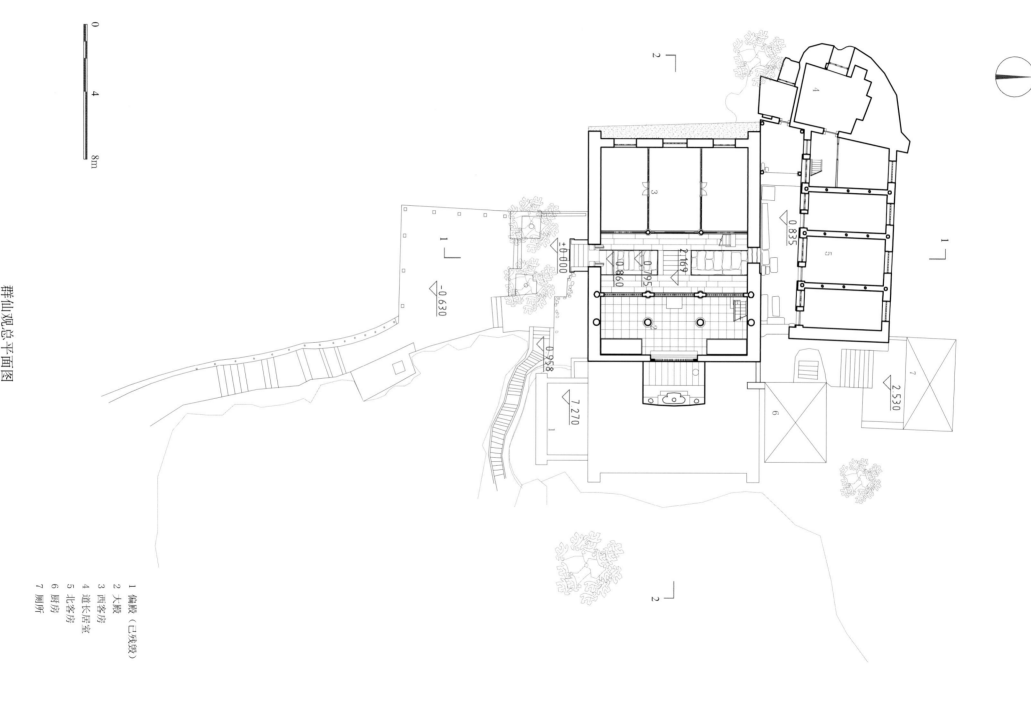

群仙观

Qunxian Taoist Temple

中国传统建筑营造技艺·武当山道教建筑——遇真宫与群仙观

群仙观总平面图
Site plan of Qunxian Taoist Temple

N

0 4 8m

1 偏殿（已残毁）
2 大殿
3 西客房
4 道长居室
5 北客房
6 厨房
7 厕所

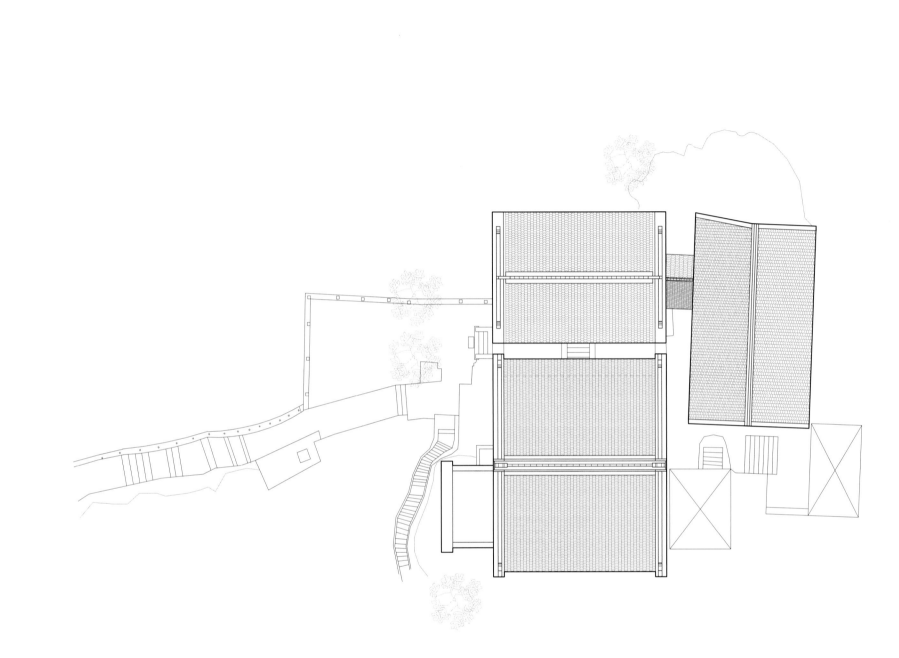

群仙观屋顶总平面图
Site roof plan of Qunxian Taoist Temple

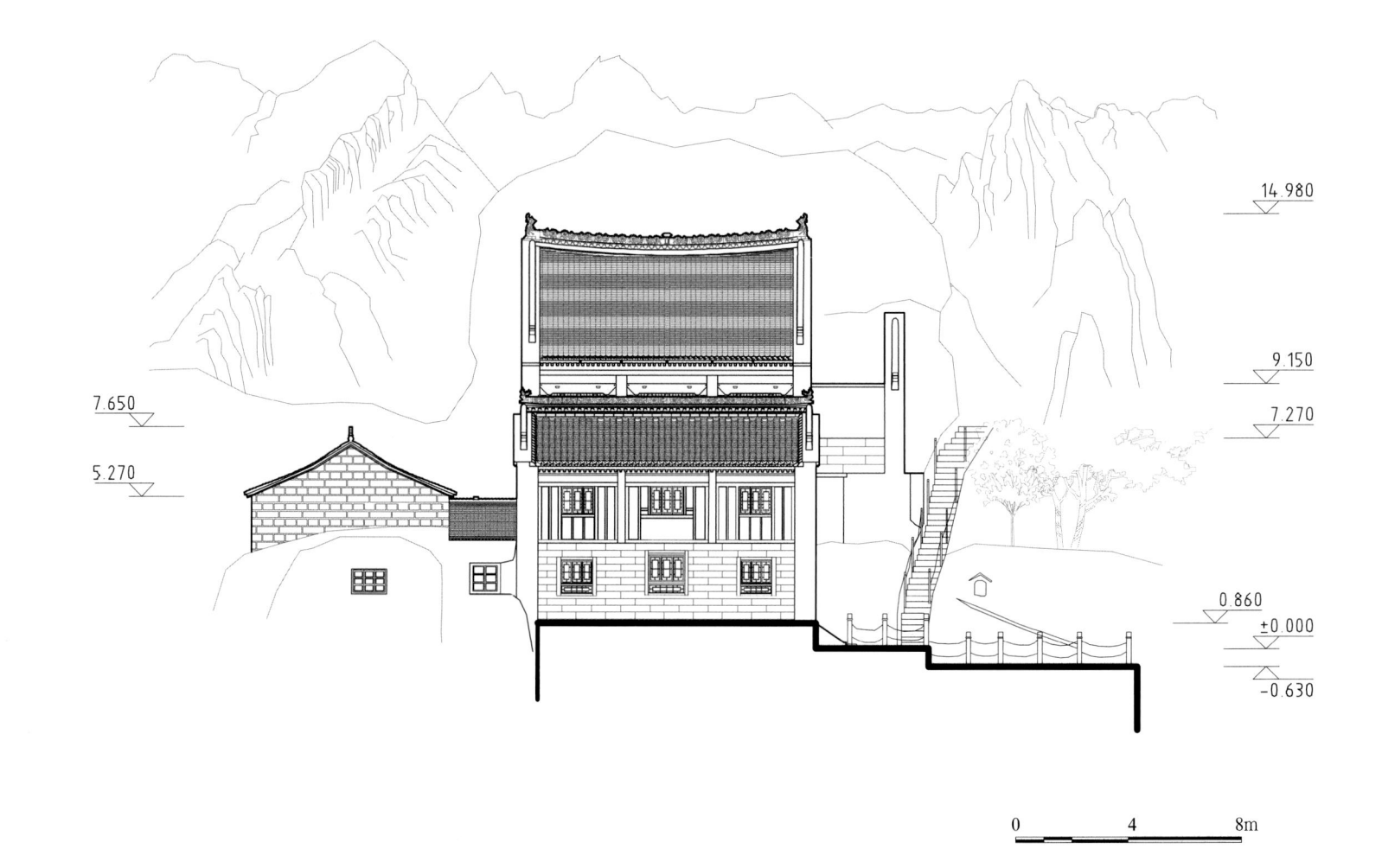

7.650

5.270

14.980

9.150

7.270

0.860

+0.000

-0.630

0   4   8m

群仙观东侧总立面图
East site elevation of Qunxian Taoist Temple

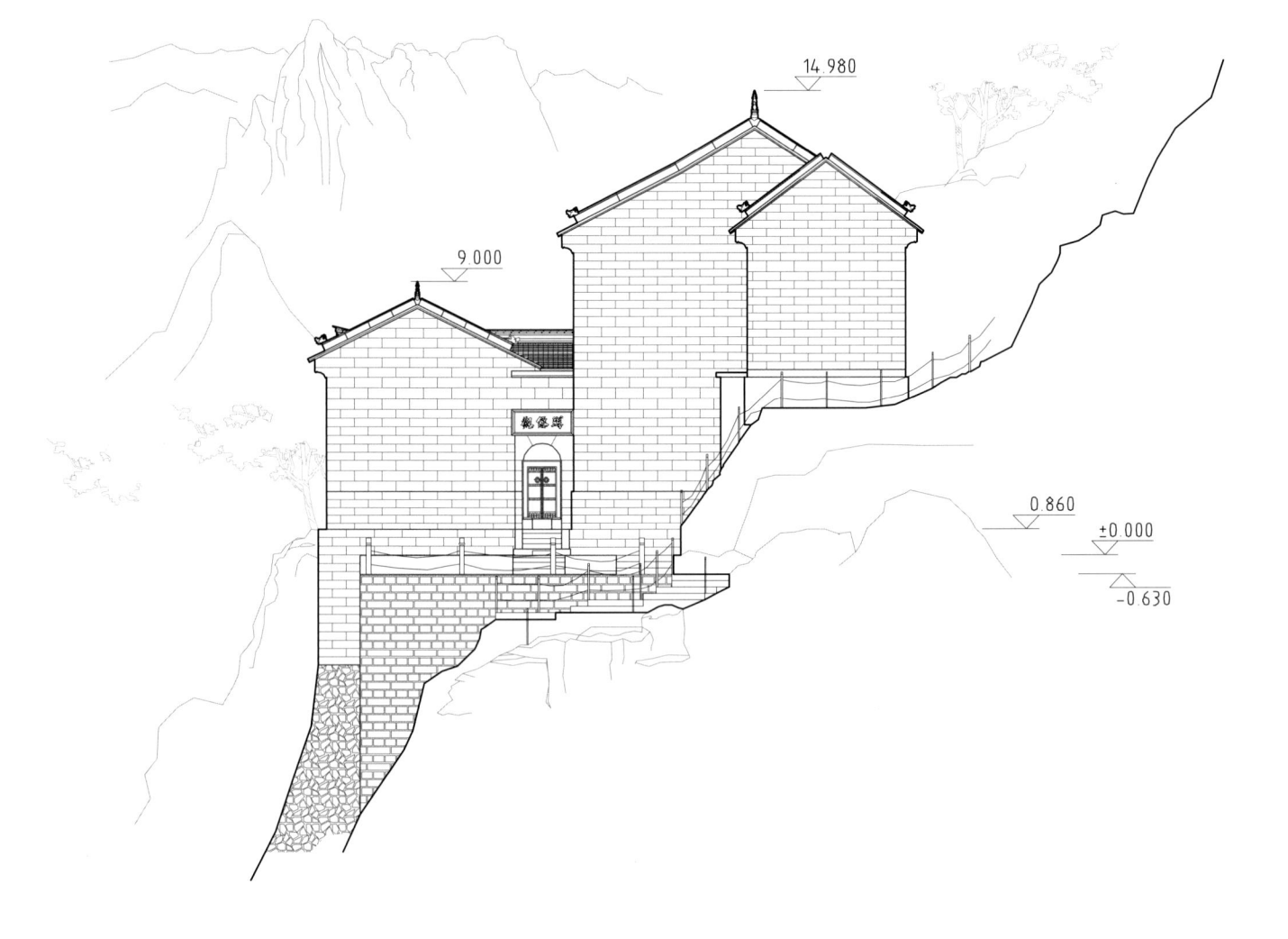

14.980

9.000

乾德坊

0.860
±0.000
-0.630

0   4   8m

群仙观南侧总立面图
South site elevation of Qunxian Taoist Temple

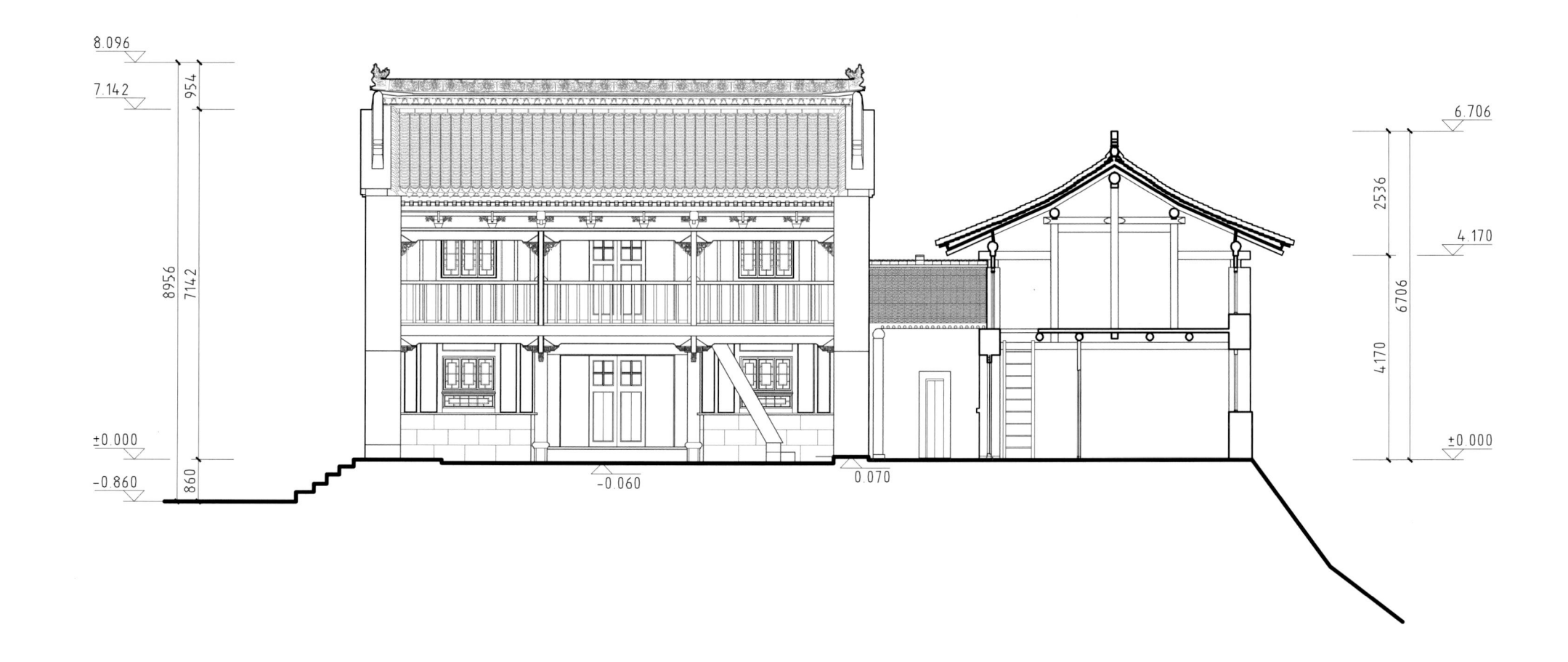

8.096

7.142

954

8956

7142

±0.000

860

-0.860

-0.060

0.070

6.706

2536

4.170

6706

4170

±0.000

群仙观 1-1 剖面图
Section 1-1 of Qunxian Taoist Temple

0　　2　　4m

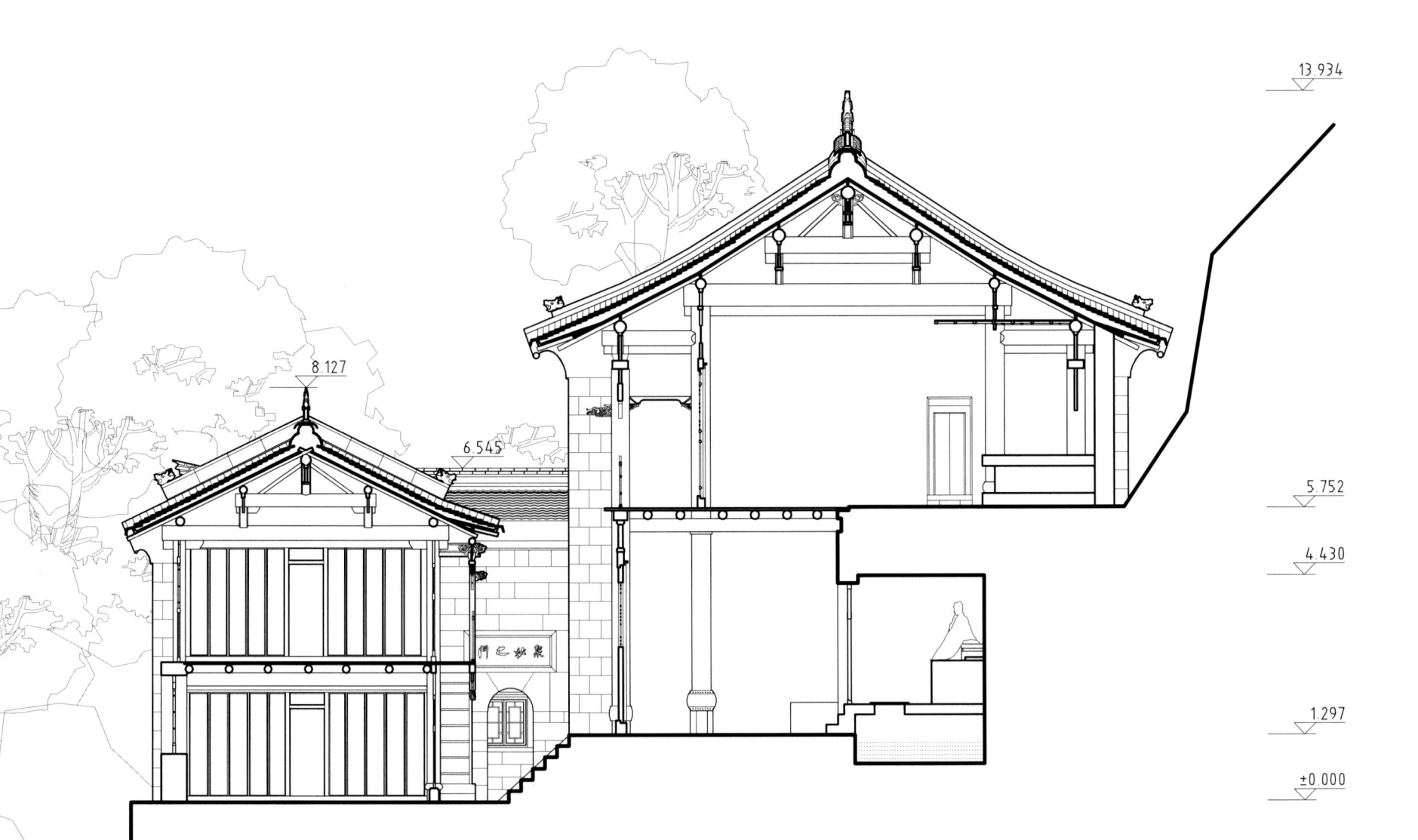

13.934

8.127

6.545

5.752

4.430

1.297

±0.000

震加记门

群仙观 2—2 剖面图
Section 2-2 of Qunxian Taoist Temple

0   2   4m

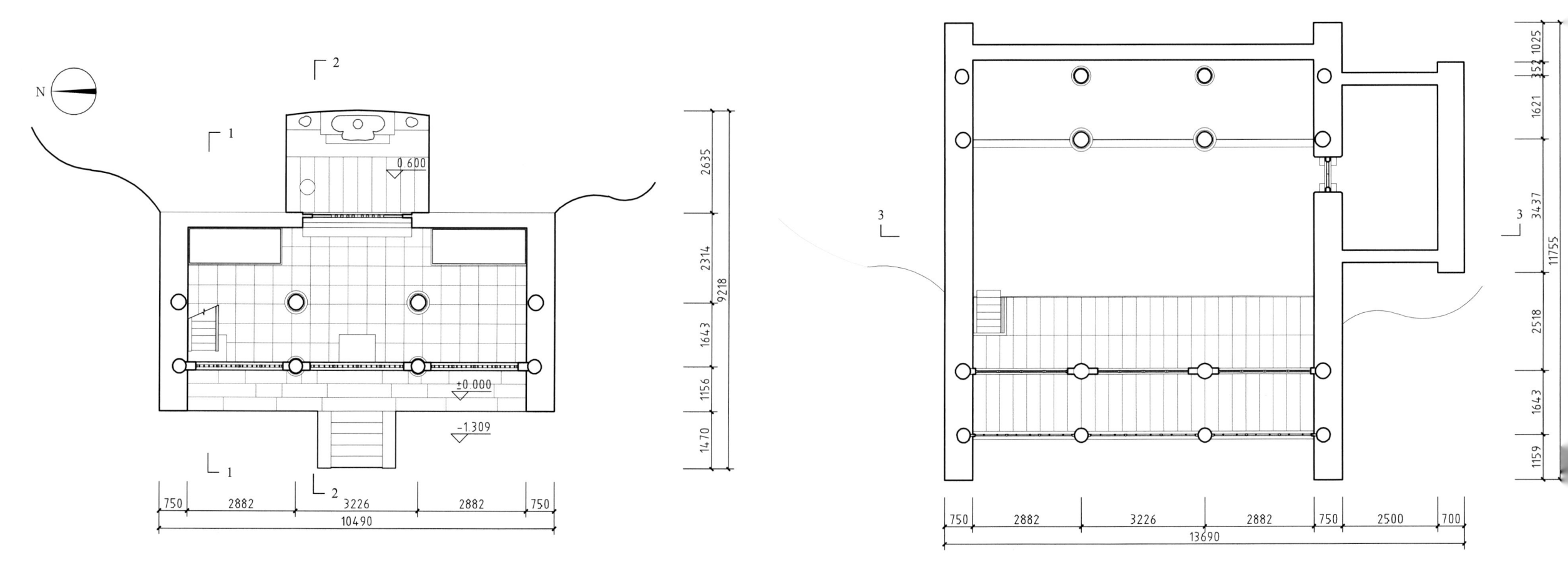

群仙观大殿一层平面图
Plan of first floor of Qunxian Taoist Temple's main hall

群仙观大殿二层平面图
Plan of second floor of Qunxian Taoist Temple's main hall

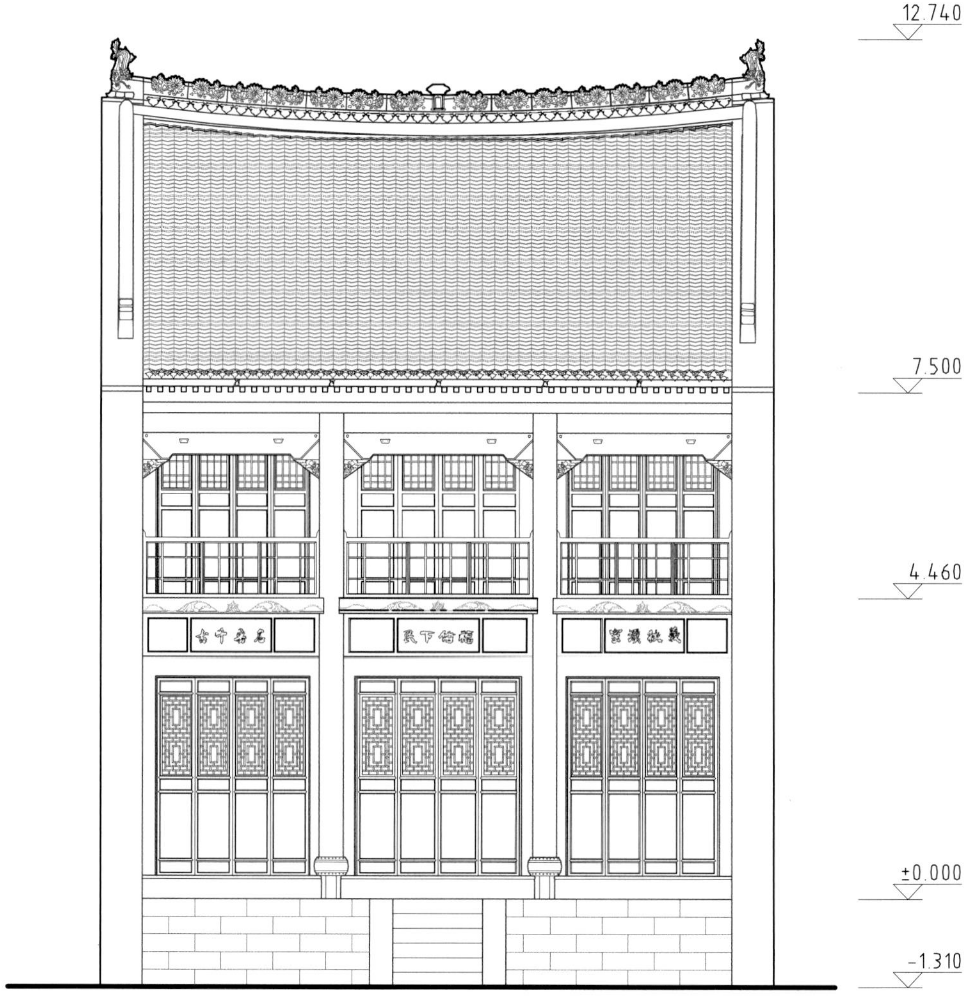

12.740

12.870

7.500

8.370

4.460

±0.000

±0.000

-1.310

-1.250

群仙观大殿正立面图
Front elevation of Qunxian Taoist Temple's main hall

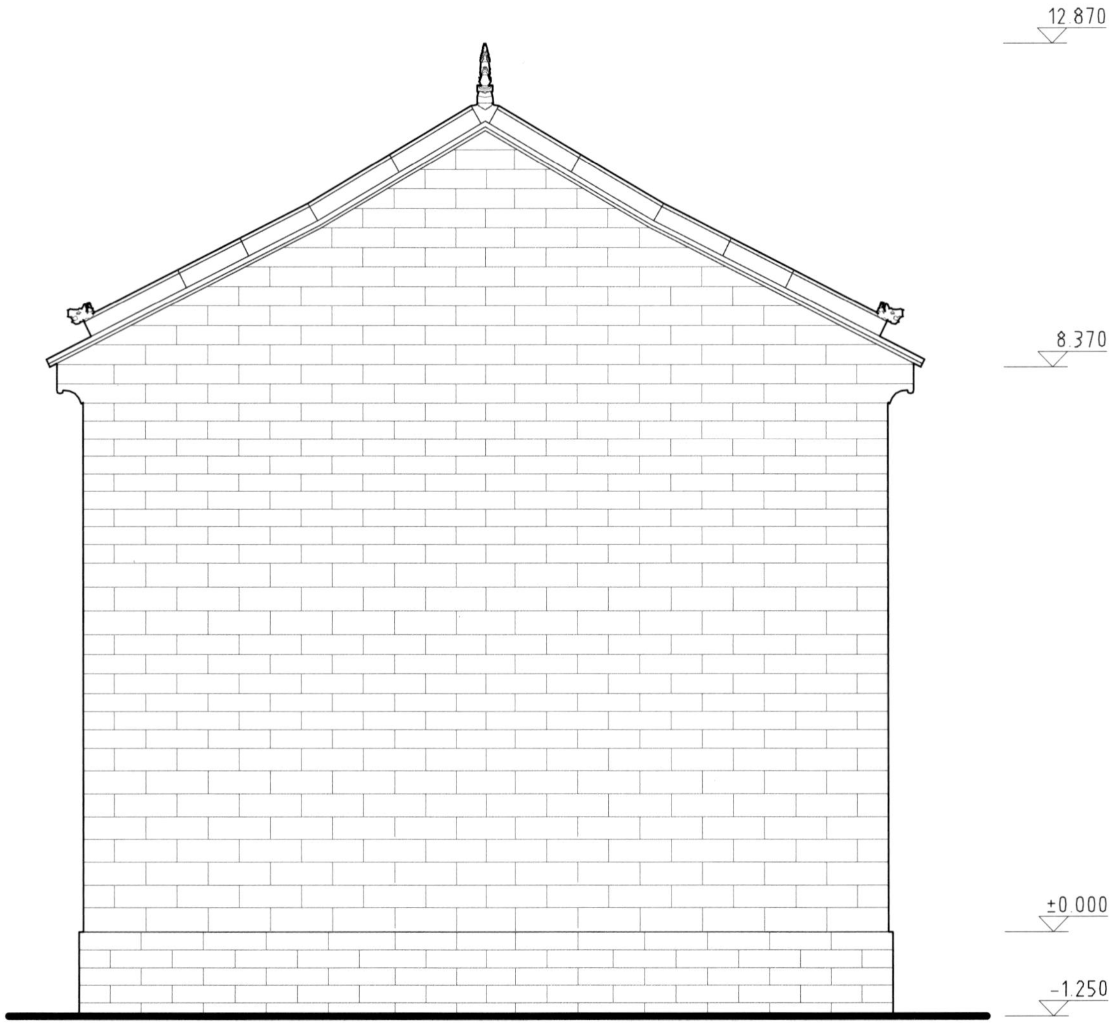

群仙观大殿侧立面图
Side elevation of Qunxian Taoist Temple's main hall

0   2   4m

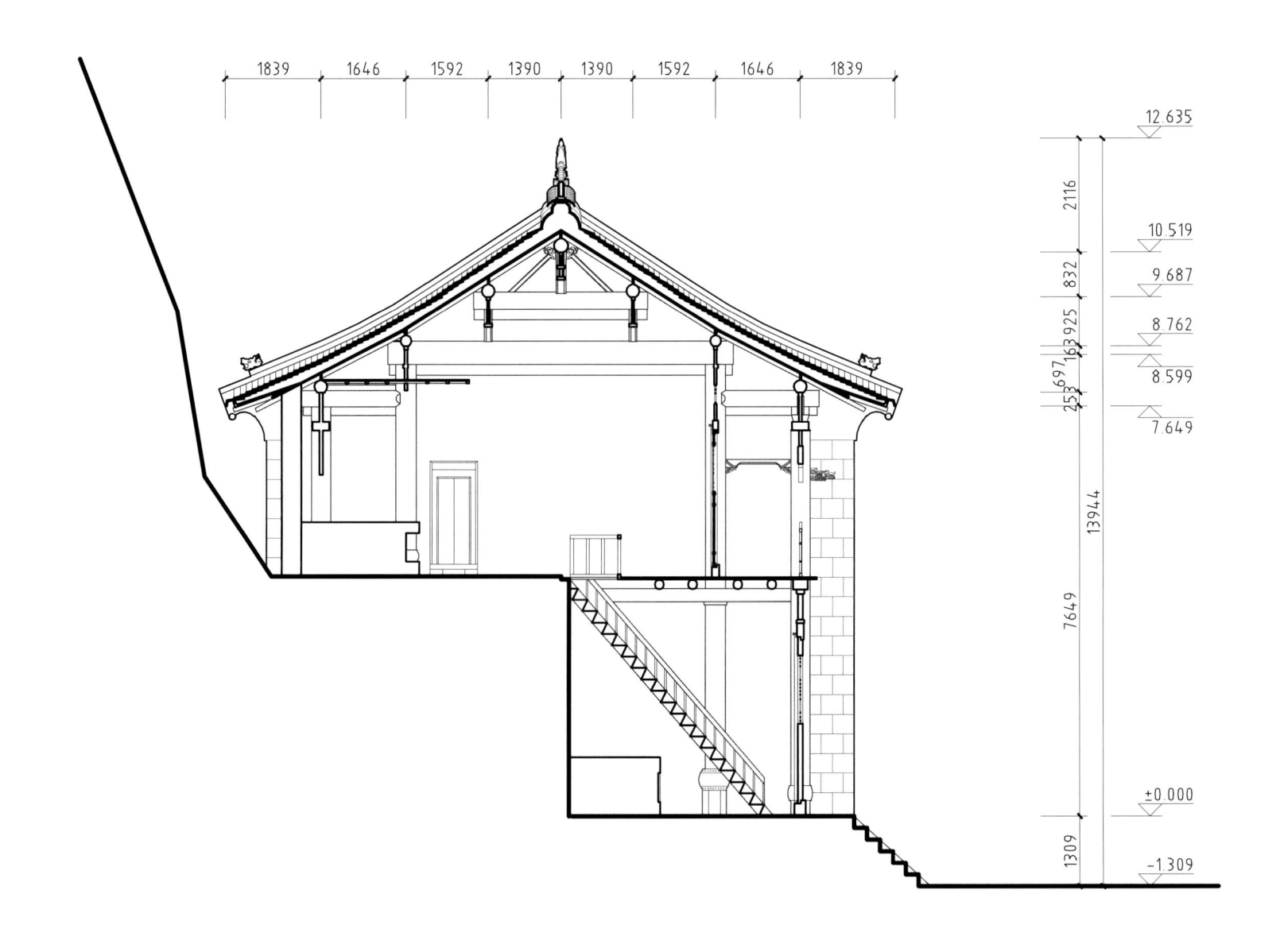

1839　1646　1592　1390　1390　1592　1646　1839

12.635

2116

10.519

832

9.687

925

8.762

163

697

8.599

253

7.649

13944

7649

±0.000

1309

-1.309

群仙观大殿 1-1 剖面图
Section 1-1 of Qunxian Taoist Temple's main hall

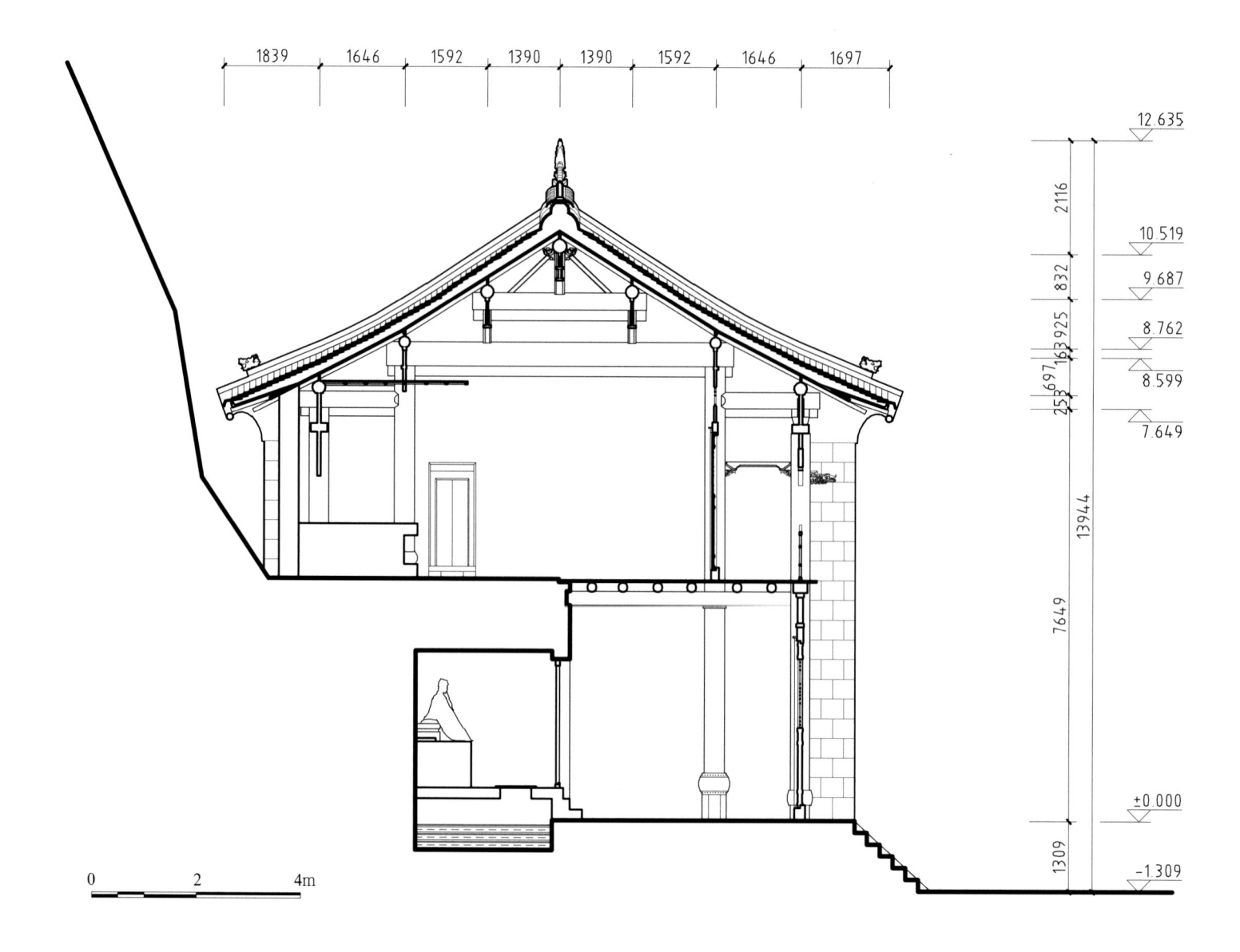

1839　1646　1592　1390　1390　1592　1646　1697

12.635

2116

10.519

832

9.687

925

8.762

163

8.599

697

7.649

253

13944

7649

±0.000

1309

-1.309

0　2　4m

群仙观大殿 2-2 剖面图
Section 2-2 of Qunxian Taoist Temple's main hall

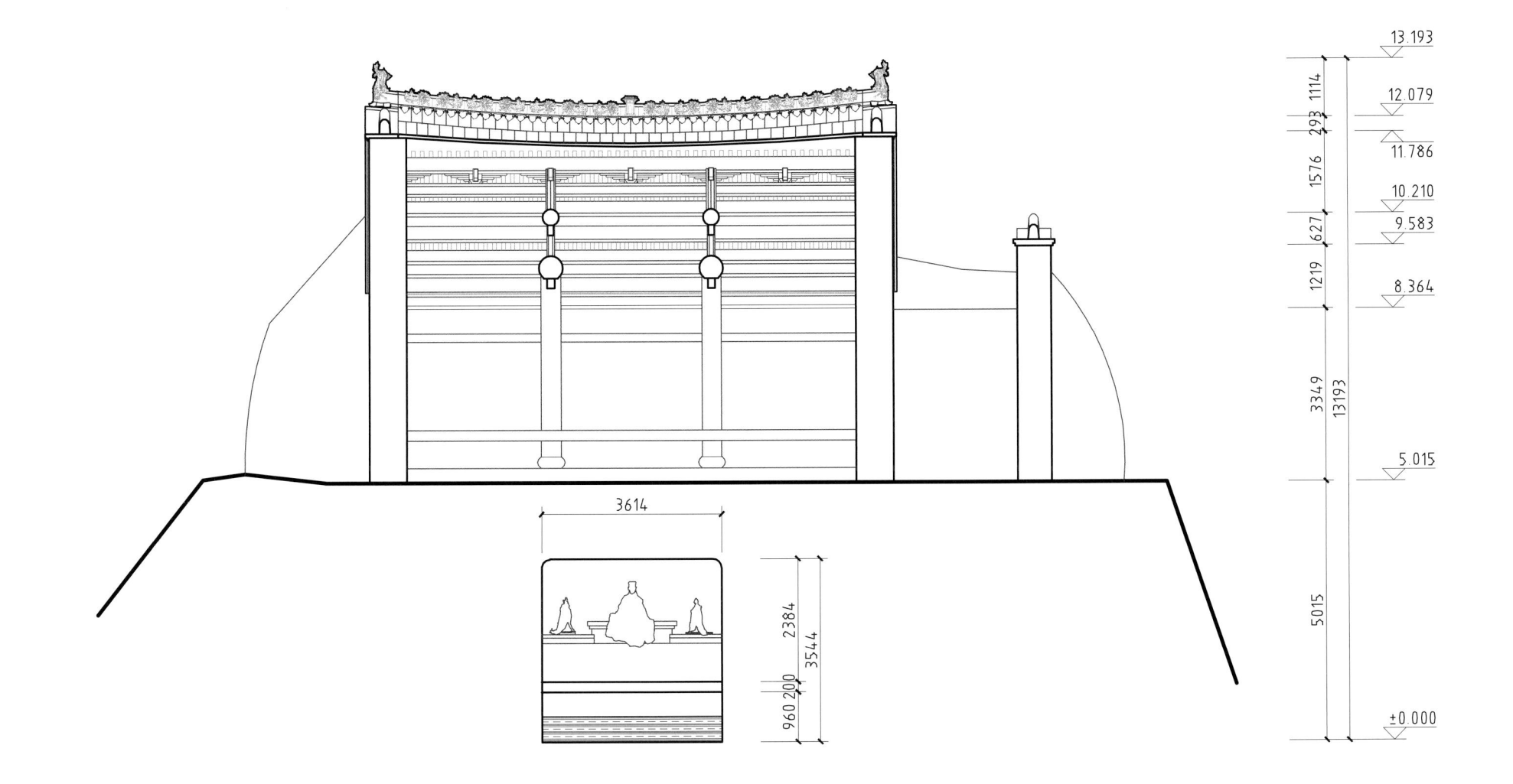

13.193

12.079

1114

293

11.786

1576

10.210

627

9.583

1219

8.364

3349

13193

5.015

5015

±0.000

3614

2384

3544

960 200

群仙观大殿 3-3 剖面图
Section 3-3 of Qunxian Taoist Temple's main hall

0   2   4m

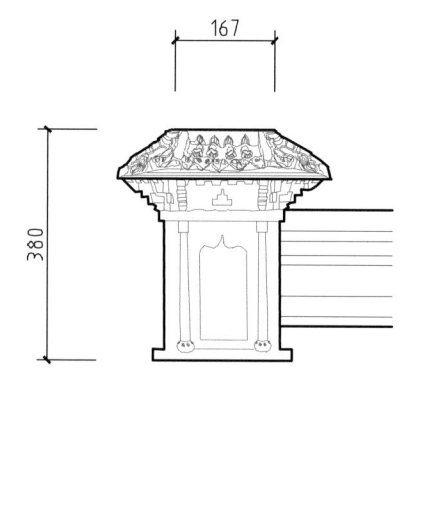

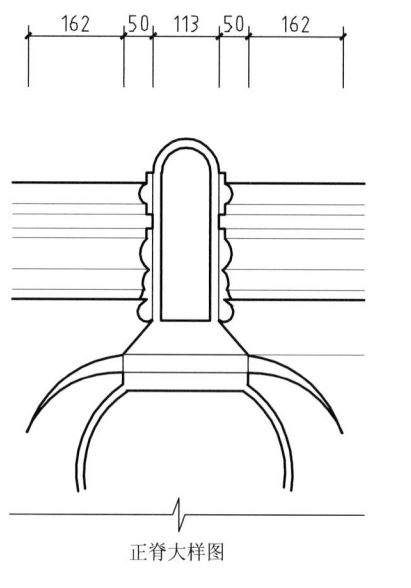

正脊大样图

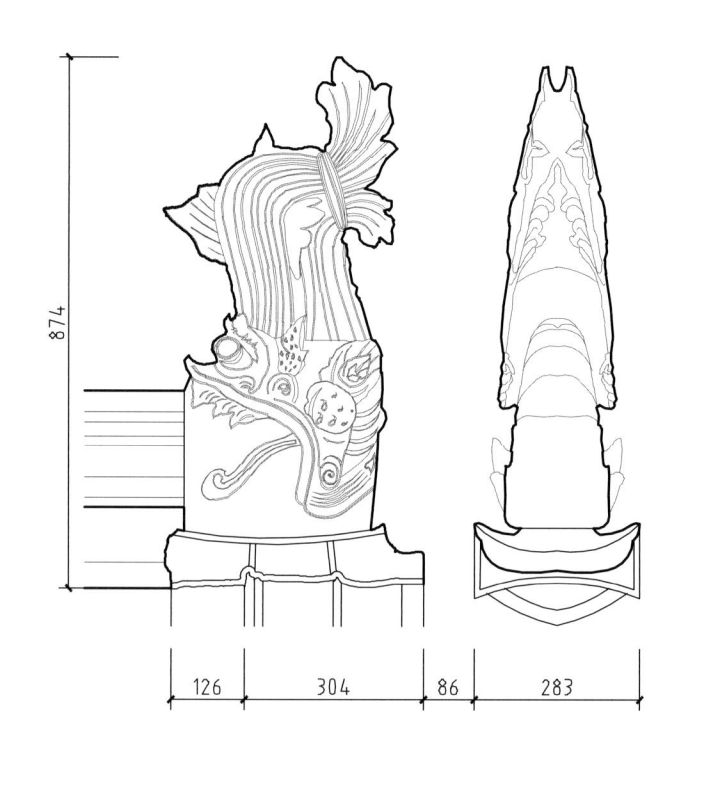

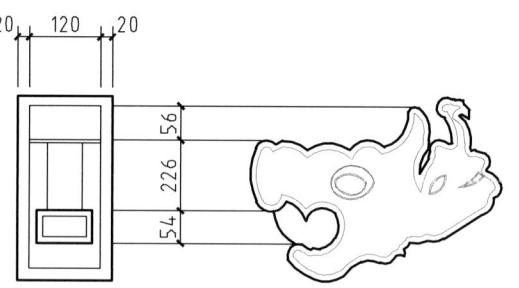

垂脊吻兽大样图

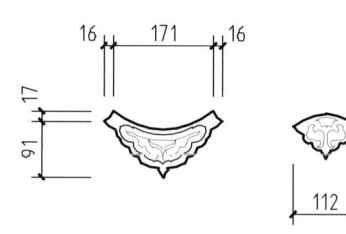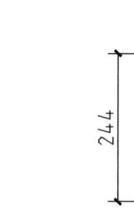

大殿檐口滴水大样

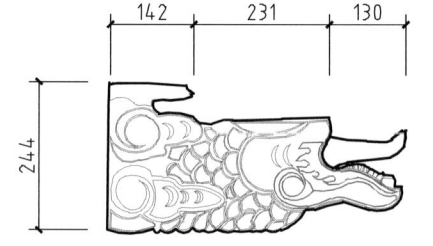

（残损）挑尖梁头大样

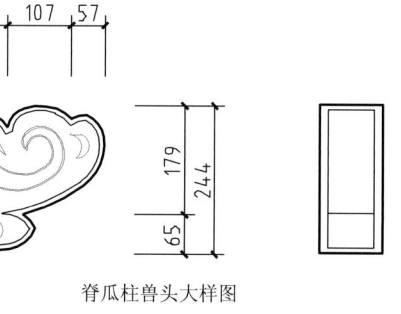

脊瓜柱兽头大样图

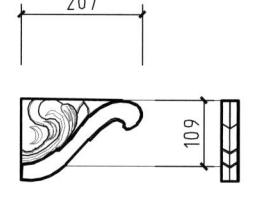

雀替大样一

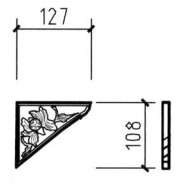

雀替大样二

群仙观大殿大样图
Qunxian Taoist Temple's main hall

0   0.25   0.5m

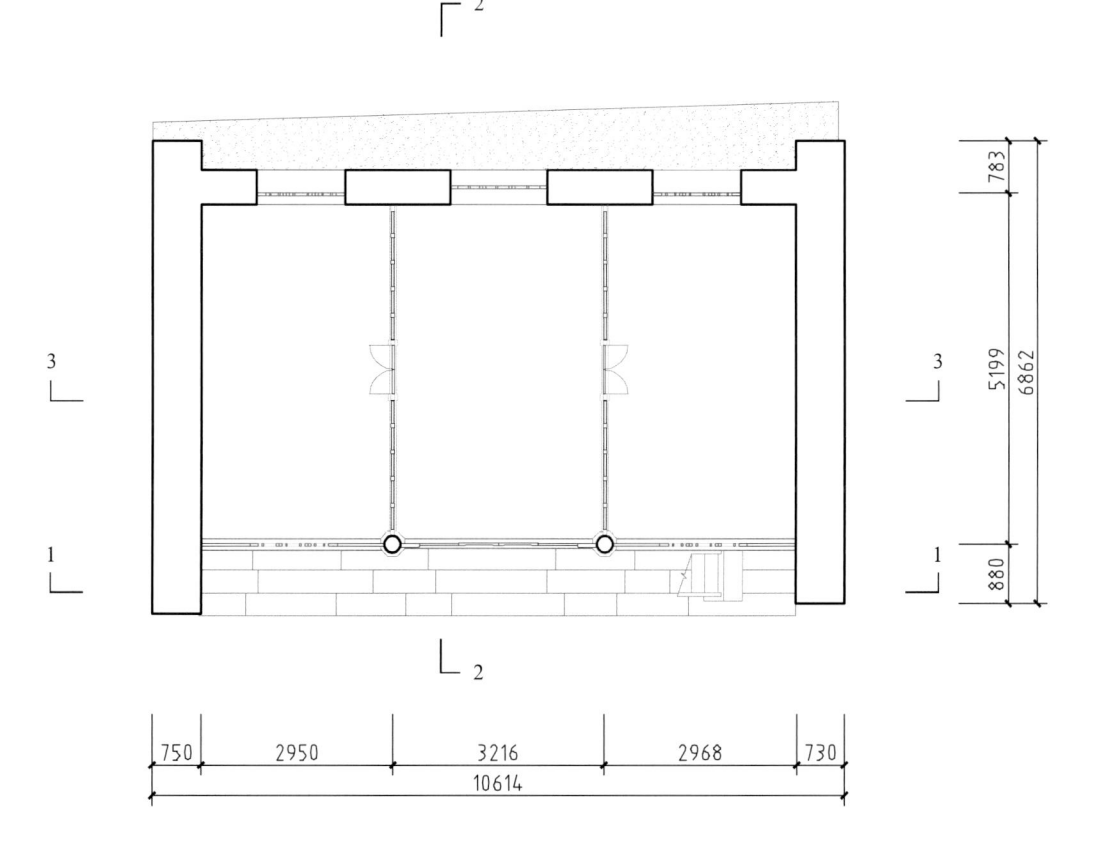

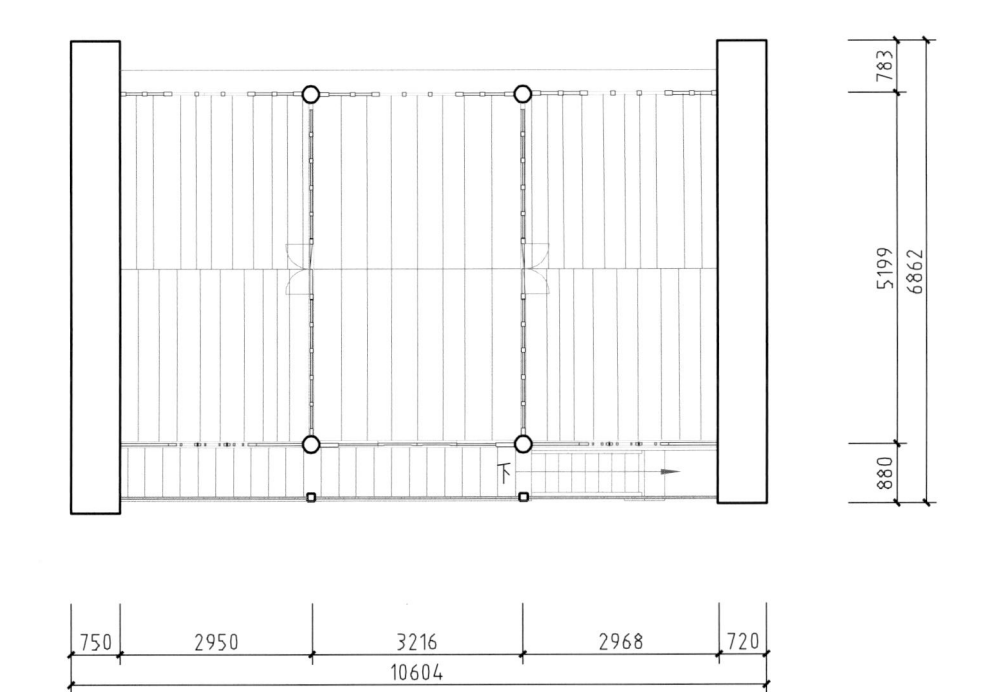

群仙观西客房一层平面图
Plan of first floor of Qunxian Taoist Temple's west guest room

群仙观西客房二层平面图
Plan of second floor of Qunxian Taoist Temple's west guest room

群仙观西客房东立面图
East elevation of Qunxian Taoist Temple's west guest room

群仙观西客房西立面图
West elevation of Qunxian Taoist Temple's west guest room

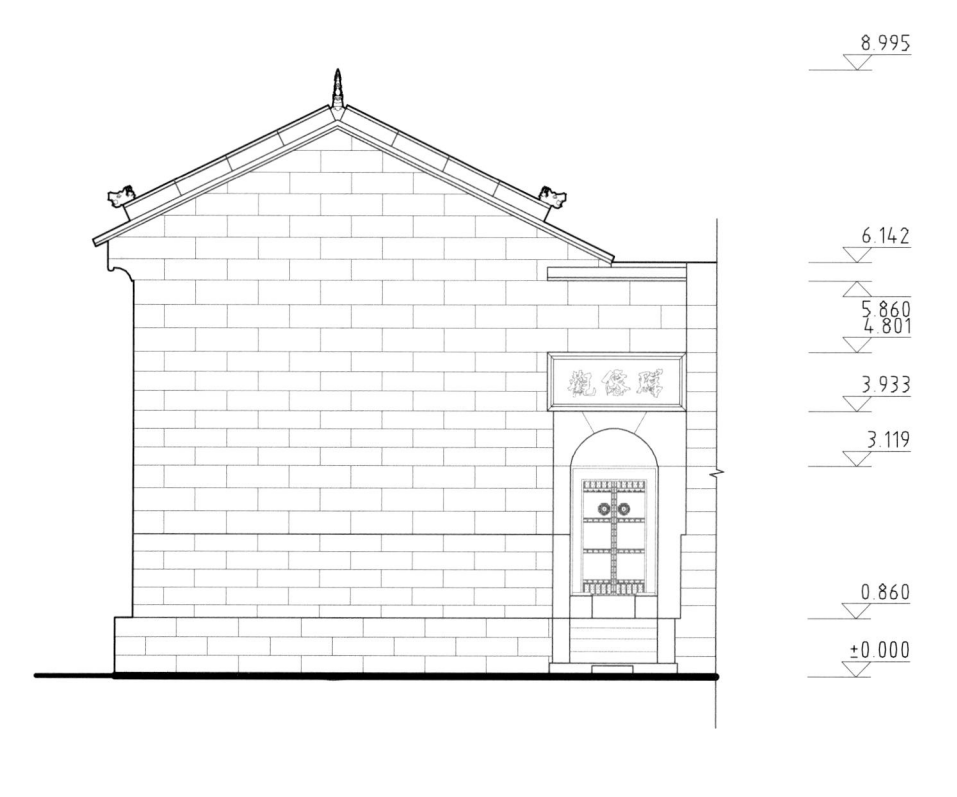

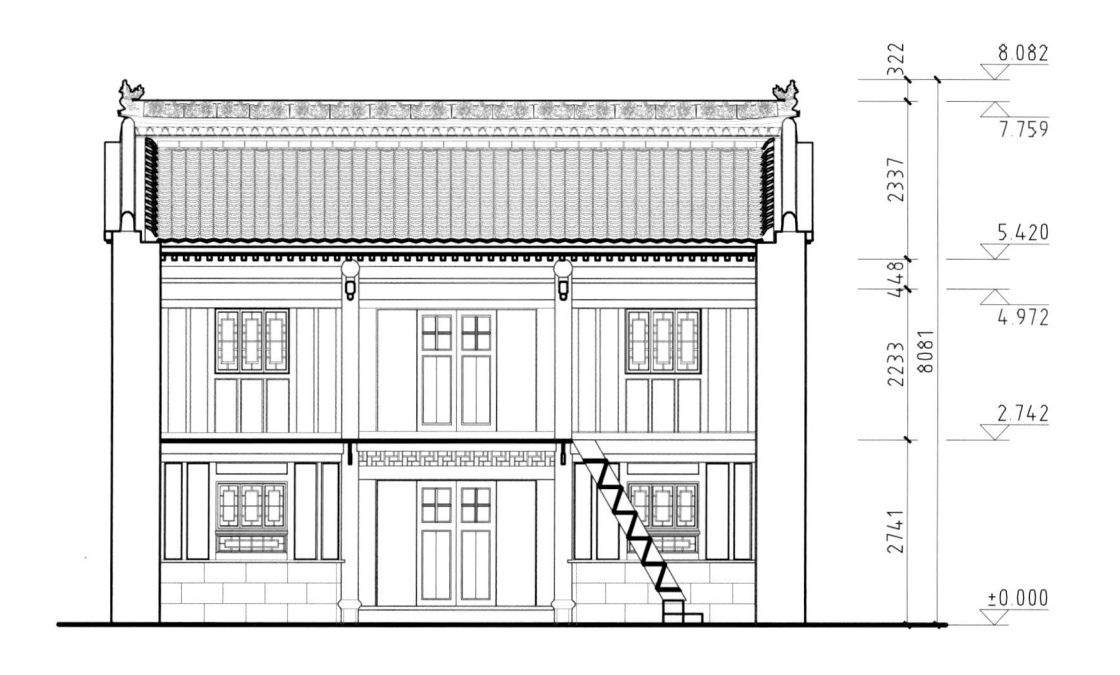

群仙观西客房南立面图
South elevation of Qunxian Taoist Temple's west guest room

群仙观西客房 1-1 剖面图
Section 1-1 of Qunxian Taoist Temple's west guest room

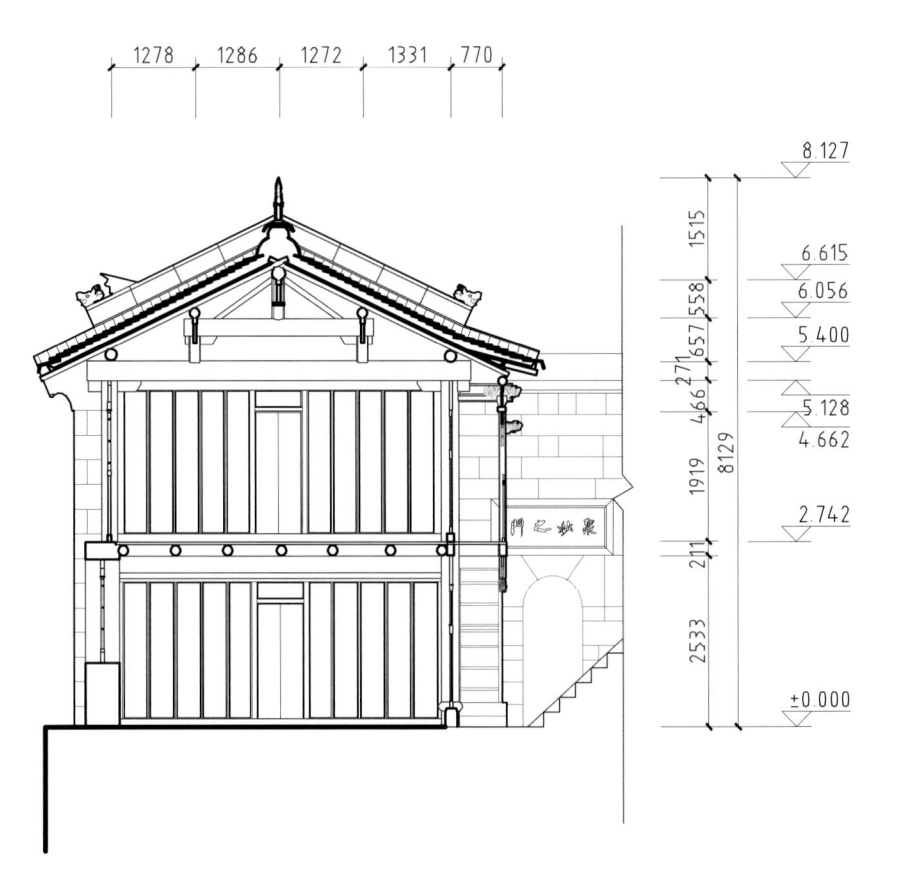

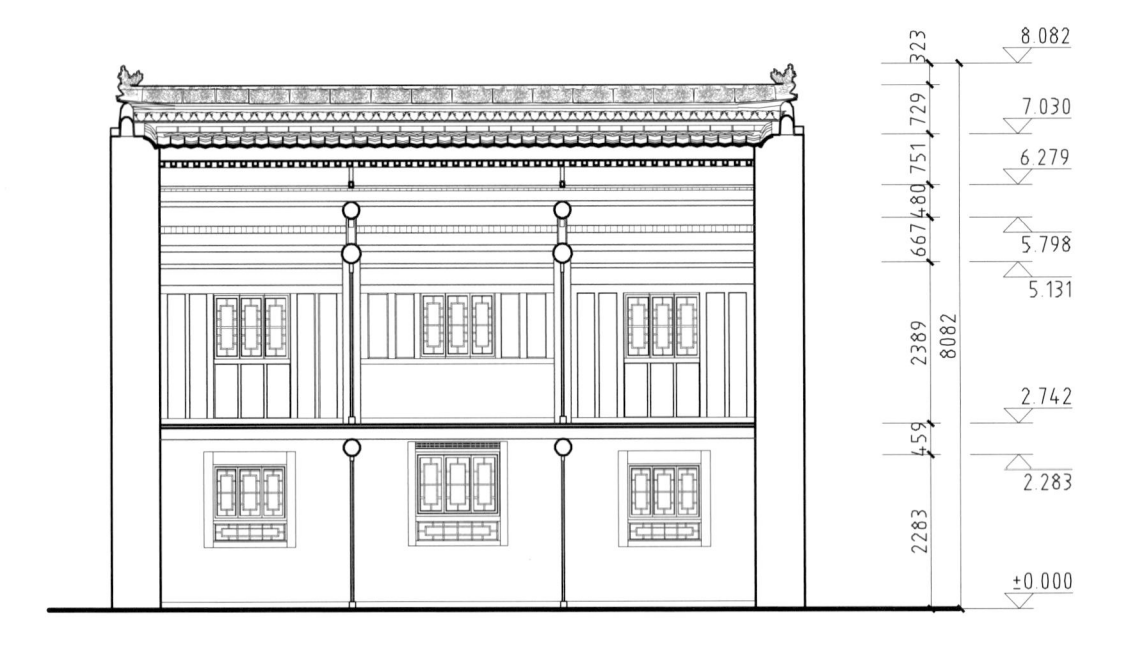

群仙观西客房 2-2 剖面图
Section 2-2 of Qunxian Taoist Temple's west guest room

群仙观西客房 3-3 剖面图
Section 3-3 of Qunxian Taoist Temple's west guest room

0   2   4m

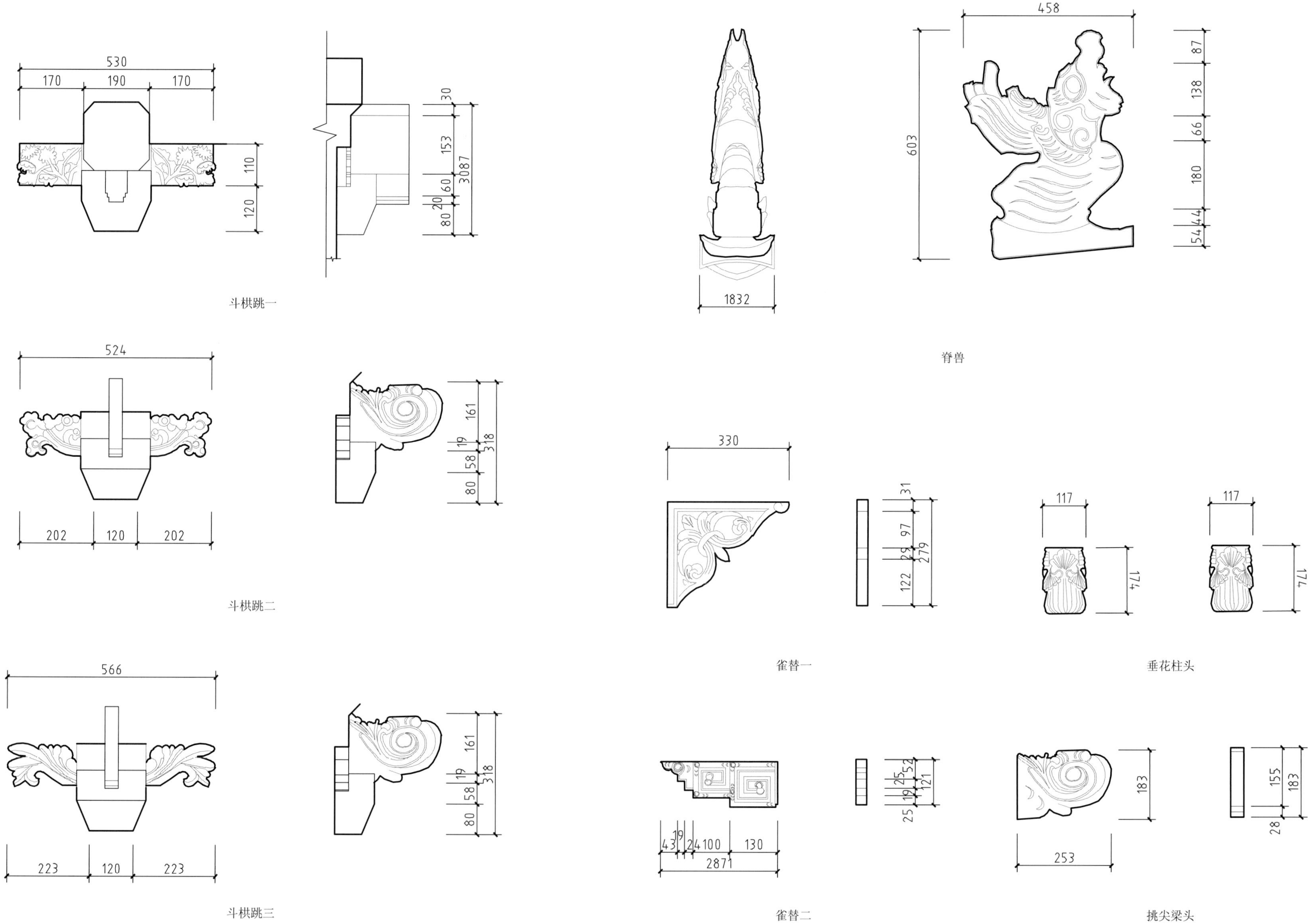

斗栱跳一

斗栱跳二

斗栱跳三

脊兽

雀替一

垂花柱头

雀替二

挑尖梁头

群仙观西客房斗栱大样图
Bracket set of Qunxian Taoist Temple's west guest room

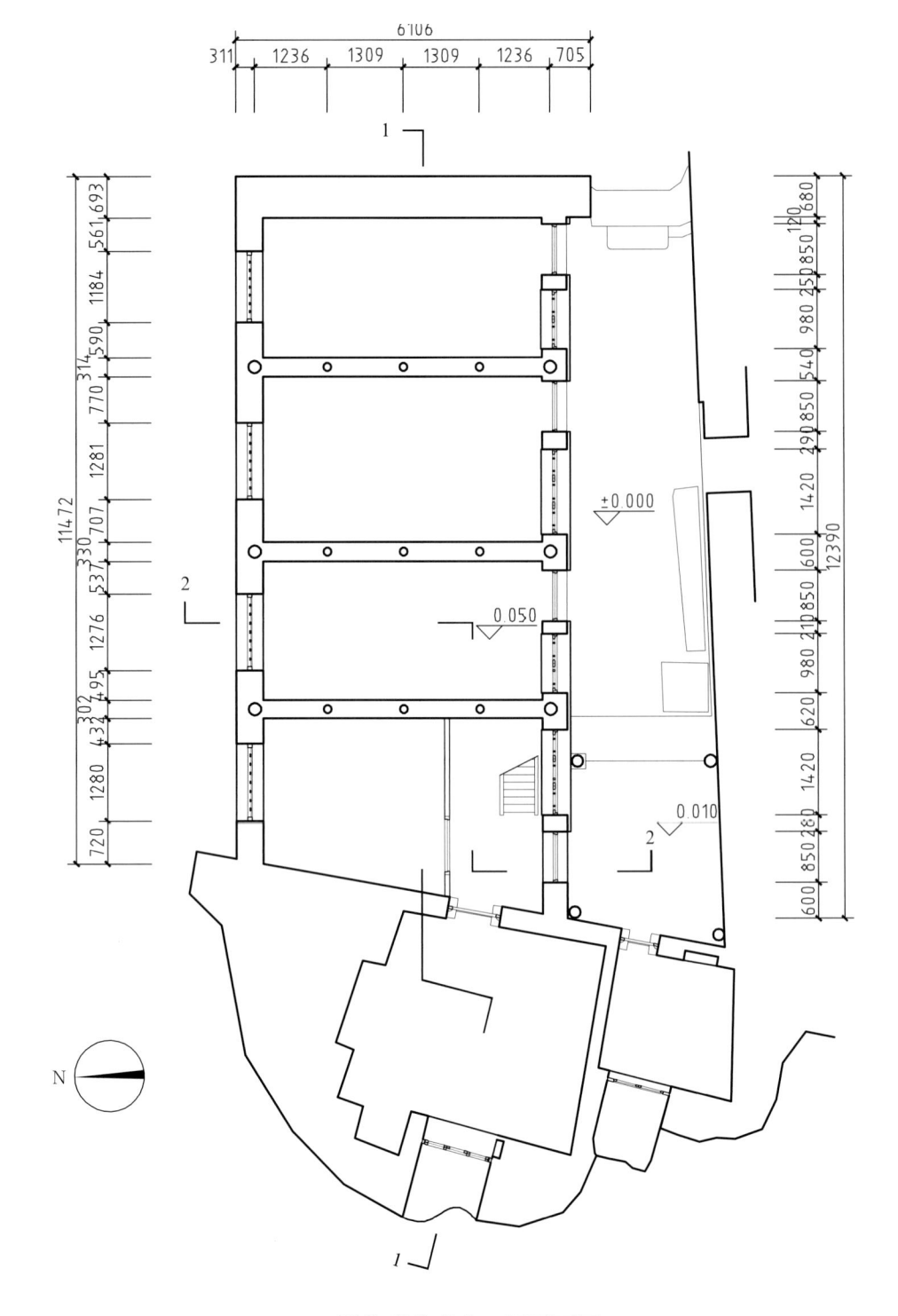

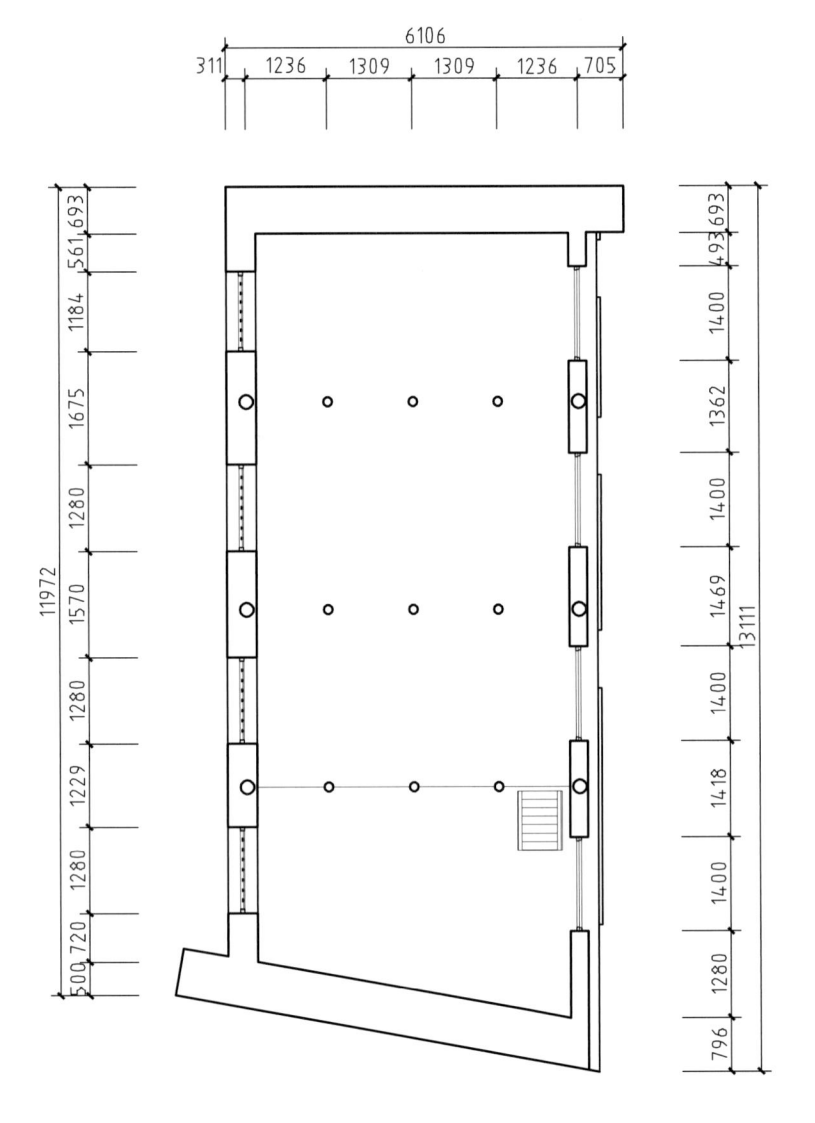

群仙观北客房一层平面图
Plan of first floor of Qunxian Taoist Temple's north guest room

群仙观北客房二层平面图
Plan of second floor of Qunxian Taoist Temple's north guest room

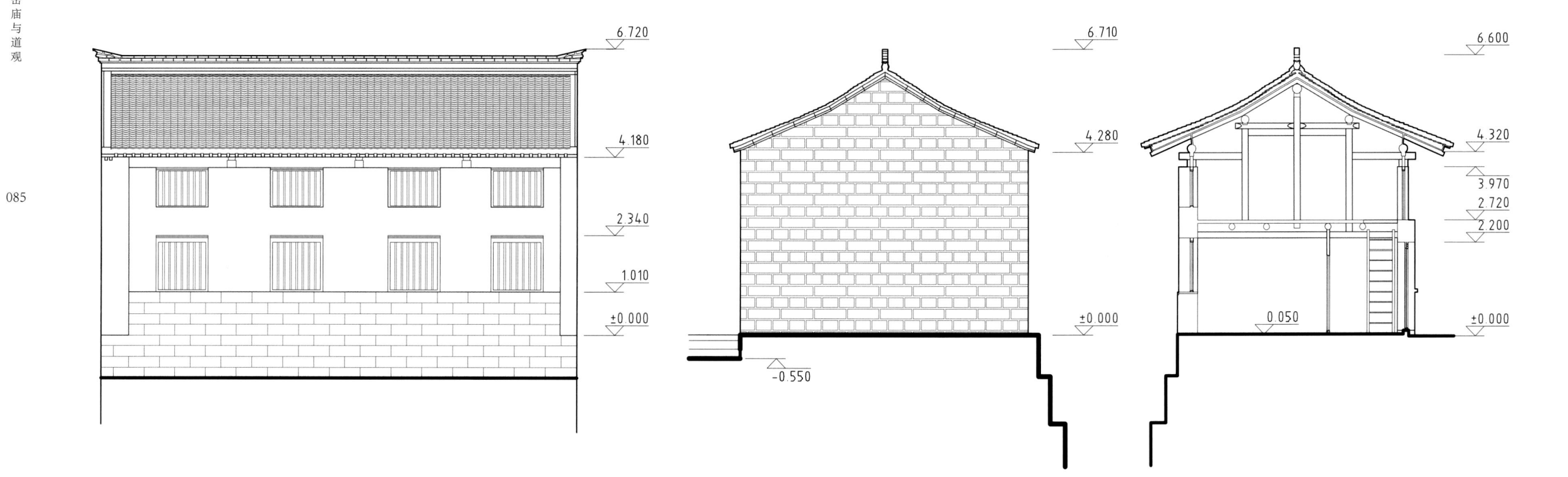

6.720

4.180

2.340

1.010

±0.000

6.710

4.280

±0.000

-0.550

6.600

4.320

3.970

2.720

2.200

0.050

±0.000

群仙观北客房背立面图
Rear elevation of Qunxian Taoist Temple's north guest room

群仙观北客房侧立面图
Side elevation of Qunxian Taoist Temple's north guest room

群仙观北客房 1-1 剖面图
Section 1-1 of Qunxian Taoist Temple's North guest room

群仙观北客房正立面图
Front elevation of Qunxian Taoist Temple's north guest room

群仙观北客房 2—2 剖面图
Section 2-2 of Qunxian Taoist Temple's North guest room

N

群仙观日月岩总平面图
Site plan of Qunxian Taoist Temple's Riyue Rock

岩洞前广场

群仙观日月岩一层平面图
Plan of first floor of Qunxian Taoist Temple's Riyue Rock

岩洞

岩洞前广场

0 2 4m

0 2 4m

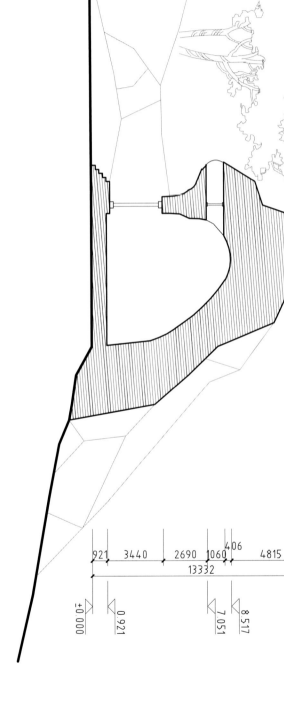

群仙观日月岩剖面图
Section of Qunxian Taoist Temple's Riyue Rock

921 3440 2690 1060 406 4815
13332

±0.000
0.921
7.051
8.517
13.332

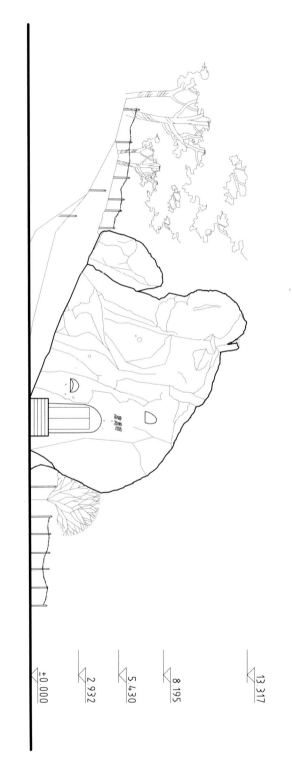

群仙观日月岩立面图
Elevation of Qunxian Taoist Temple's Riyue Rock

±0.000
2.932
5.430
8.195
13.317

0 2 4m

0 2 4m

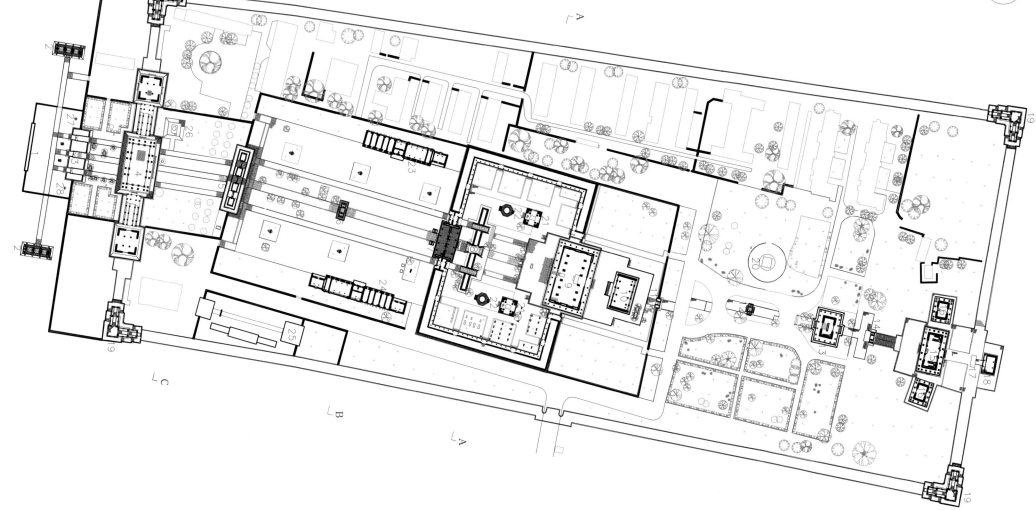

西岳庙总平面图
Site plan of Xiyue Temple

西岳庙
Xiyue Temple

N

0　40　80m

1　影壁
2　牌楼基址
3　灝灵门
4　五凤楼
5　棂星门
6　天威咫尺坊
7　金城门
8　金水桥
9　灝灵殿
10　寝宫
11　后宰门
12　少昊之都
13　御书楼
14　藏收之府
15　万寿阁
16　藏经阁
17　石牌楼基座
18　游岳坊
19　角楼
20　放生池
21　禅亭
22　玉春亭
23　冥王庄
24　灵官殿
25　三圣母殿
26　天下第一碑
27　乾隆四十二年庙貌碑
28　全国重点文物保护单位碑

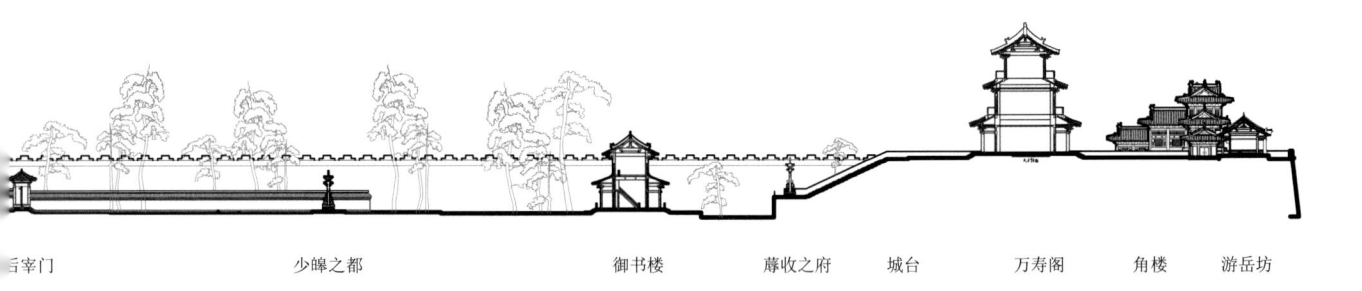

三宰门　　　　　少昊之都　　　　　御书楼　蓐收之府　城台　万寿阁　角楼　游岳坊

西岳庙总剖面图
Site section of Xiyue Temple

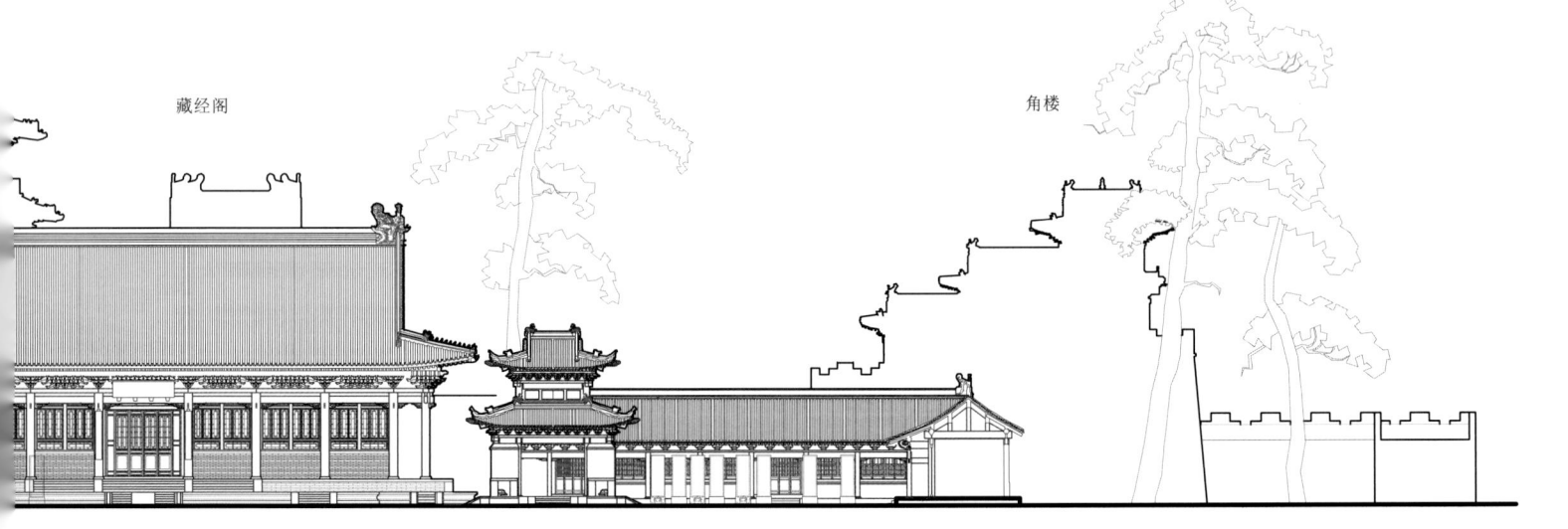

藏经阁　　　　　　　　　　　　　　　　　角楼

灏灵殿　　　　碑亭　　　长廊

西岳庙 A-A 剖面图
Section A-A of Xiyue Temple

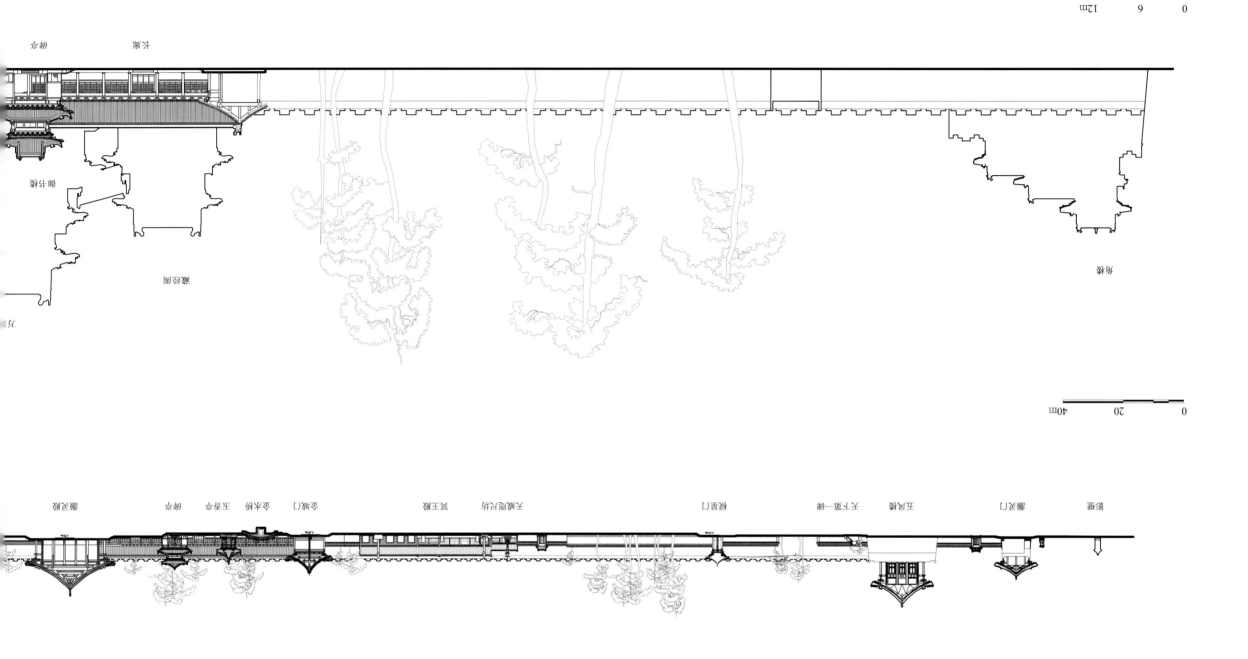

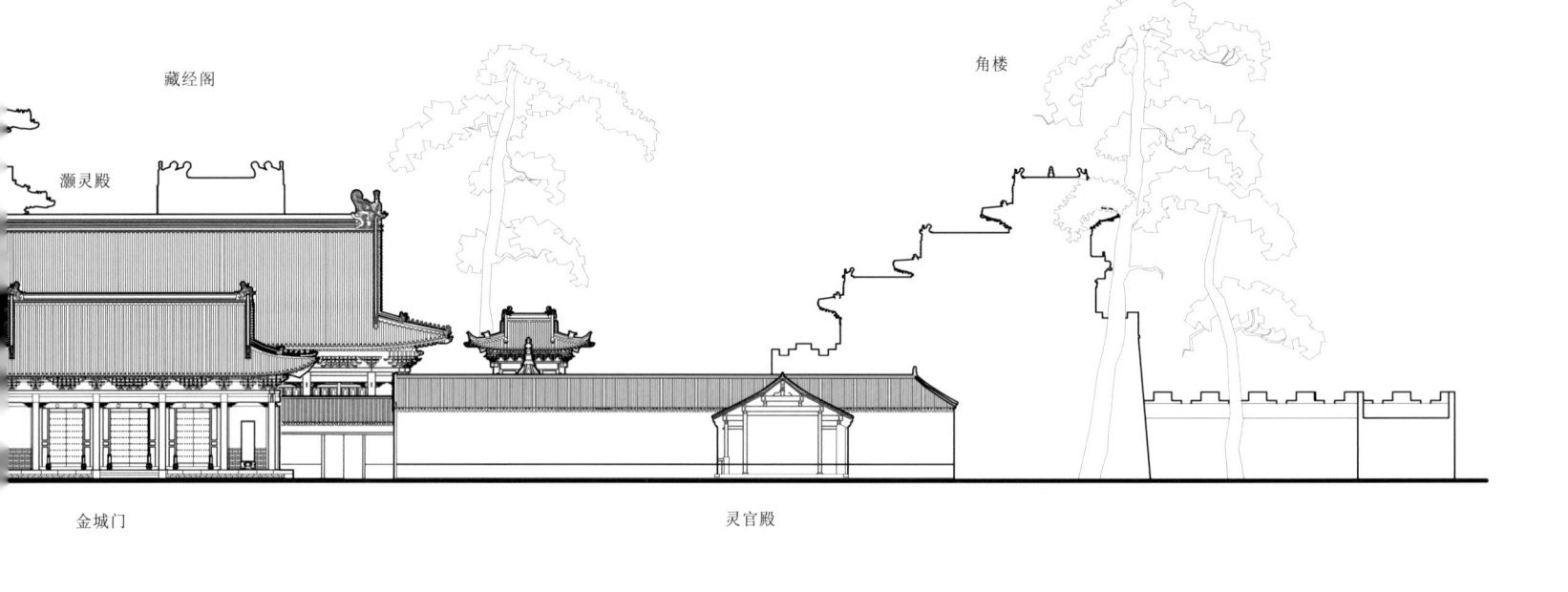

藏经图

灏灵殿

角楼

金城门

灵官殿

西岳庙 B—B 剖面图
Section B-B of Xiyue Temple

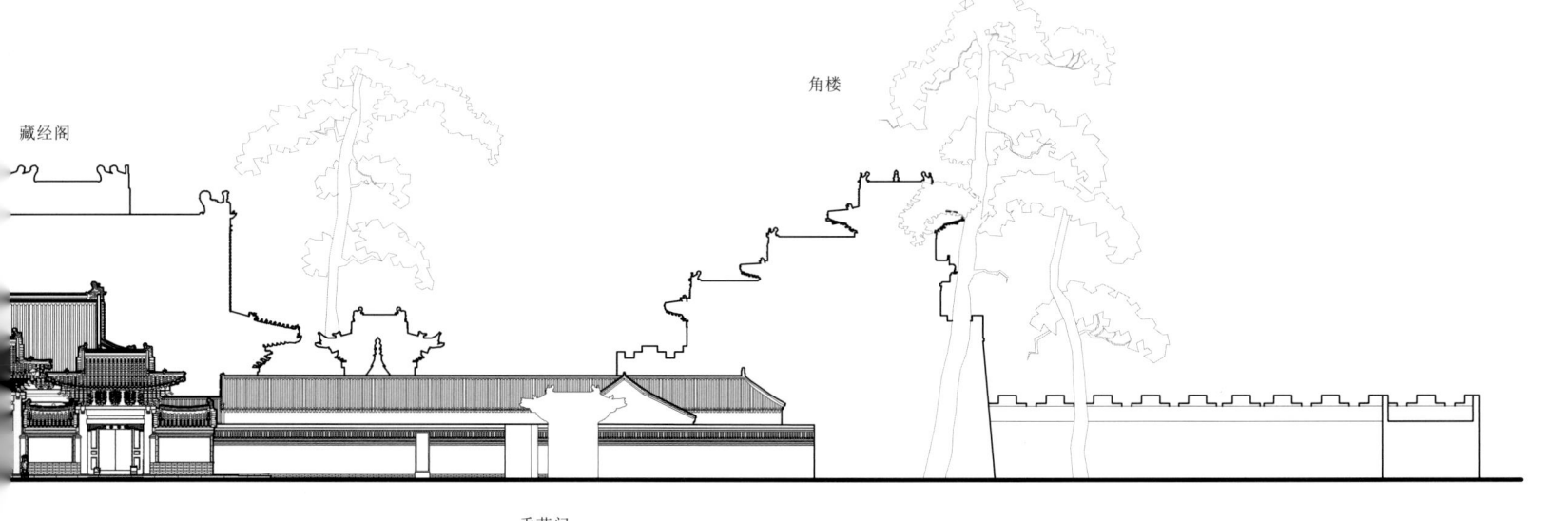

藏经图

角楼

垂花门

西岳庙 C—C 剖面图
Section C-C of Xiyue Temple

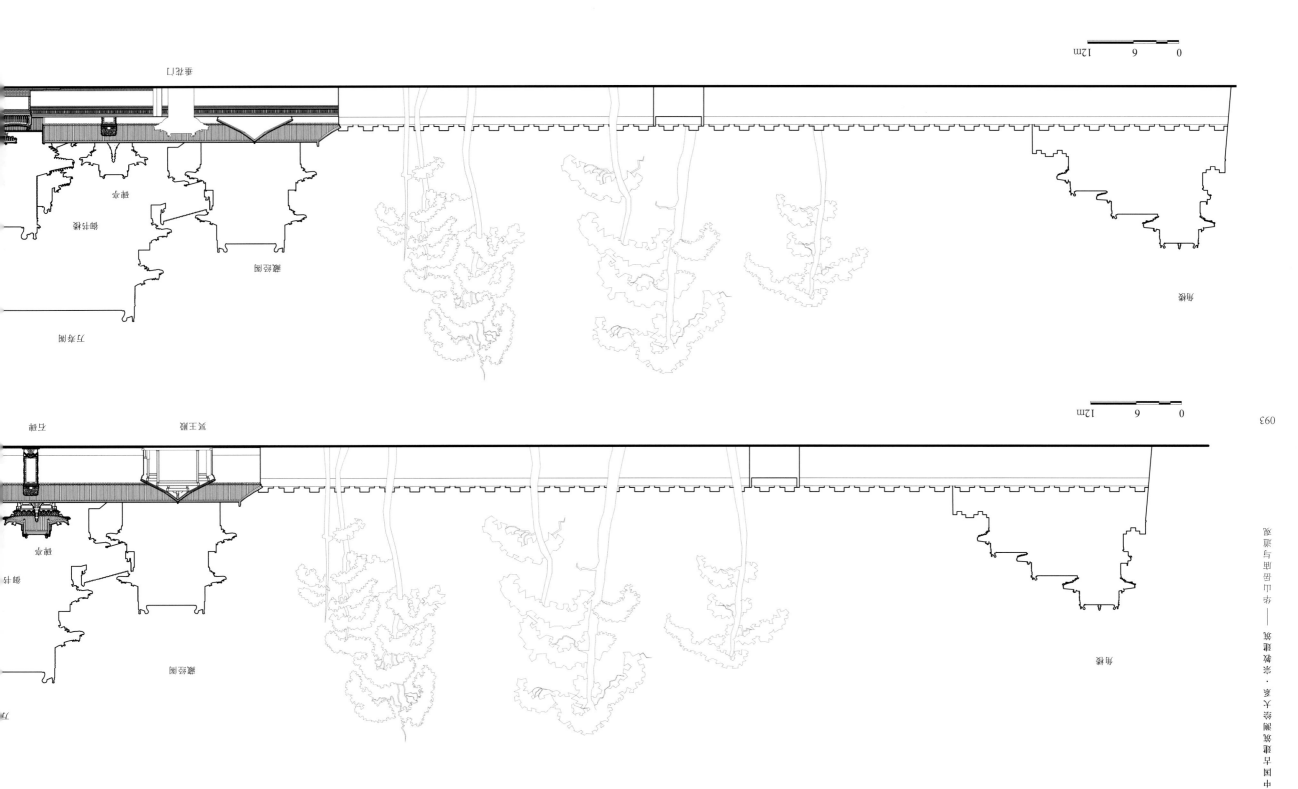

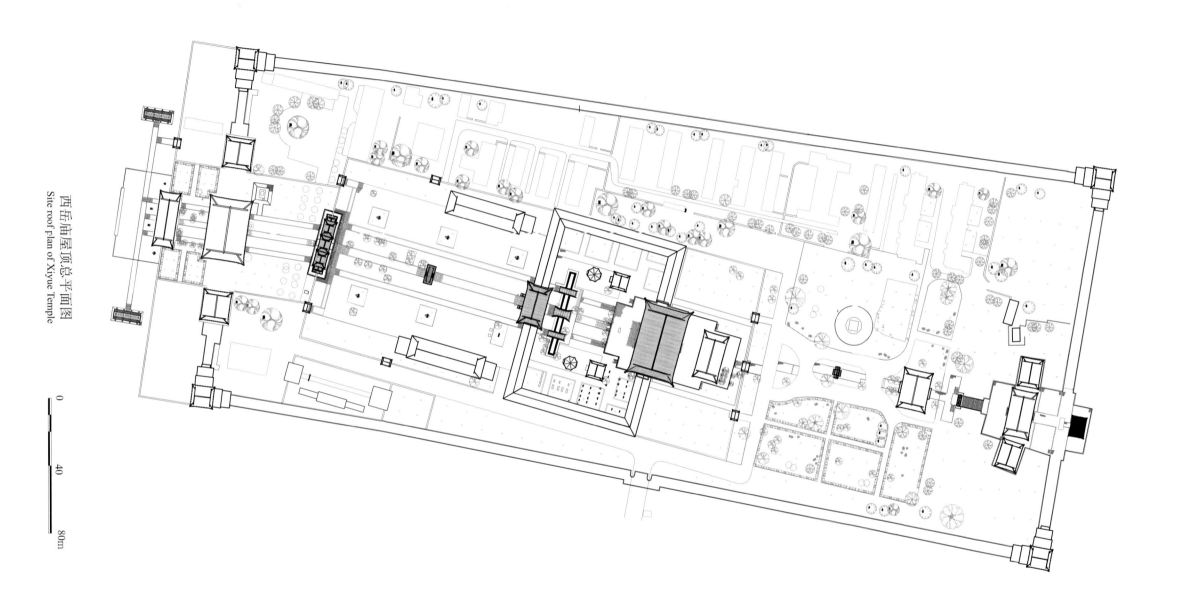

西岳庙屋顶总平面图
Site roof Plan of Xiyue Temple

0   40   80m

临漳与思想中华 —— 陈谦陈道·南天顶陈陈陈早国中

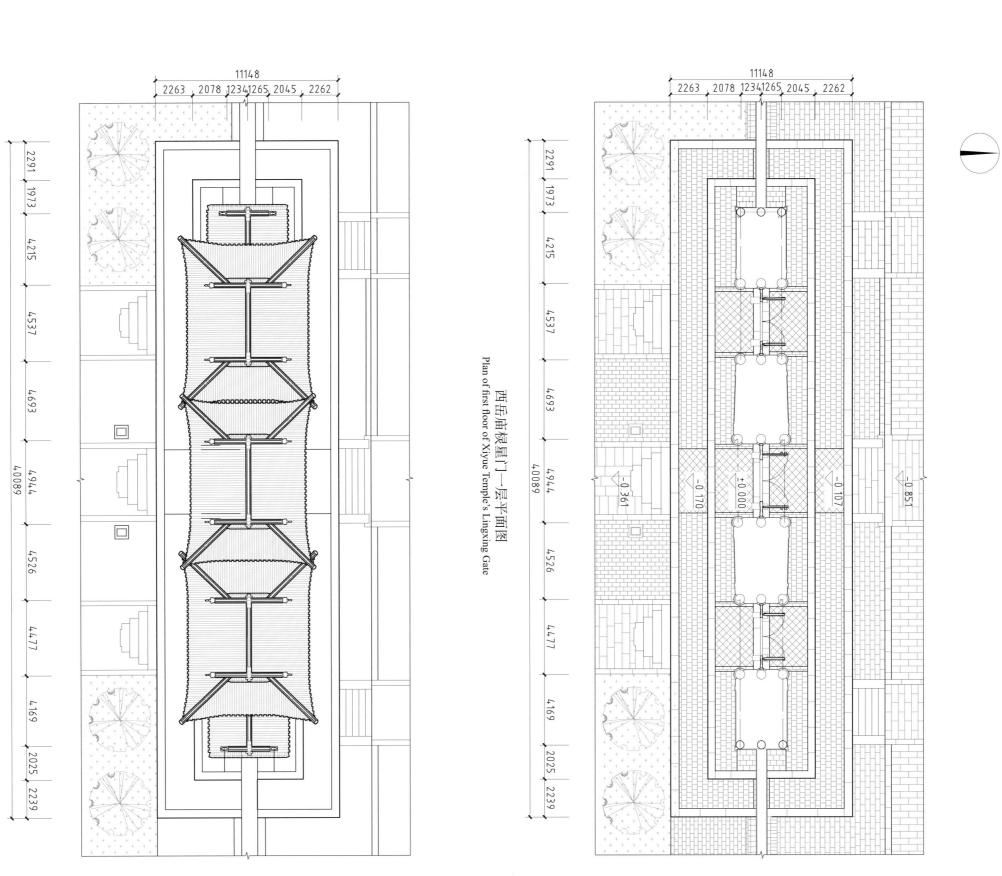

西岳庙棂星门屋顶平面图
Roof plan of Xiyue Temple's Lingxing Gate

西岳庙棂星门一层平面图
Plan of first floor of Xiyue Temple's Lingxing Gate

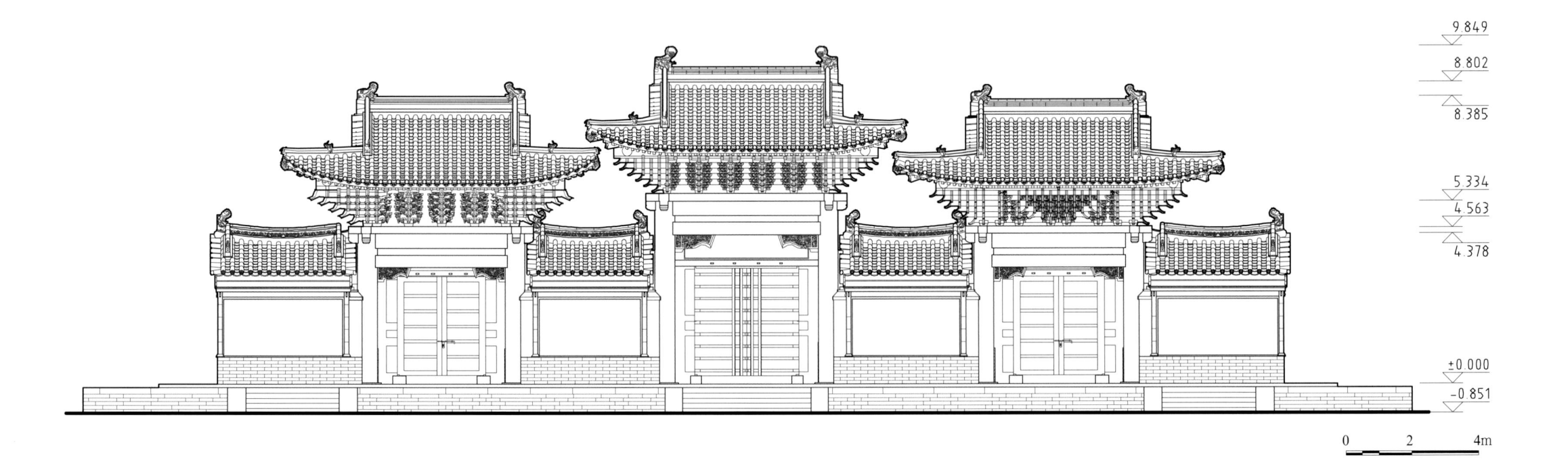

9.849

8.802

8.385

5.334

4.563

4.378

±0.000

-0.851

0　　2　　4m

西岳庙棂星门北立面图
North elevation of Xiyue Temple's Lingxing Gate

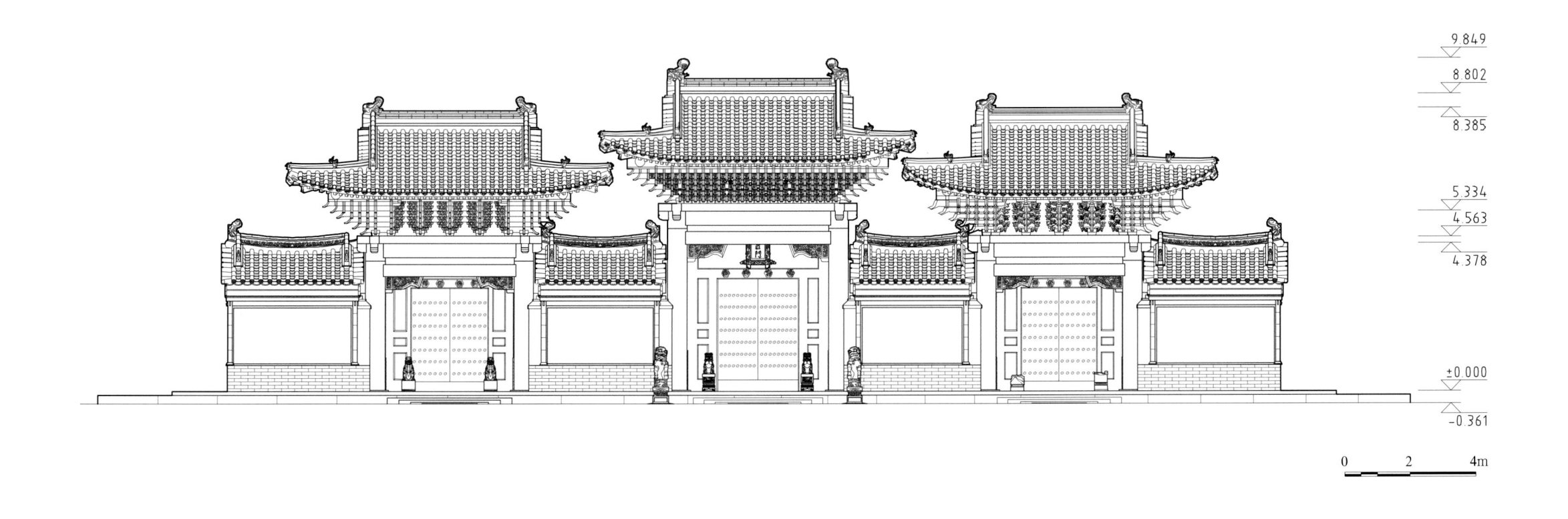

9 849

8 802

8 385

5 334

4 563

4 378

±0.000

-0.361

0　　2　　4m

西岳庙棂星门南立面图
South elevation of Xiyue Temple's Lingxing Gate

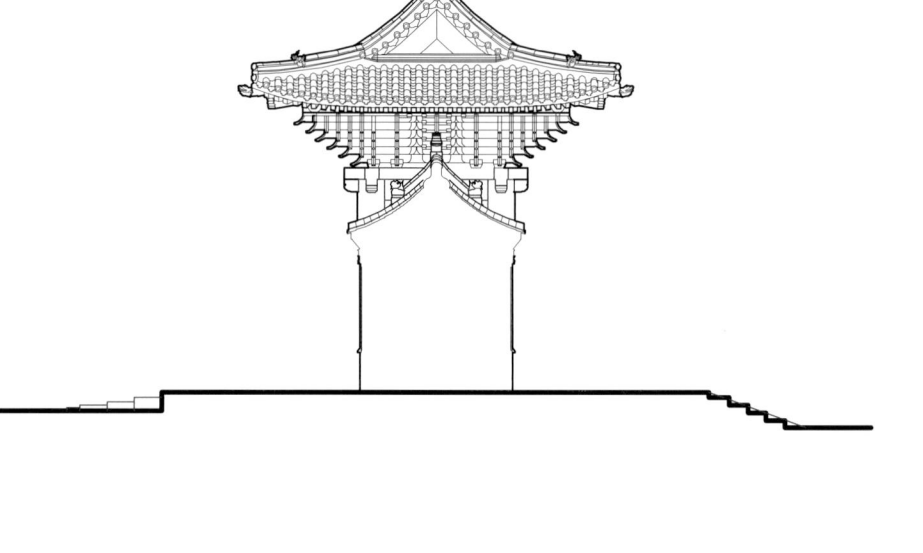

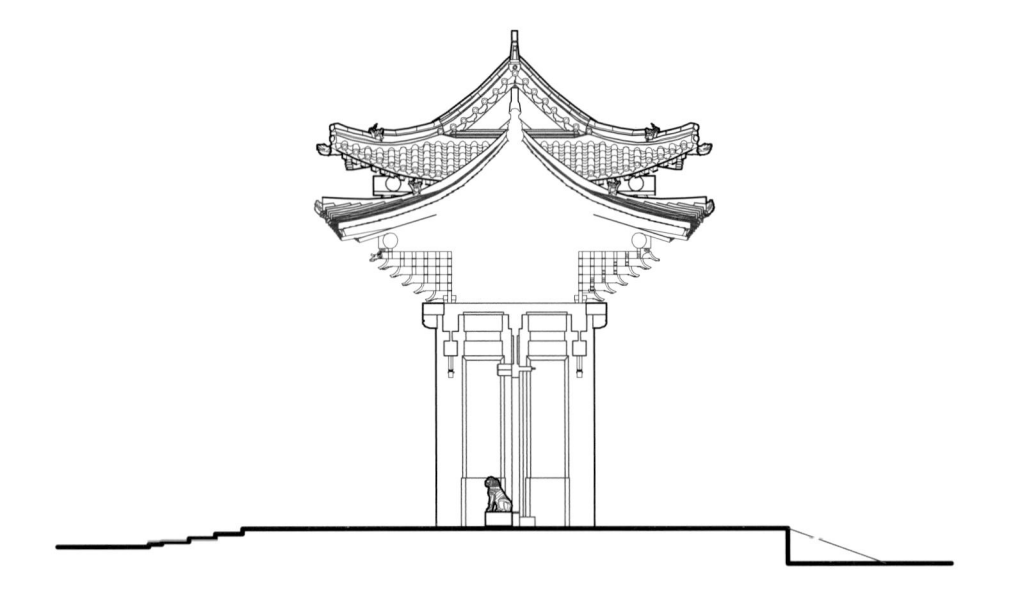

0　　2　　4m

西岳庙棂星门横剖面图（一）
Cross-section of Xiyue Temple's Lingxing Gate (1)

西岳庙棂星门横剖面图（二）
Cross-section of Xiyue Temple's Lingxing Gate (2)

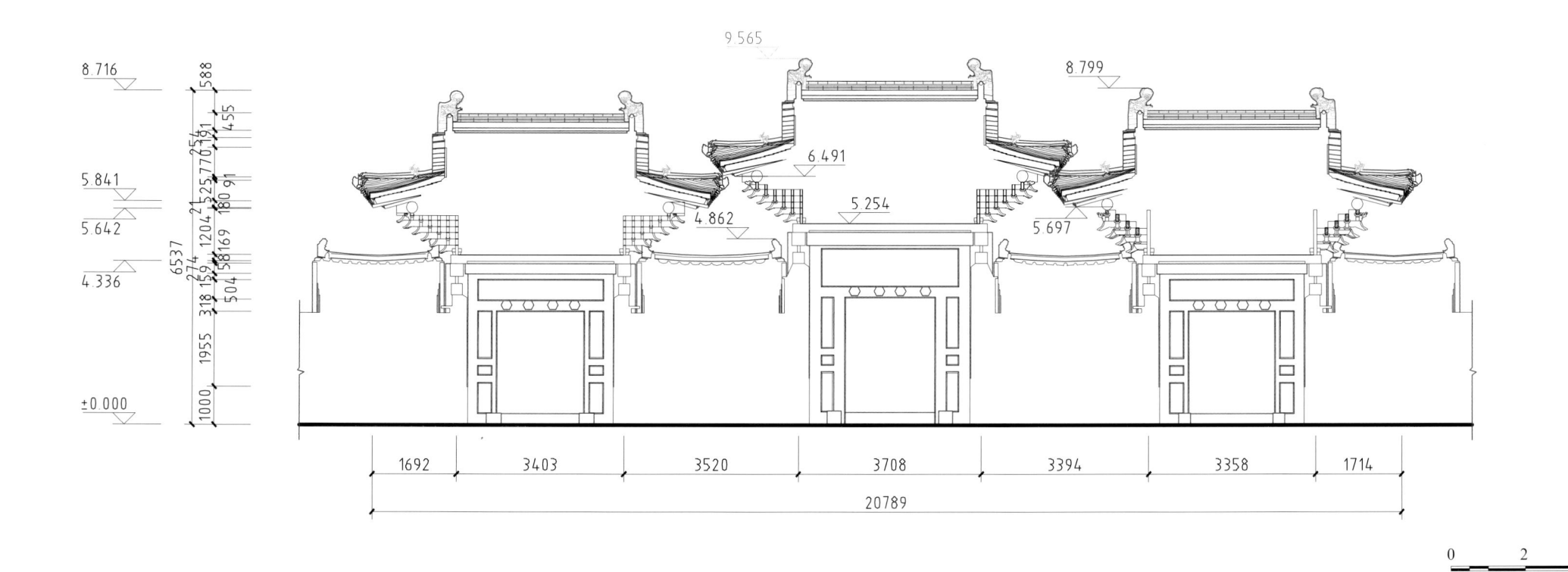

8.716

9.565

8.799

6.491

5.841

5.254

5.642

4.862

5.697

4.336

±0.000

588
256
1
21
577
91
1204
91
274
159
180
58169
504
318
1955
6537
1000
455

1692   3403   3520   3708   3394   3358   1714

20789

0   2   4m

西岳庙棂星门纵剖面图
Longitudinal section of Xiyue Temple's Lingxing Gate

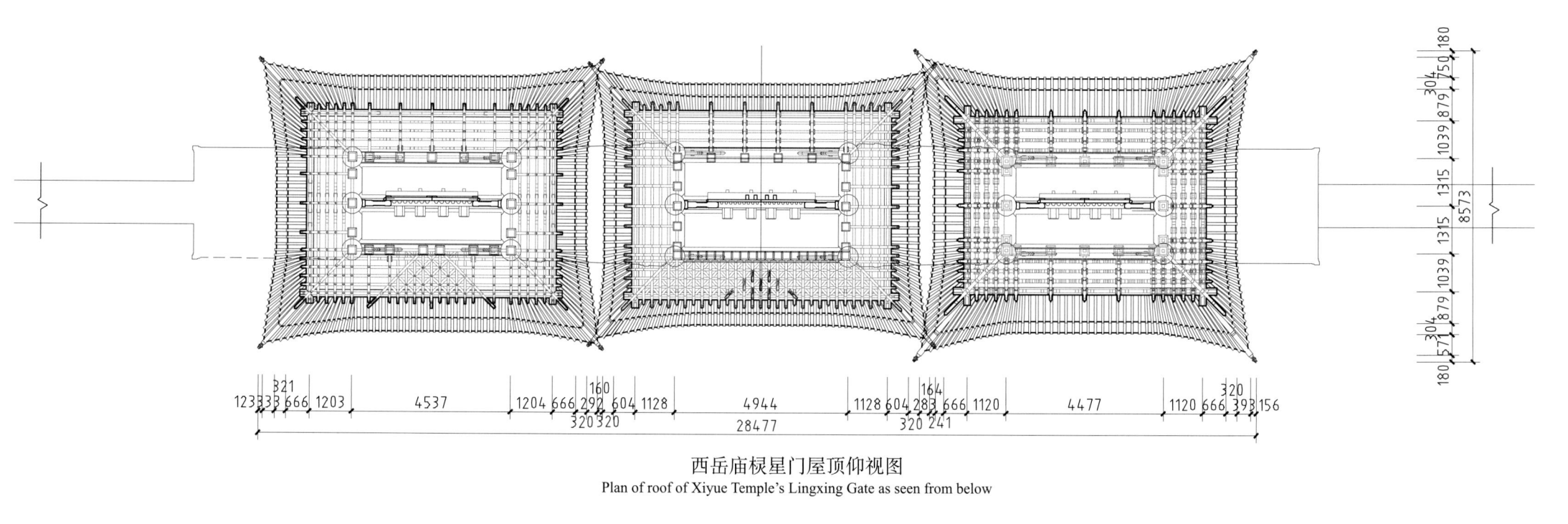

西岳庙棂星门屋顶仰视图
Plan of roof of Xiyue Temple's Lingxing Gate as seen from below

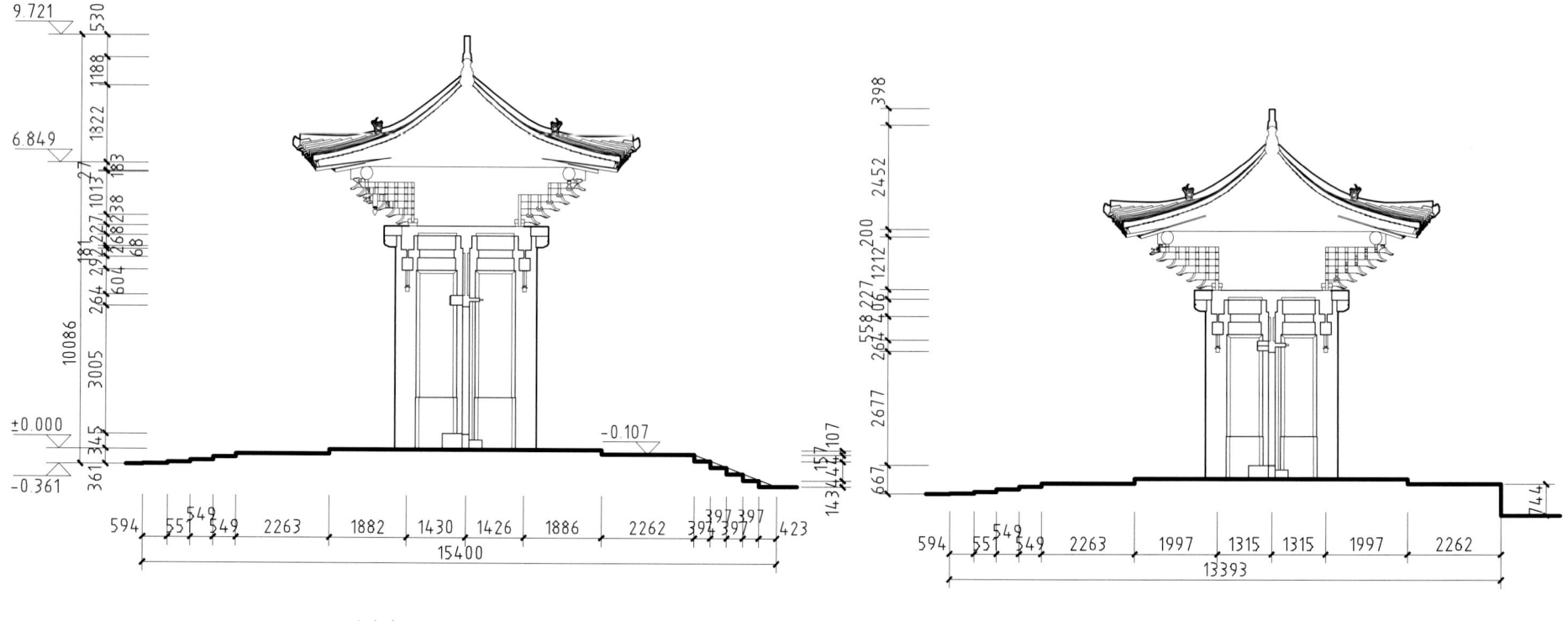

西岳庙棂星门中门横剖面图
Cross-section of Xiyue Temple's Lingxing Gate's central gate

西岳庙棂星门边门横剖面图
Cross-section of Xiyue Temple's Lingxing Gate's side gate

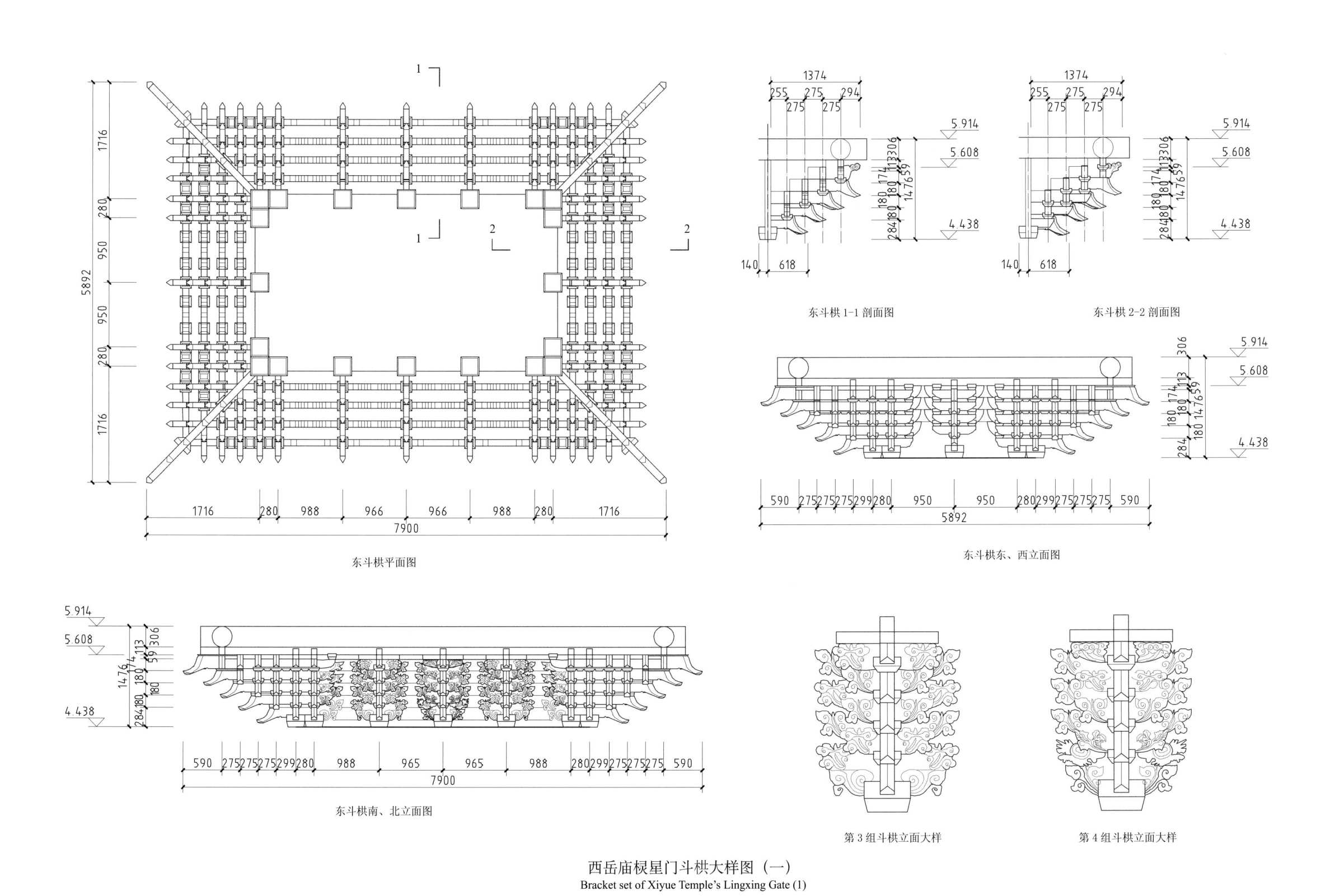

东斗栱 1-1 剖面图

东斗栱 2-2 剖面图

东斗栱平面图

东斗栱东、西立面图

东斗栱南、北立面图

第 3 组斗栱立面大样

第 4 组斗栱立面大样

西岳庙棂星门斗栱大样图（一）
Bracket set of Xiyue Temple's Lingxing Gate (1)

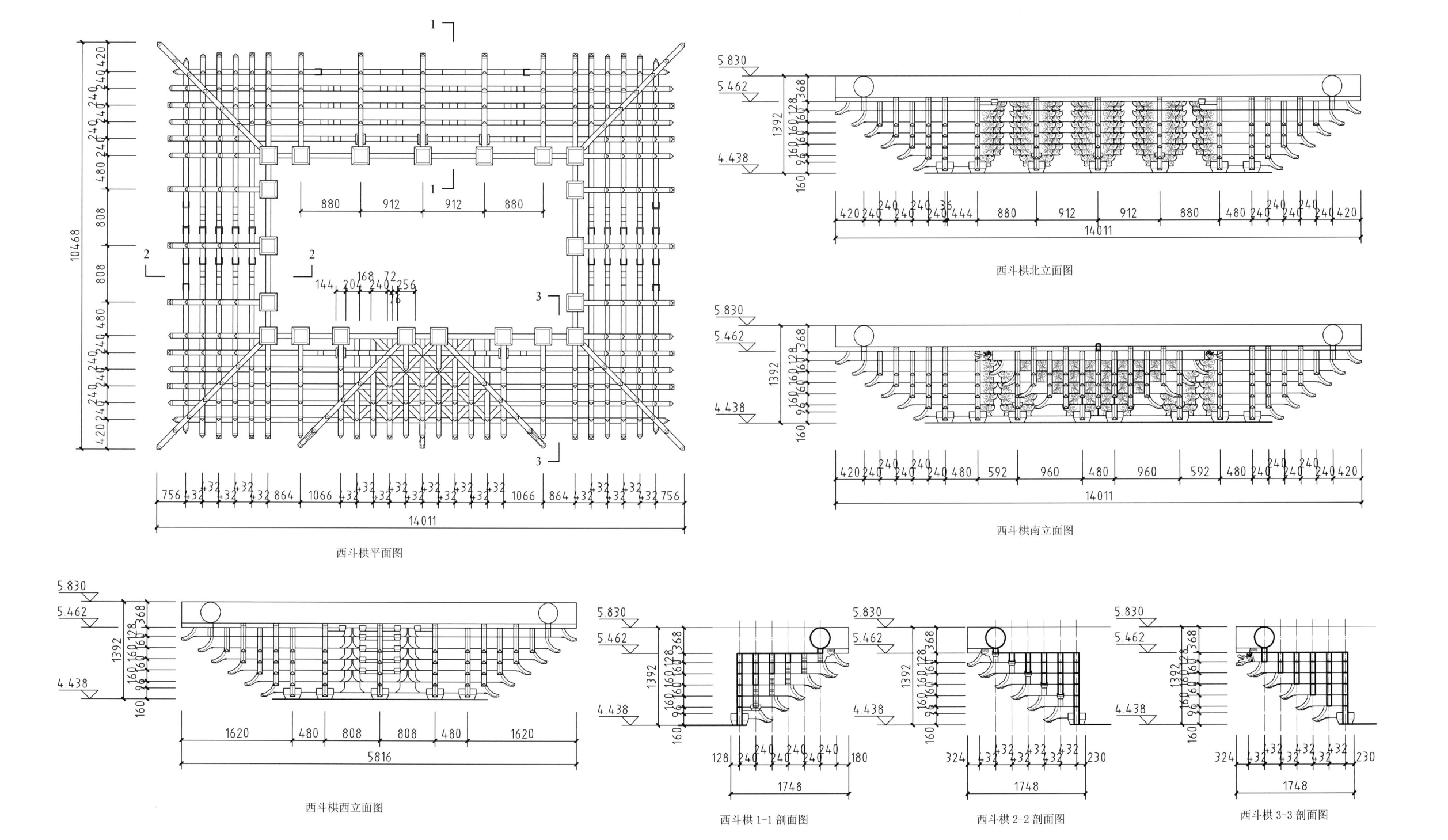

西斗栱平面图

西斗栱北立面图

西斗栱南立面图

西斗栱西立面图

西斗栱 1-1 剖面图

西斗栱 2-2 剖面图

西斗栱 3-3 剖面图

西岳庙棂星门斗栱大样图（二）
Bracket set of Xiyue Temple's Lingxing Gate (2)

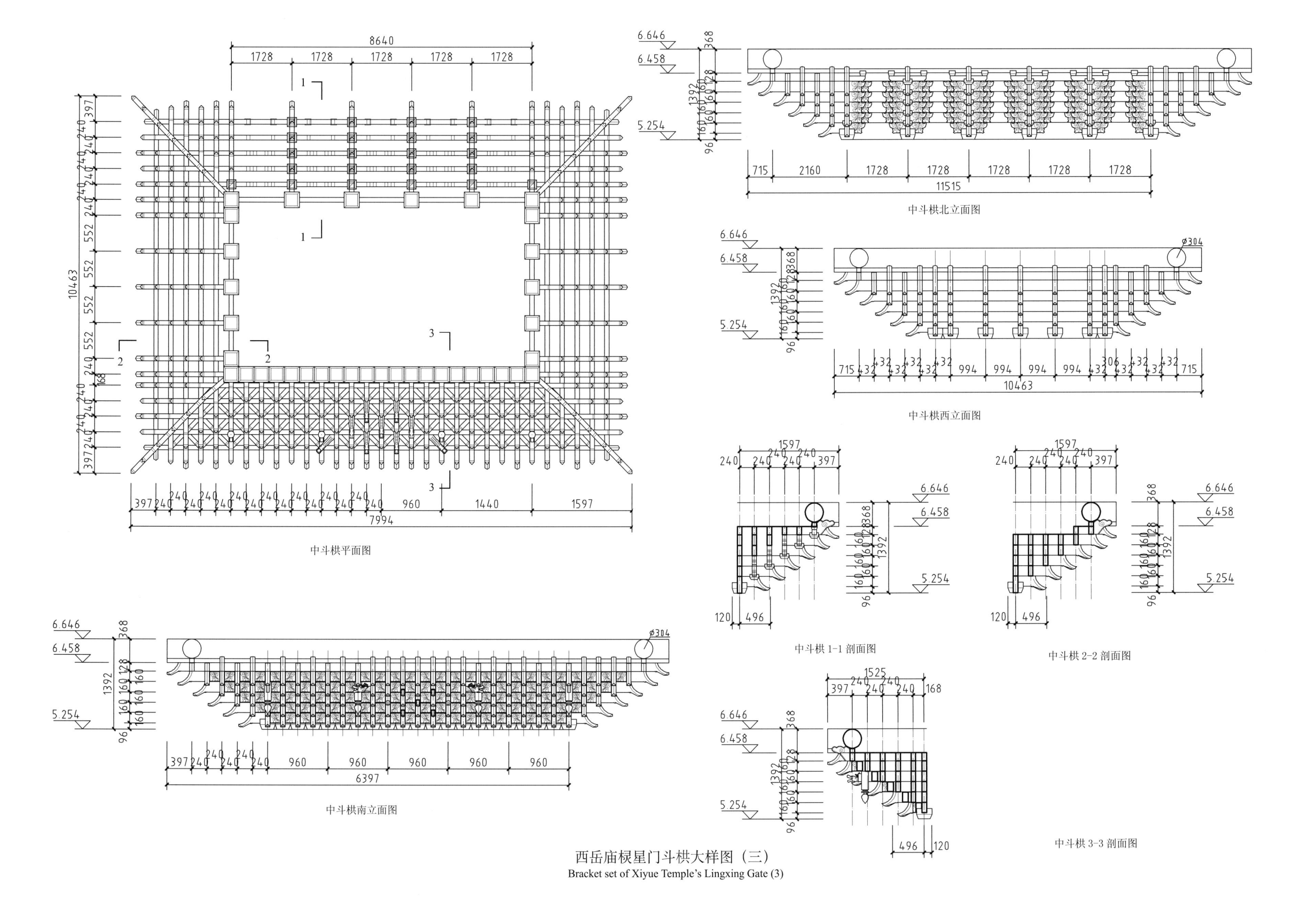

中斗栱平面图

中斗栱北立面图

中斗栱西立面图

中斗栱南立面图

中斗栱 1-1 剖面图

中斗栱 2-2 剖面图

中斗栱 3-3 剖面图

西岳庙棂星门斗栱大样图（三）
Bracket set of Xiyue Temple's Lingxing Gate (3)

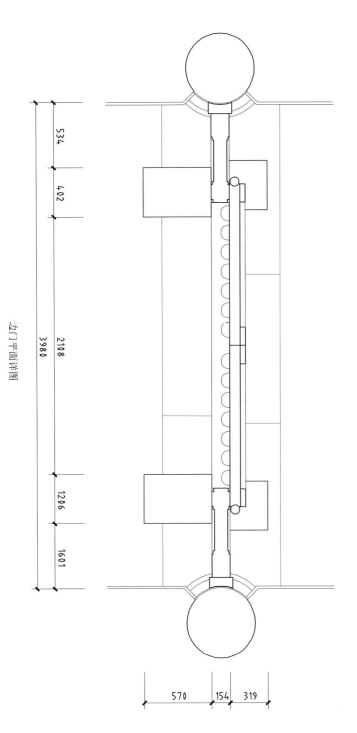

边门平面详图

西岳庙棂星门构件大样图（一）
Structural component of Xiyue Temple's Lingxing Gate (1)

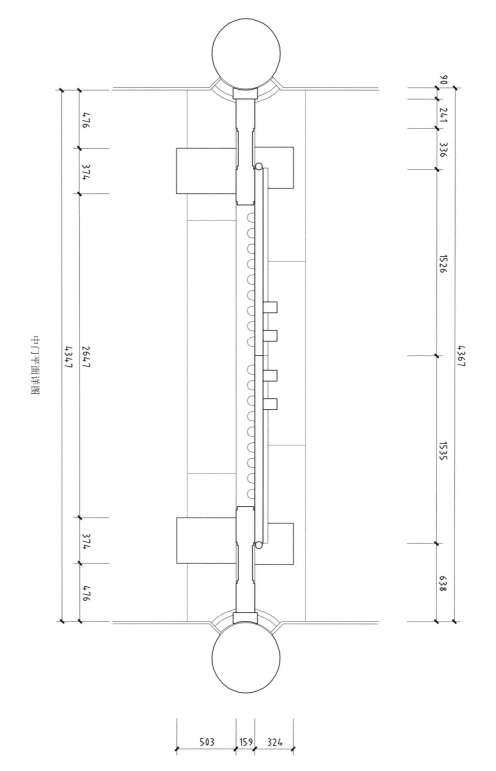

中门平面详图

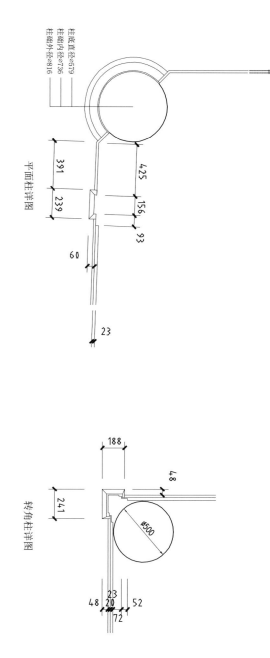

平面柱详图

柱底直径ø579
柱础内径ø736
柱础外径ø816

转角柱详图

ø500

中国古建筑营造技术·陕西韩城——党家村与周边地区

西门右狮正立面（残损状态）

西门右狮侧立面（残损状态）

西门右狮平面（残损状态）

东门石狮侧立面（残损状态）

东门石狮正立面（残损状态）

东门石狮平面（残损状态）

中门石狮

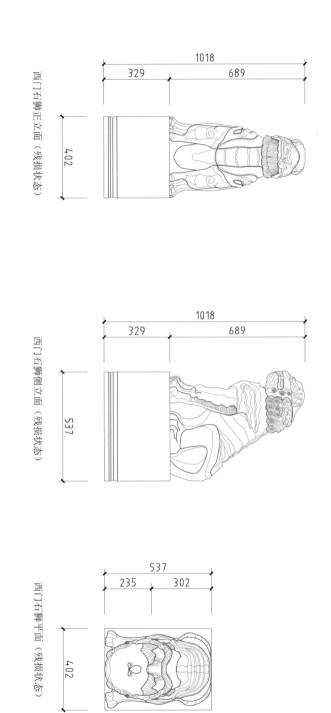

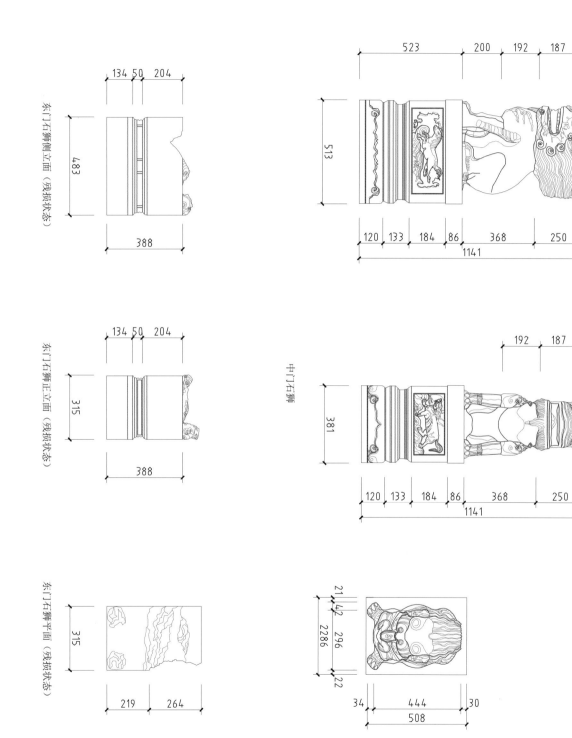

西岳庙棂星门构件大样图（二）
Structural component of Xiyue Temple's Lingxing Gate (2)

中国古代建筑知识普及与传承系列丛书·故宫营造

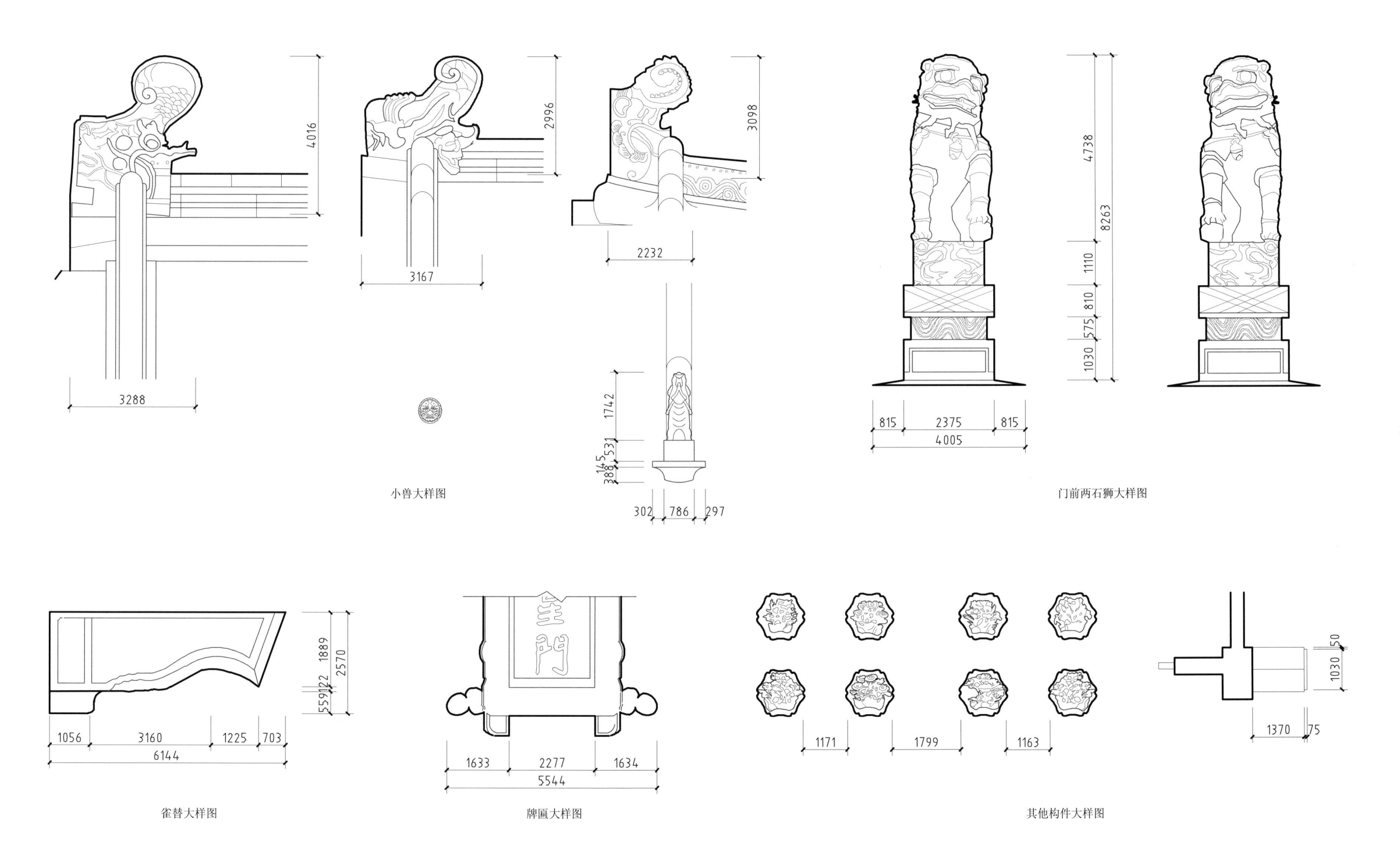

小兽大样图

门前两石狮大样图

雀替大样图

牌匾大样图

其他构件大样图

西岳庙棂星门构件大样图（三）
Structural component of Xiyue Temple's Lingxing Gate (3)

0   0.4   0.8m

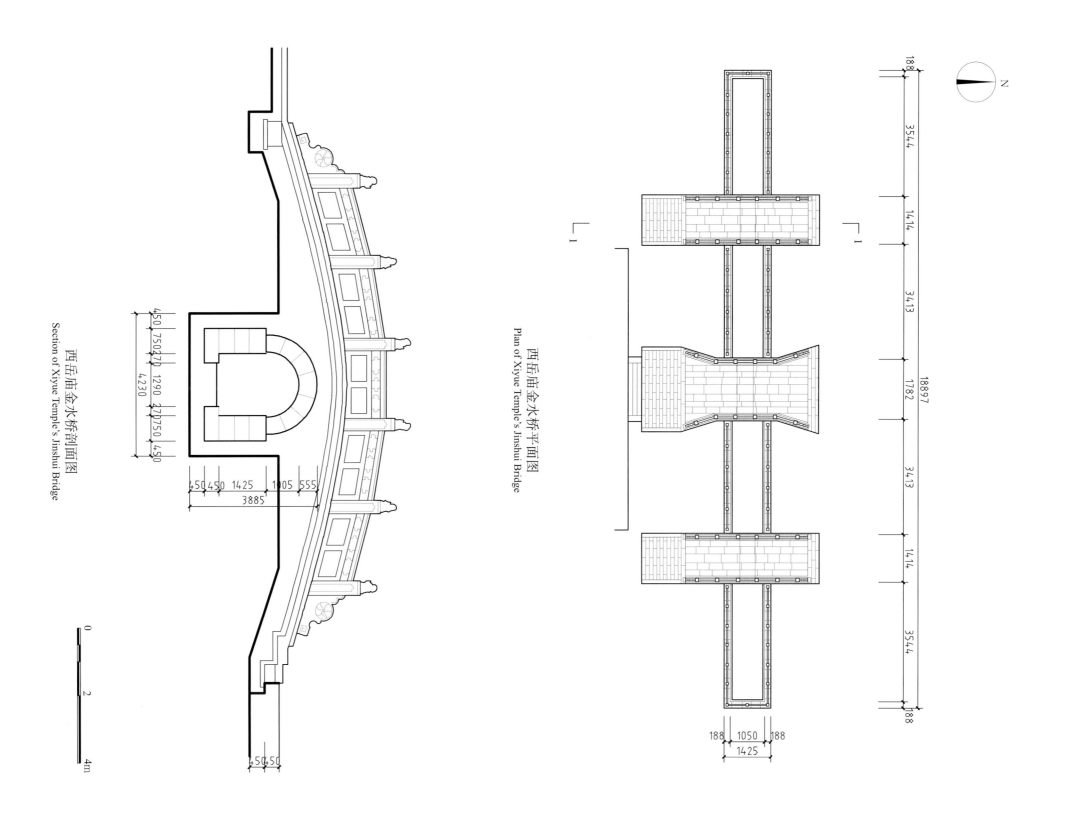

西岳庙金水桥剖面图
Section of Xiyue Temple's Jinshui Bridge

西岳庙金水桥平面图
Plan of Xiyue Temple's Jinshui Bridge

N

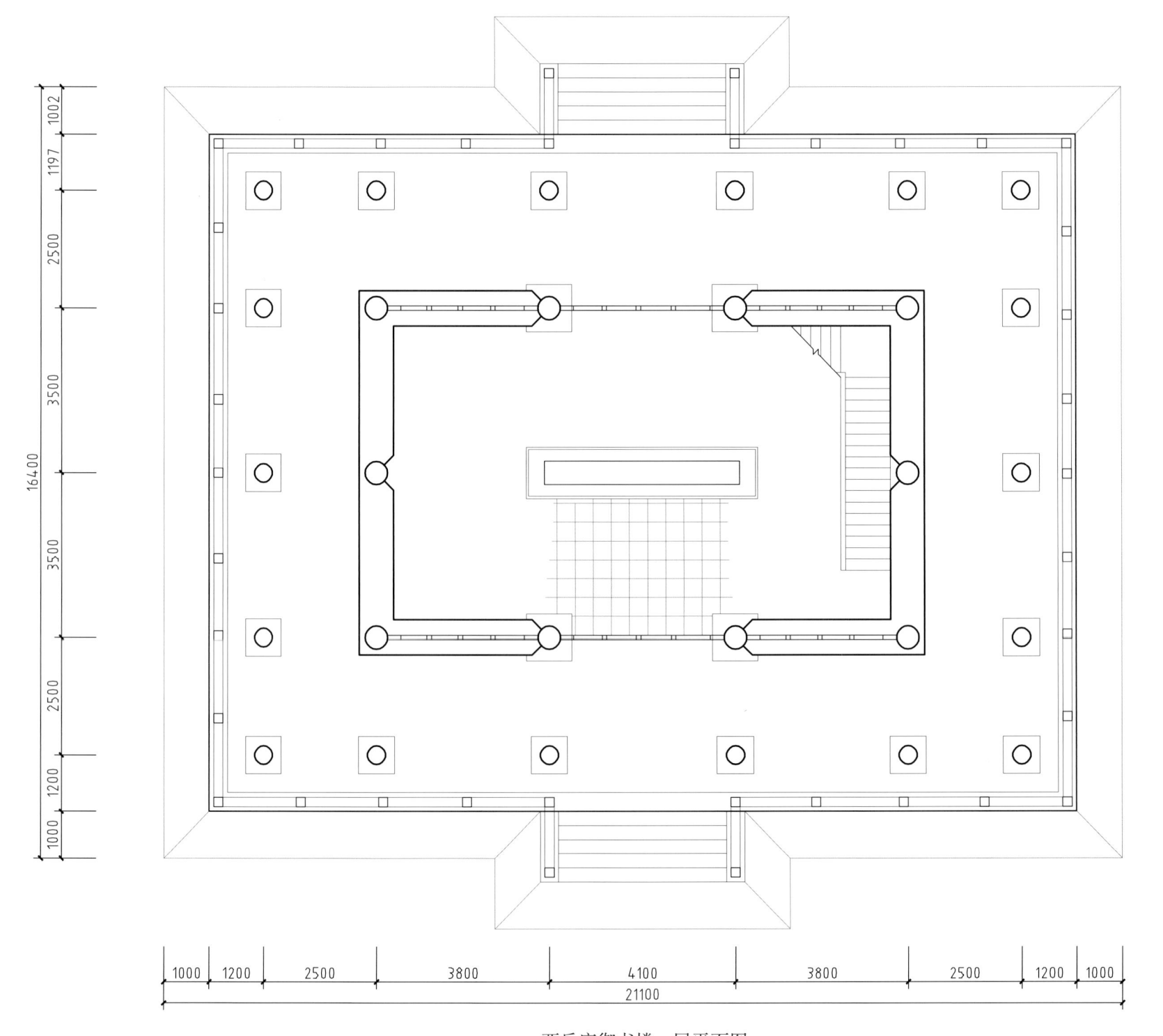

西岳庙御书楼一层平面图
Plan of first floor of Xiyue Temple's Yushulou

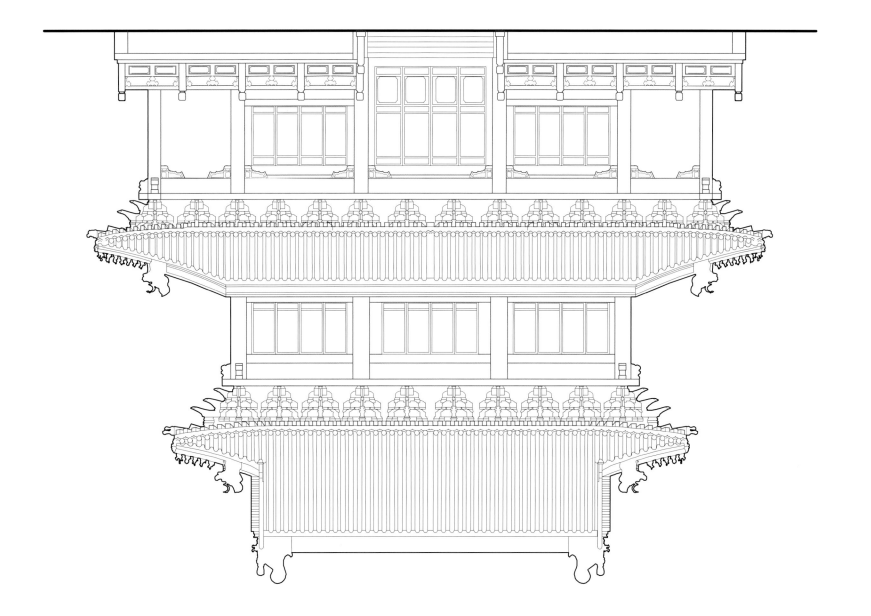

中国古建筑测绘大系·宗教建筑——西岳庙与耀州文庙

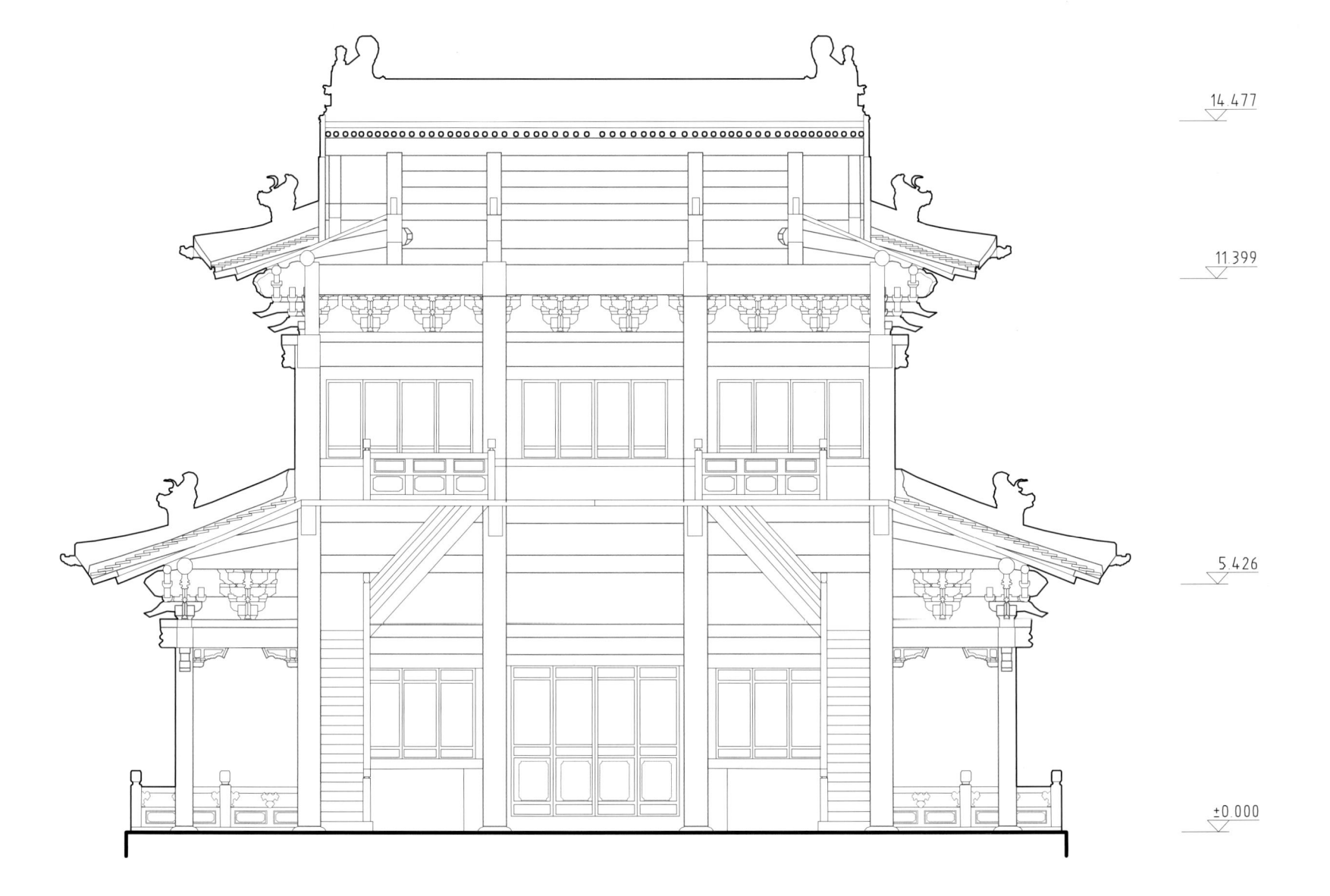

14.477

11.399

5.426

±0.000

西岳庙御书楼横剖面图
Cross-section of Xiyue Temple's Yushulou

0   2   4m

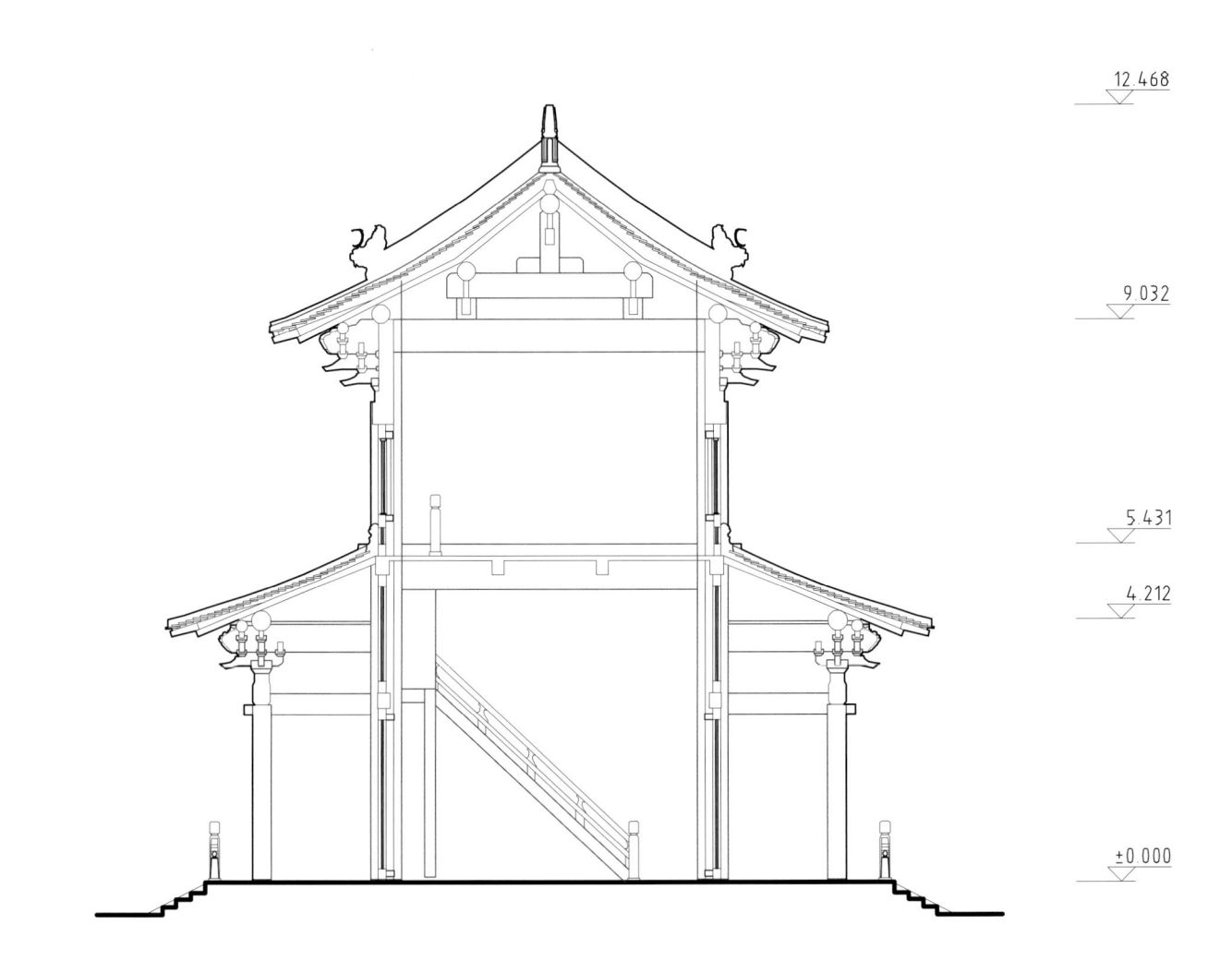

12.468

9.032

5.431

4.212

±0.000

西岳庙御书楼纵剖面图
Longitudinal section of Xiyue Temple's Yushulou

0　　2　　4m

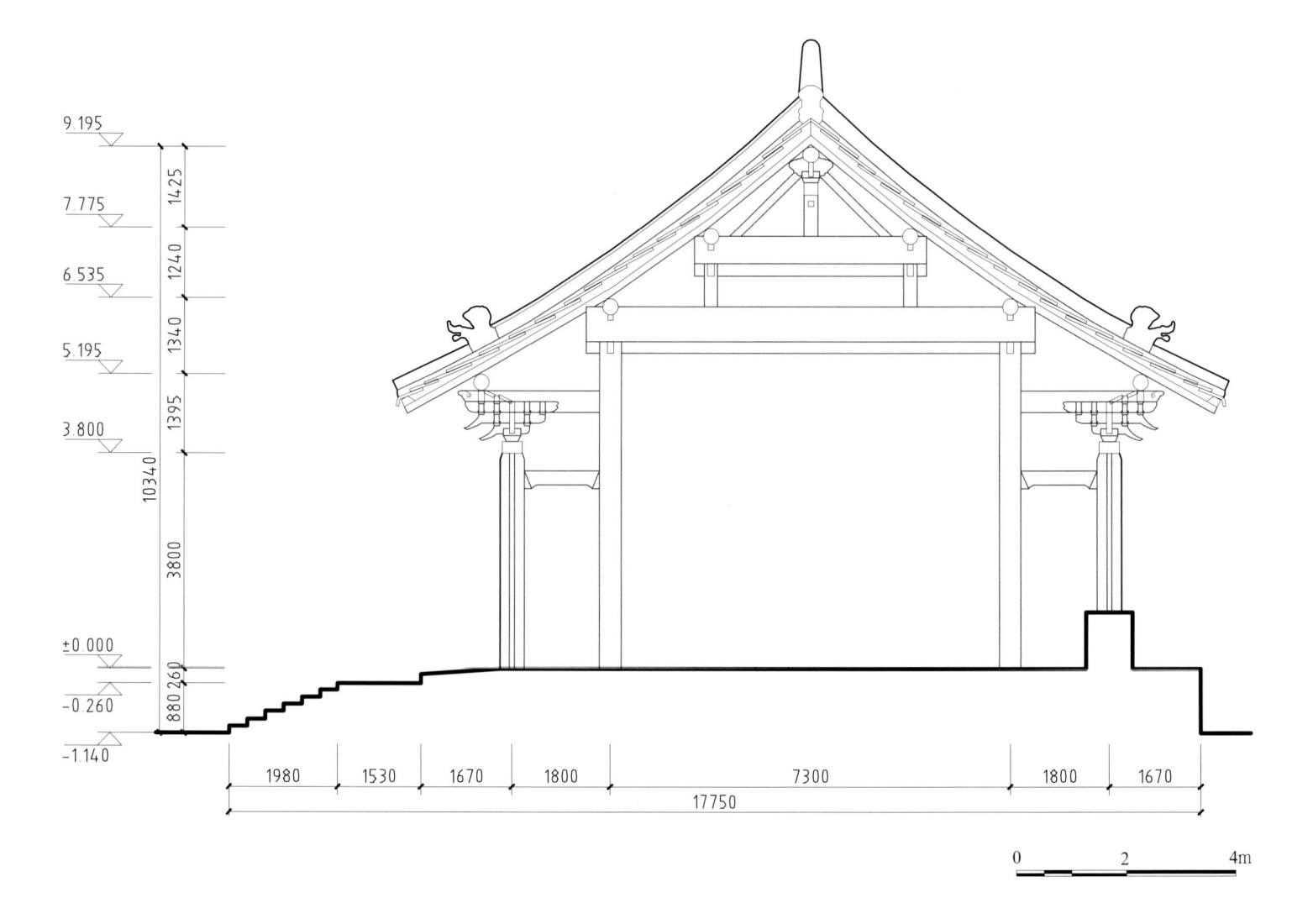

9.195

7.775

6.535

5.195

3.800

±0.000

-0.260

-1.140

1425

1240

1340

1395

10340

3800

880 260

1980 1530 1670 1800 7300 1800 1670

17750

0   2   4m

西岳庙寝宫剖面图
Section of Xiyue Temple's *qingong*

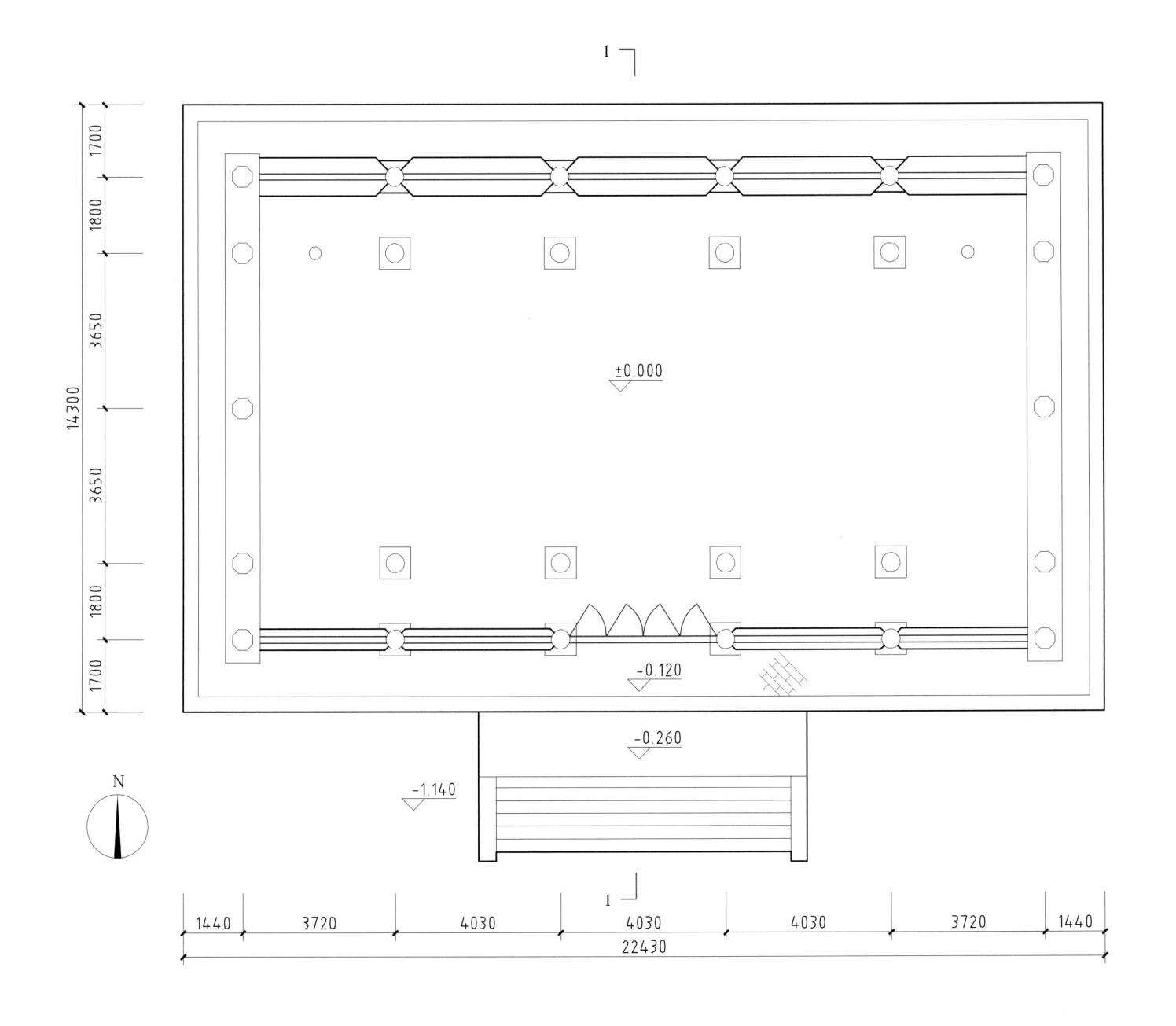

西岳庙寝宫平面图
Plan of Xiyue Temple's *qingong*

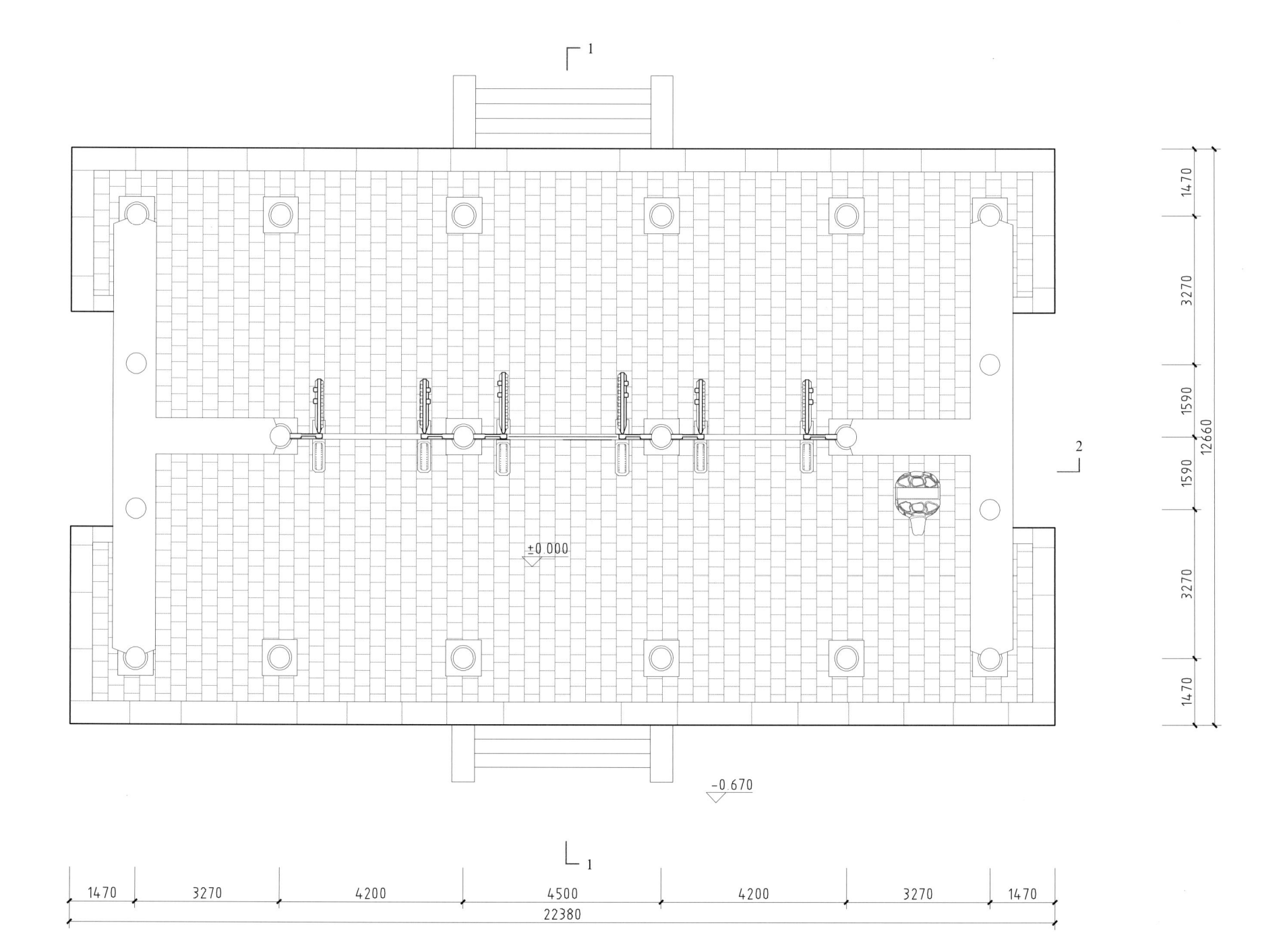

N

±0 000

−0 670

1470 3270 4200 4500 4200 3270 1470
22380

1470 3270 1590 1590 3270 1470
12660

西岳庙金城门平面图
Plan of Xiyue Temple's Jincheng Gate

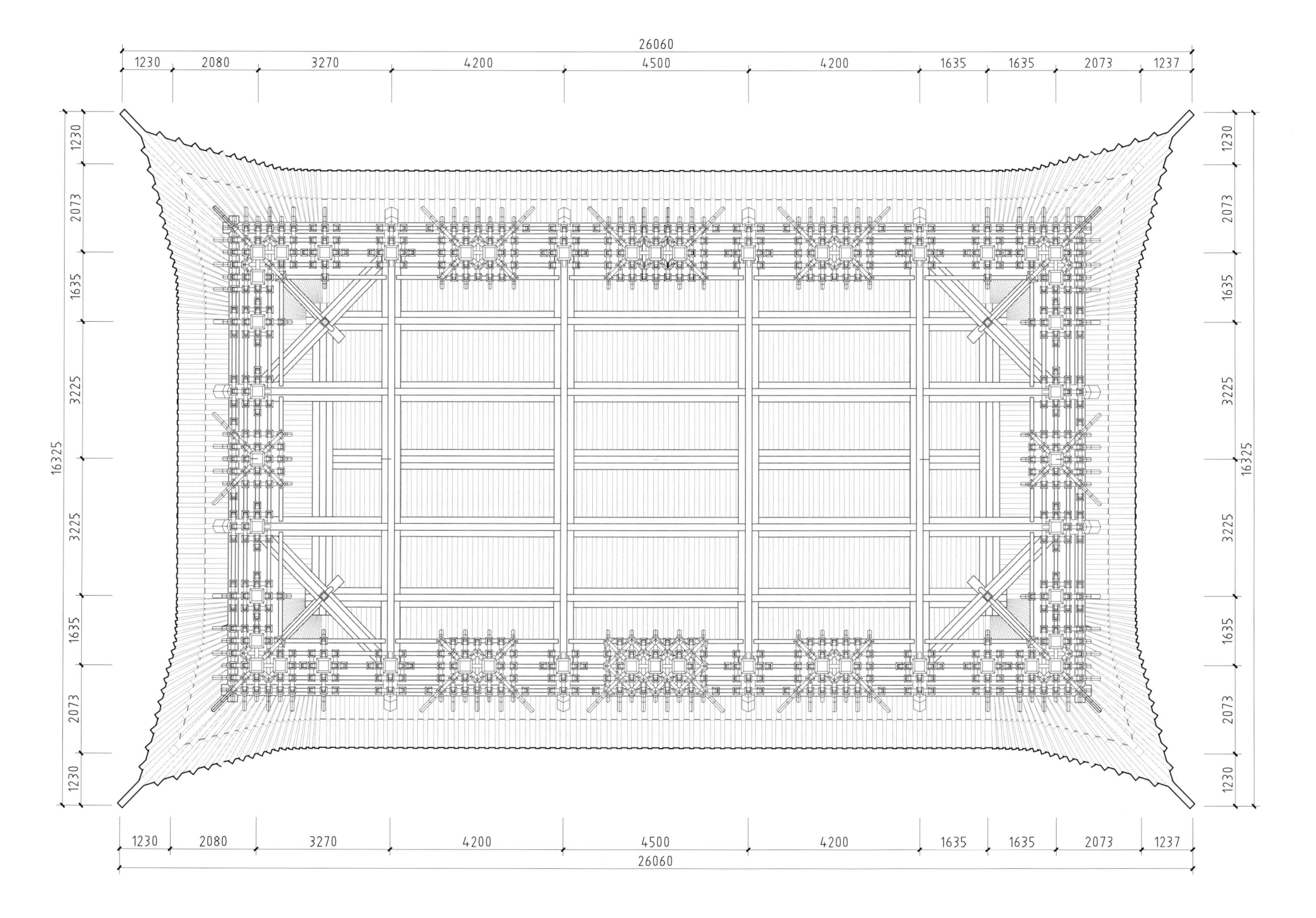

西岳庙金城门梁架仰视平面图
Plan of framework of Xiyue Temple's Jincheng Gate as seen from below

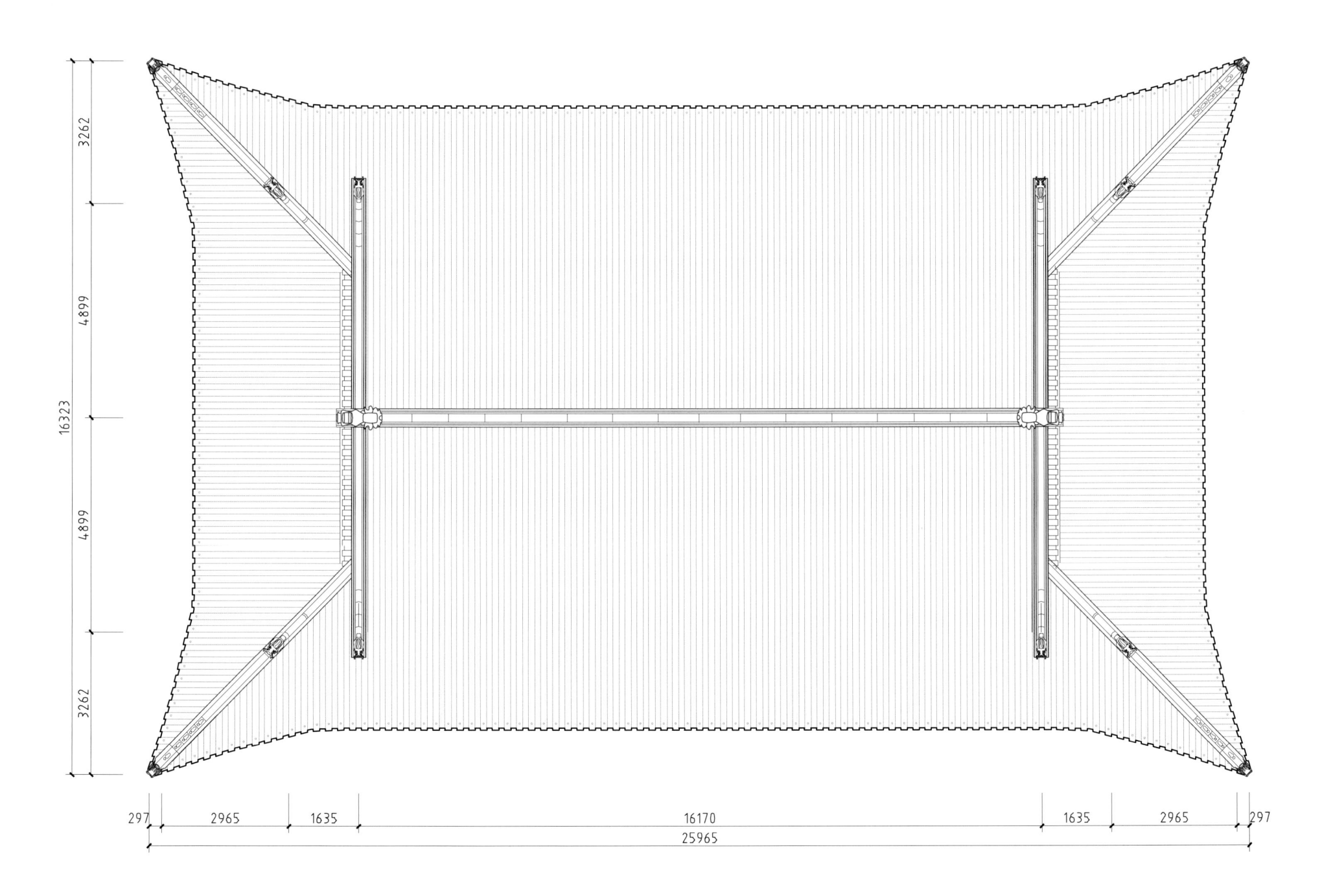

3262

4899

16323

4899

3262

297　2965　1635　16170　1635　2965　297

25965

西岳庙金城门屋顶平面图
Roof plan of Xiyue Temple's Jincheng Gate

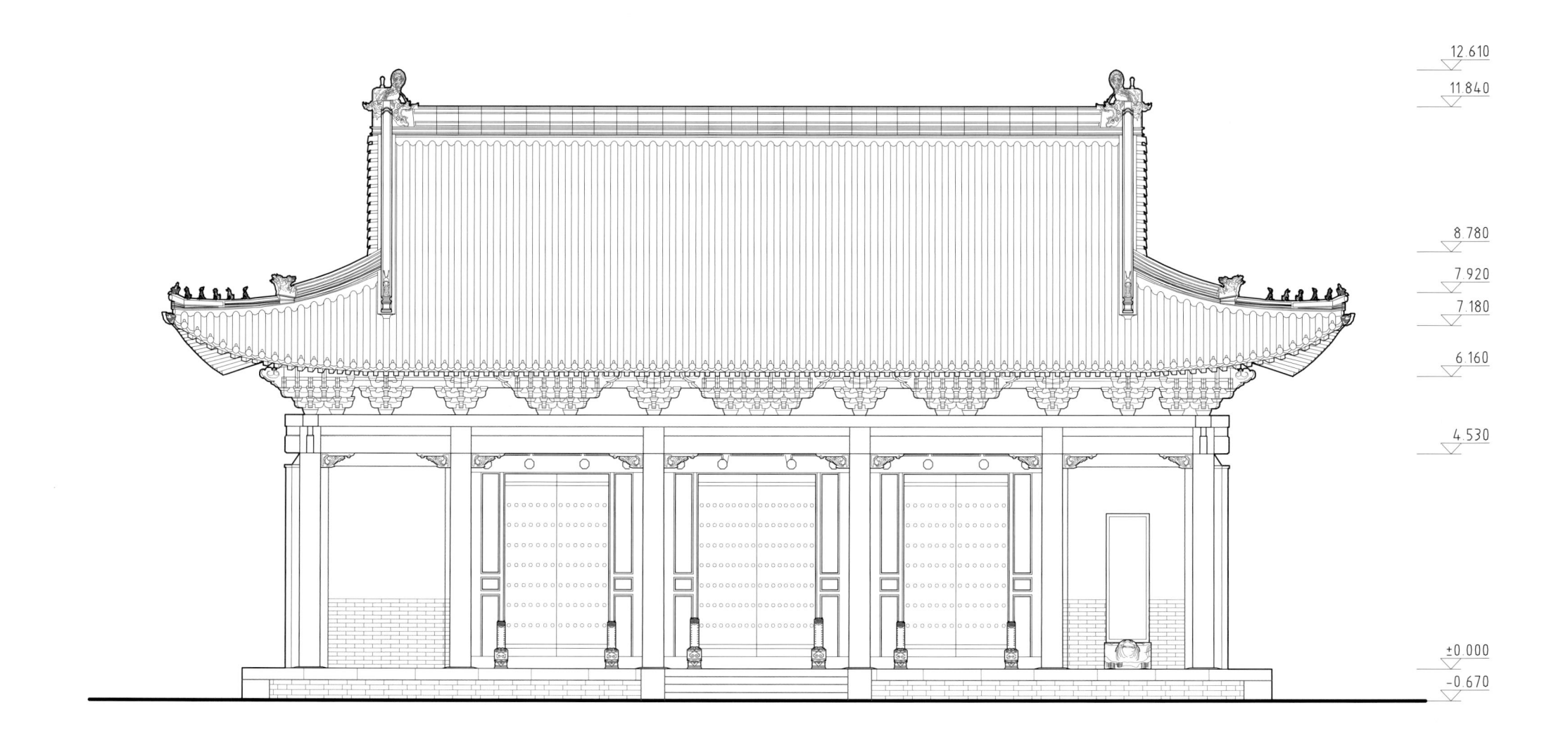

12 610

11 840

8 780

7 920

7 180

6 160

4 530

±0 000

-0 670

西岳庙金城门南立面图
South elevation of Xiyue Temple's Jincheng Gate

0   2   4m

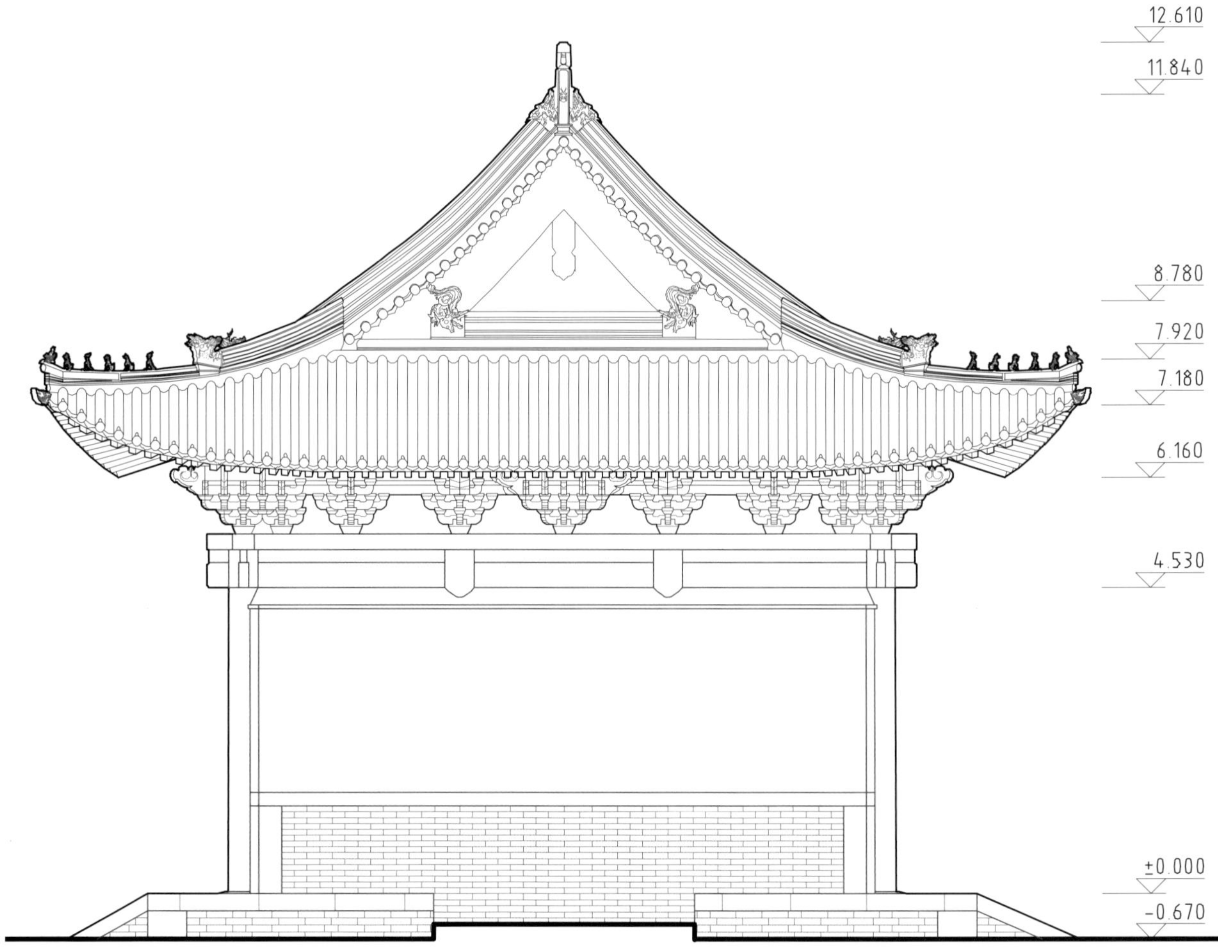

12.610

11.840

8.780

7.920

7.180

6.160

4.530

±0.000

-0.670

1470 1635 1635 1590 1590 1635 1635 1470

12660

西岳庙金城门东立面图
East elevation of Xiyue Temple's Jincheng Gate

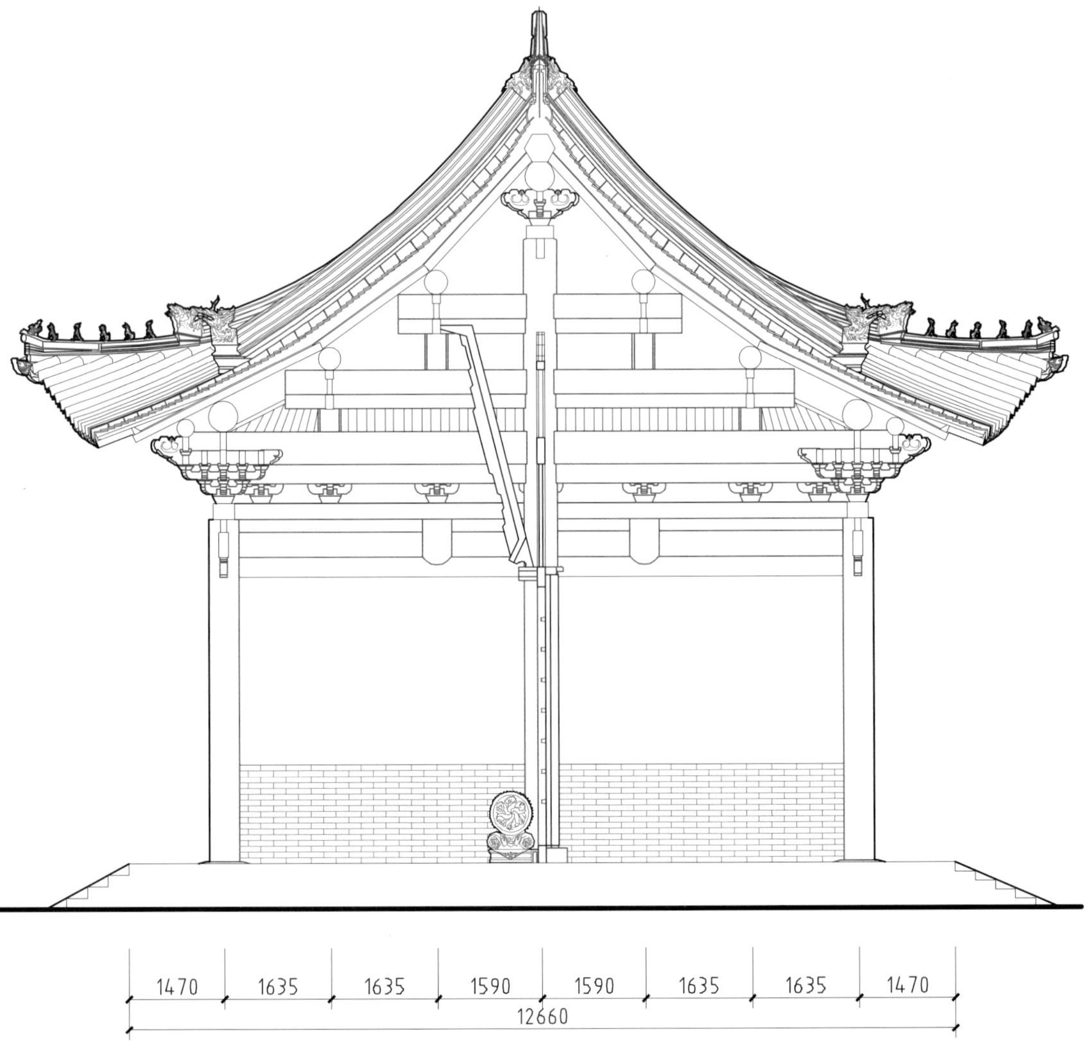

西岳庙金城门 1—1 剖面图
Section 1-1 of Xiyue Temple's Jincheng Gate

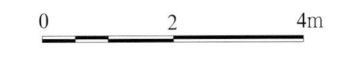

0   2   4m

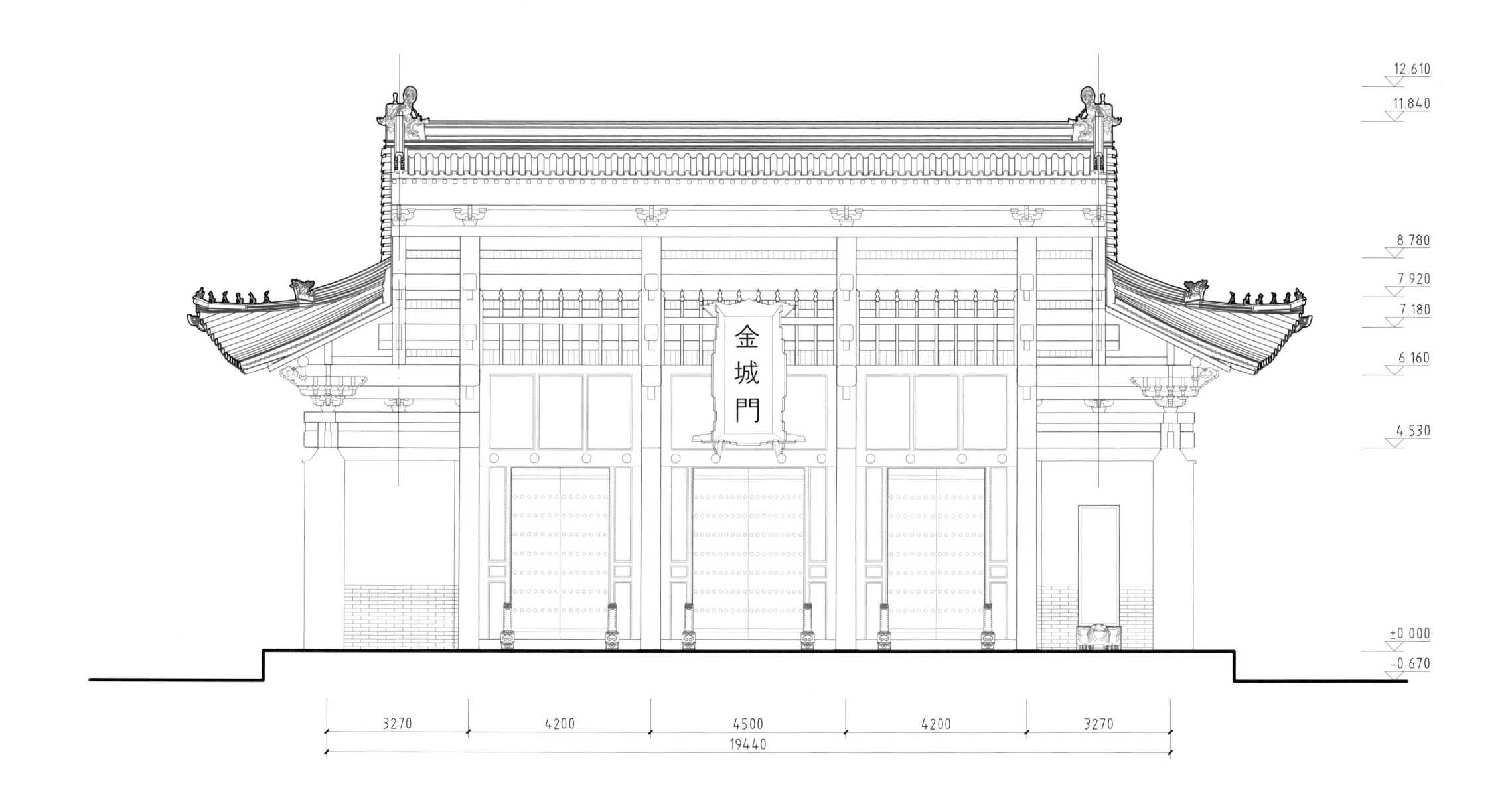

12.610

11.840

8.780

7.920

7.180

6.160

4.530

金
城
門

±0.000

-0.670

| 3270 | 4200 | 4500 | 4200 | 3270 |

19440

西岳庙金城门 2-2 剖面图
Section 2-2 of Xiyue Temple's Jincheng Gate

0   2   4m

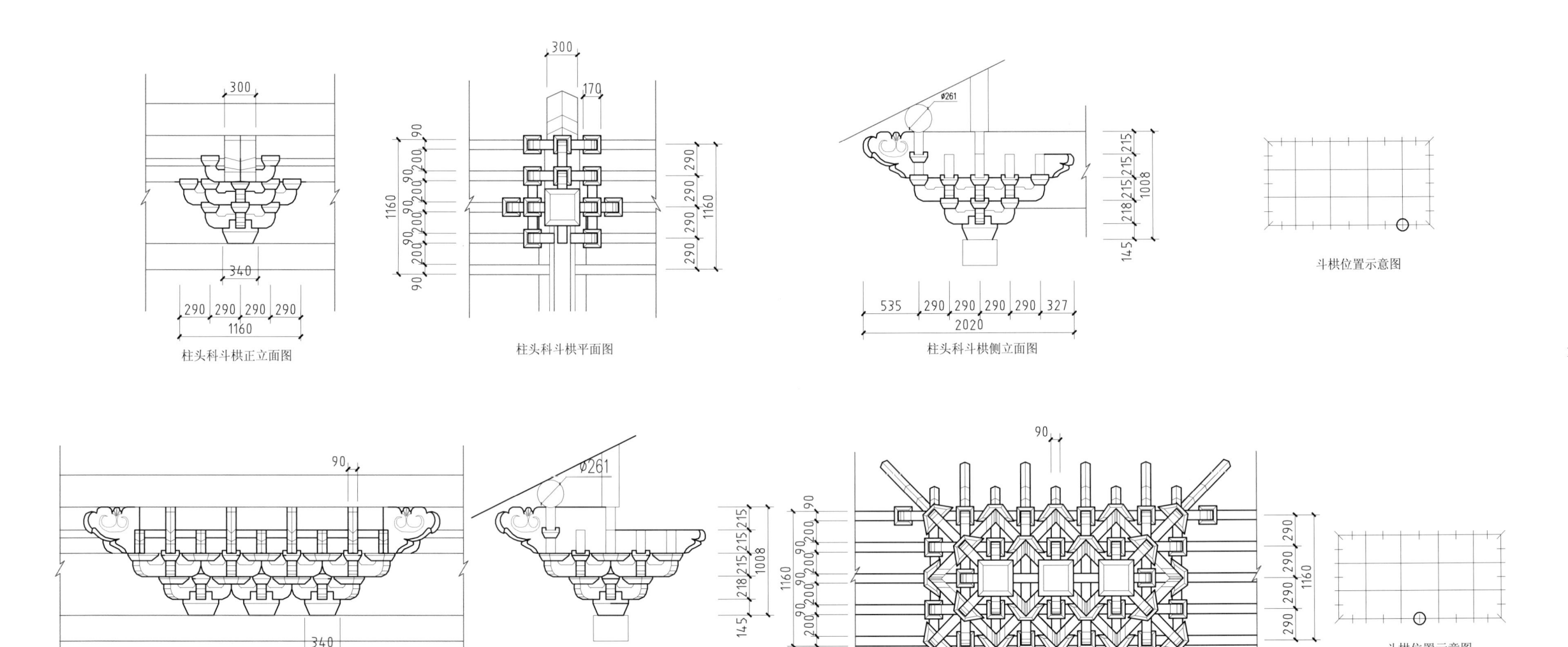

柱头科斗栱正立面图

柱头科斗栱平面图

柱头科斗栱侧立面图

斗栱位置示意图

平身科斗栱 1 正立面图

平身科斗栱 1 侧立面图

平身科斗栱 1 平面图

斗栱位置示意图

西岳庙金城门斗栱大样图（一）

Bracket set of Xiyue Temple's Jincheng Gate (1)

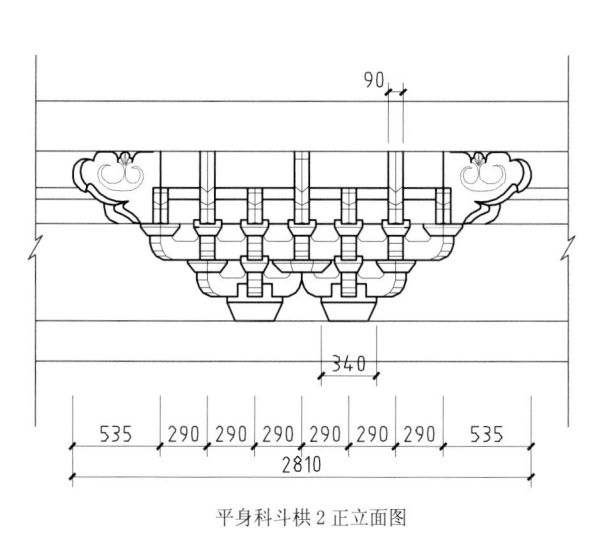

平身科斗栱 2 正立面图

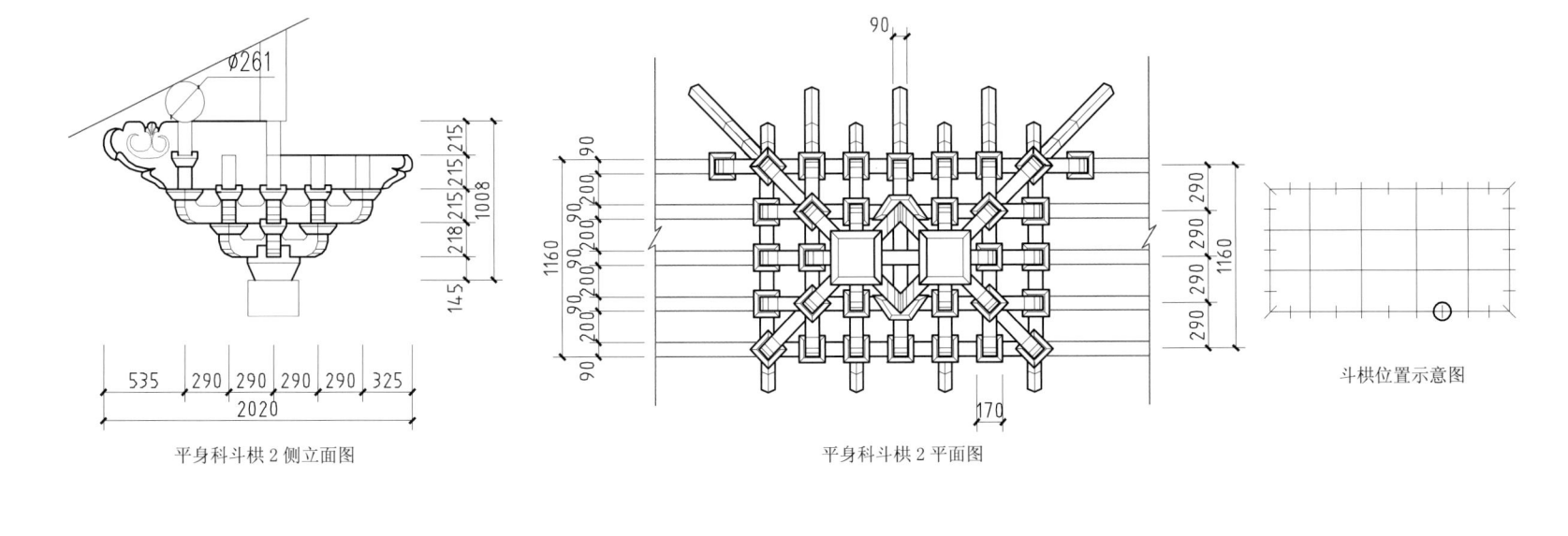

平身科斗栱 2 侧立面图

平身科斗栱 2 平面图

斗栱位置示意图

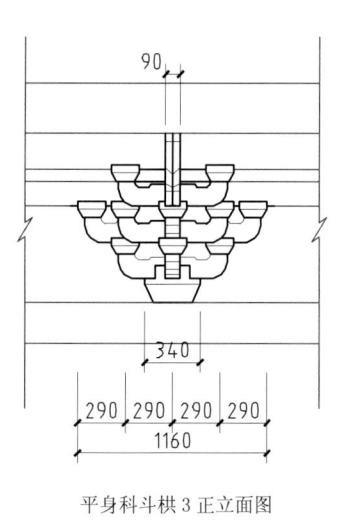

平身科斗栱 3 正立面图

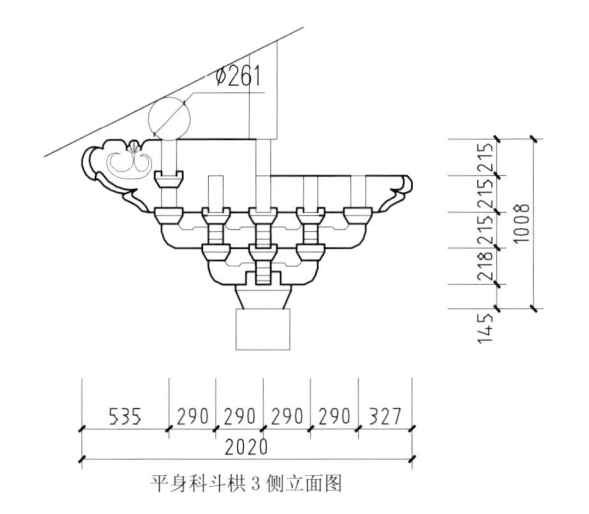

平身科斗栱 3 侧立面图

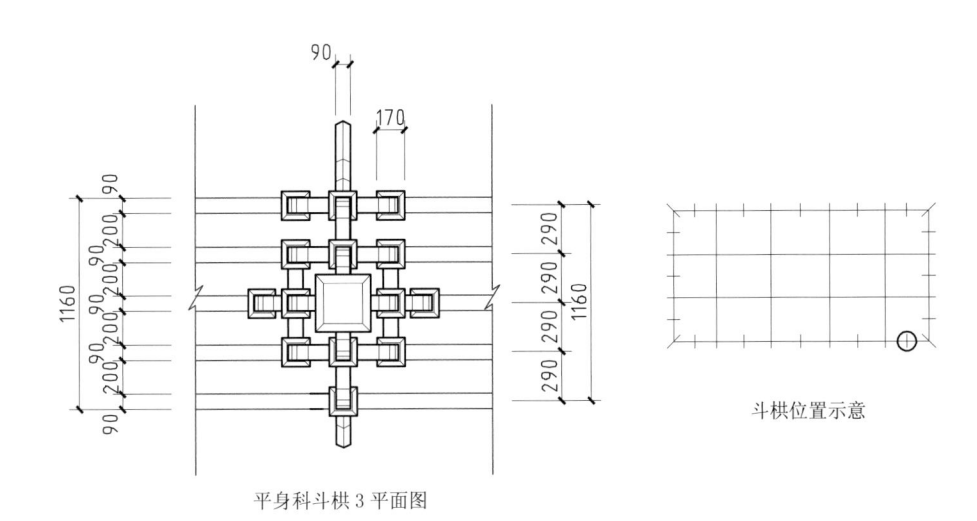

平身科斗栱 3 平面图

斗栱位置示意

西岳庙金城门斗栱大样图（二）

Bracket set of Xiyue Temple's Jincheng Gate (2)

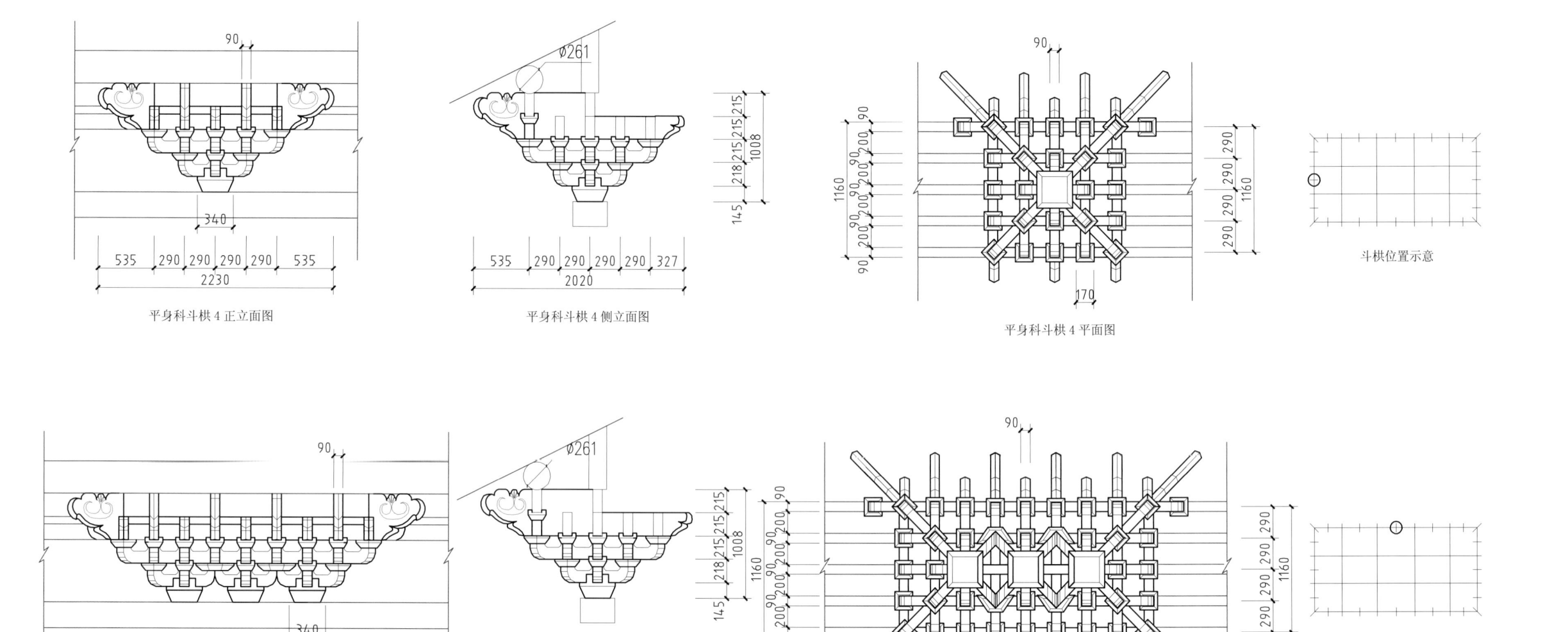

平身科斗栱 4 正立面图

平身科斗栱 4 侧立面图

平身科斗栱 4 平面图

斗栱位置示意

平身科斗栱 5 正立面图

平身科斗栱 5 侧立面图

平身科斗栱 5 平面图

斗栱位置示意

西岳庙金城门斗栱大样图（三）
Bracket set of Xiyue Temple's Jincheng Gate (3)

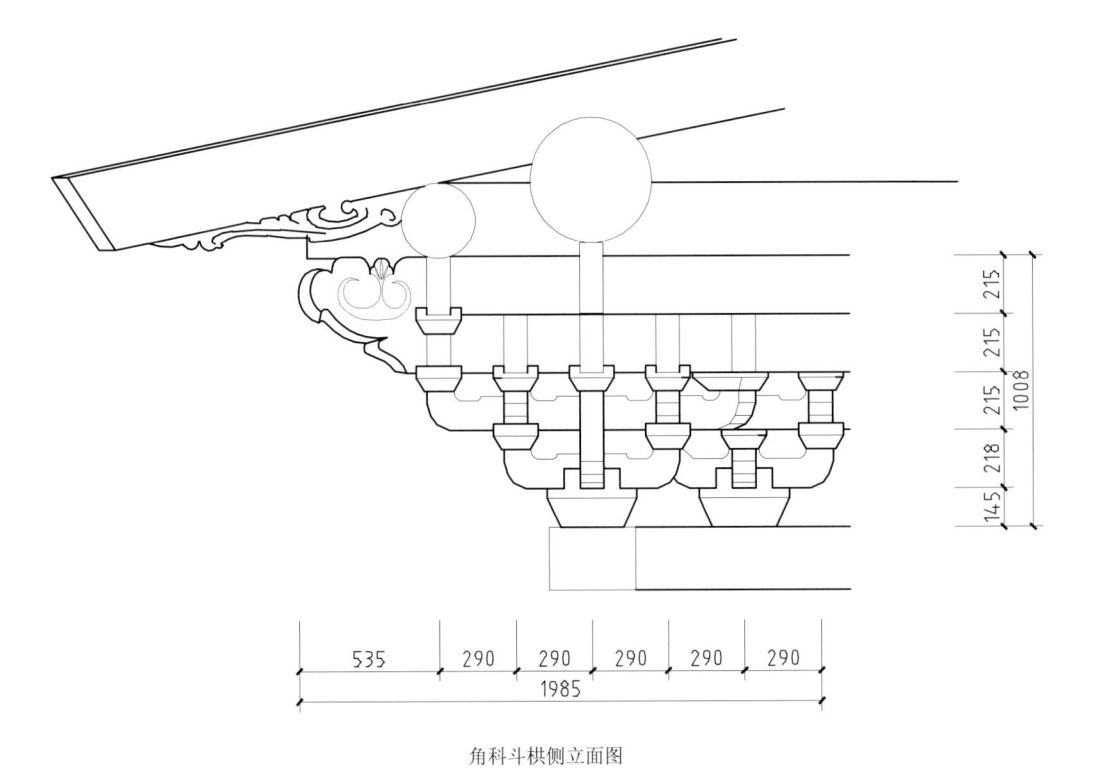

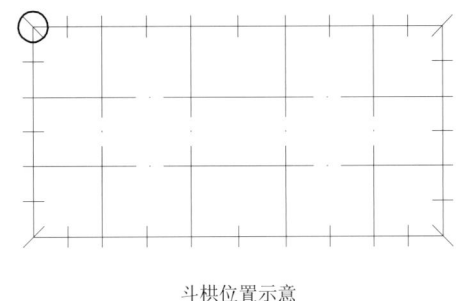

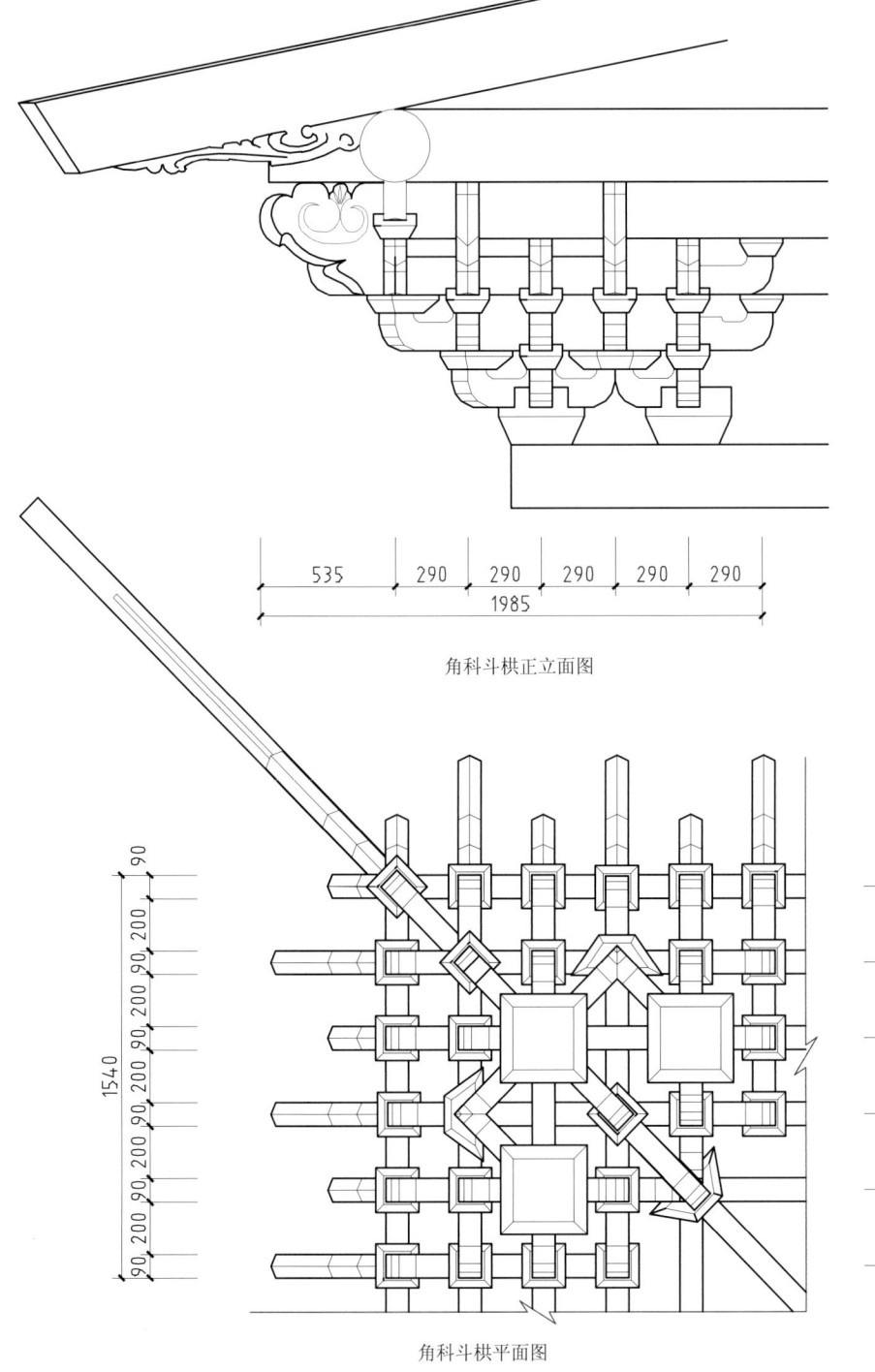

角科斗栱正立面图

角科斗栱侧立面图

斗栱位置示意

角科斗栱平面图

西岳庙金城门斗栱大样图（四）
Bracket set of Xiyue Temple's Jincheng Gate (4)

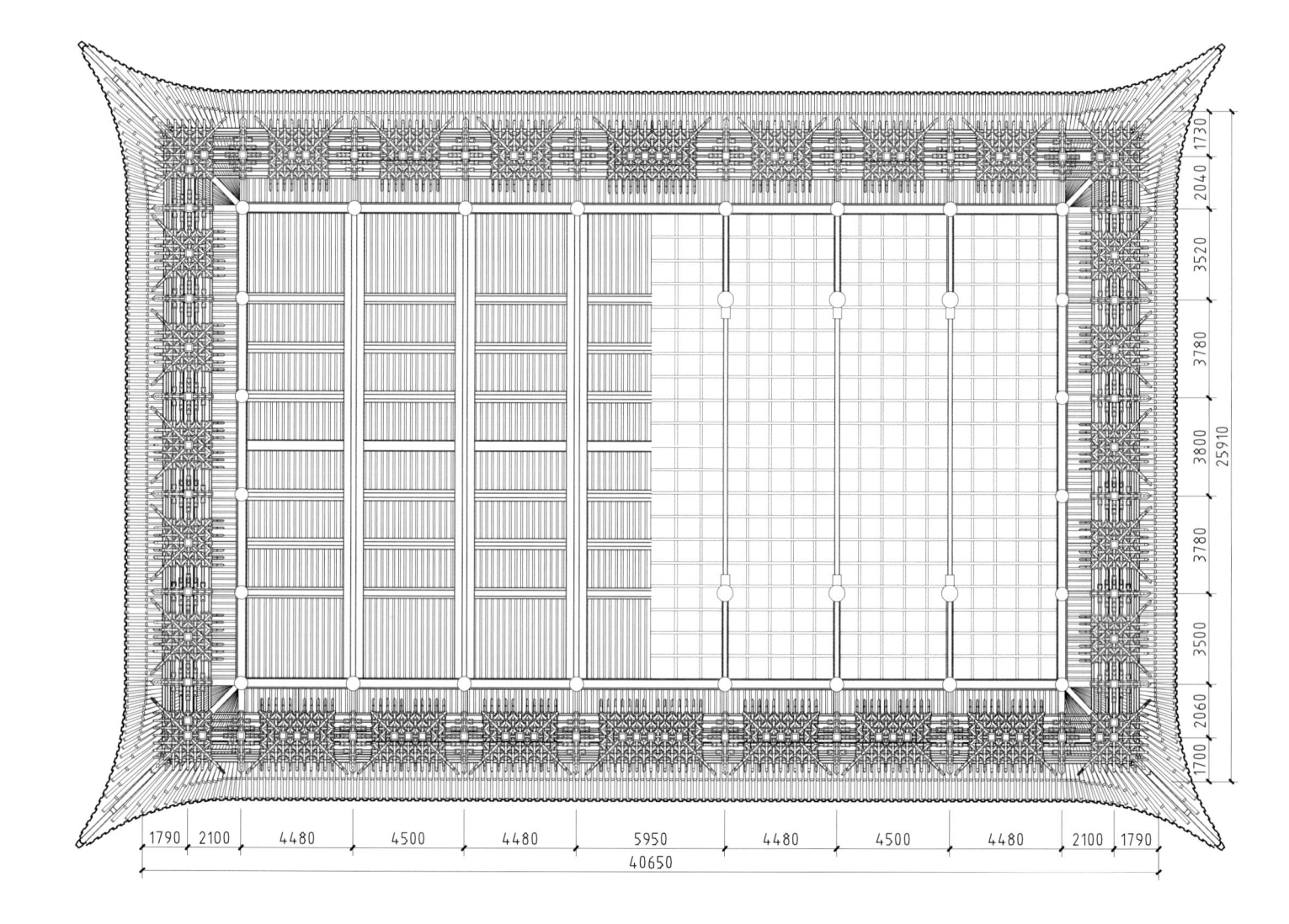

西岳庙灏灵殿梁架仰视平面图
Plan of framework of Xiyue Temple's Haoling Hall as seen from below

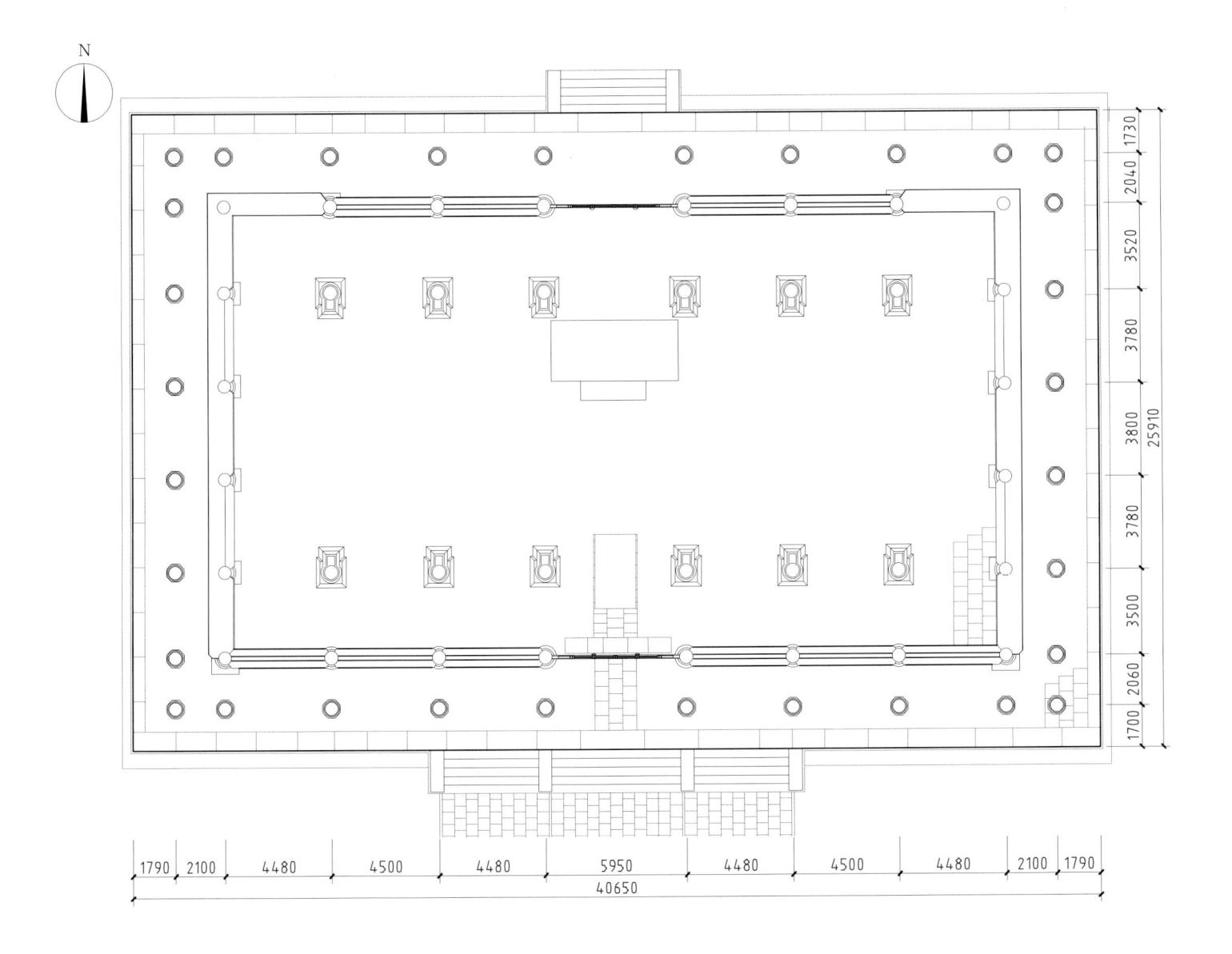

N

| 1730 | 2040 | 3520 | 3780 | 3800 | 3780 | 3500 | 2060 | 1700 |

25910

| 1790 | 2100 | 4480 | 4500 | 4480 | 5950 | 4480 | 4500 | 4480 | 2100 | 1790 |

40650

西岳庙灏灵殿平面图
Plan of Xiyue Temple's Haoling Hall

西岳庙灏灵殿屋顶平面图
Roof plan of Xiyue Temple's Haoling Hall

0   4   8m

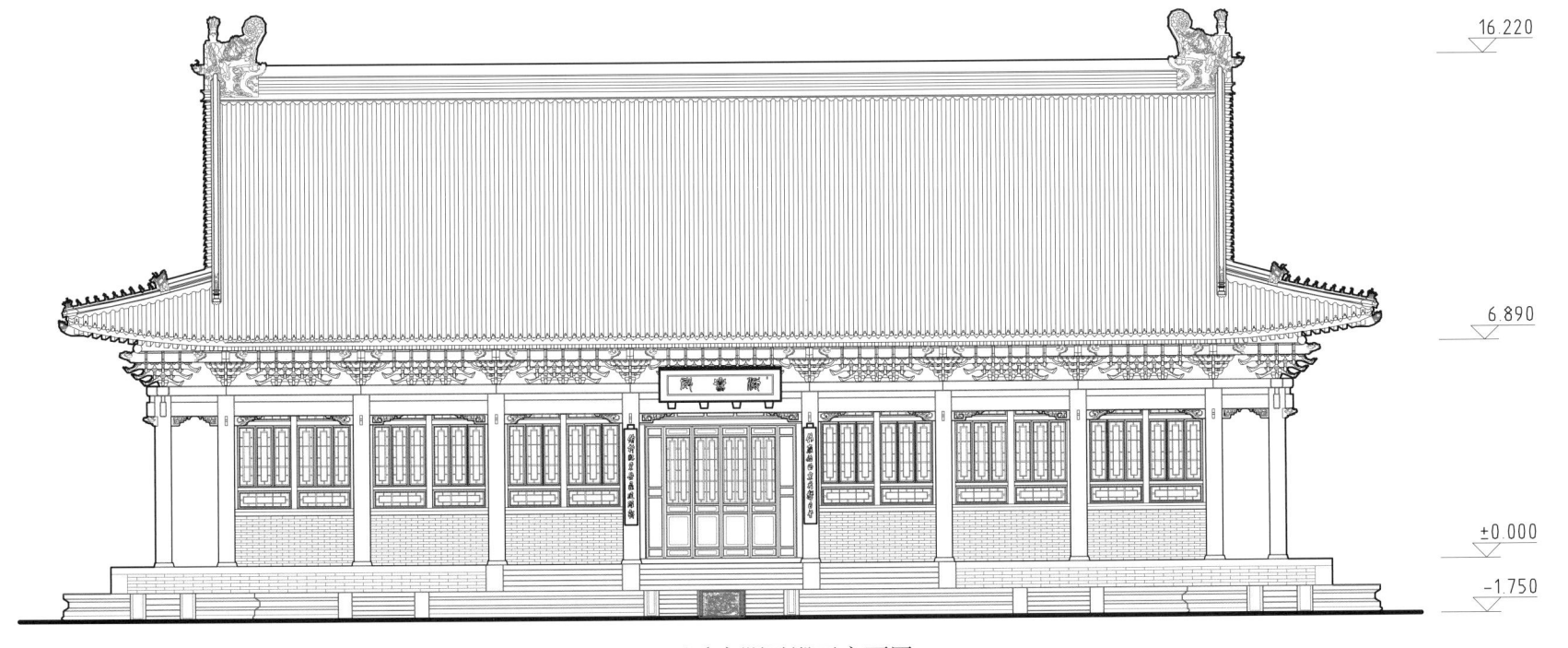

16.220

6.890

±0.000

−1.750

西岳庙灏灵殿正立面图
Front elevation of Xiyue Temple's Haoling Hall

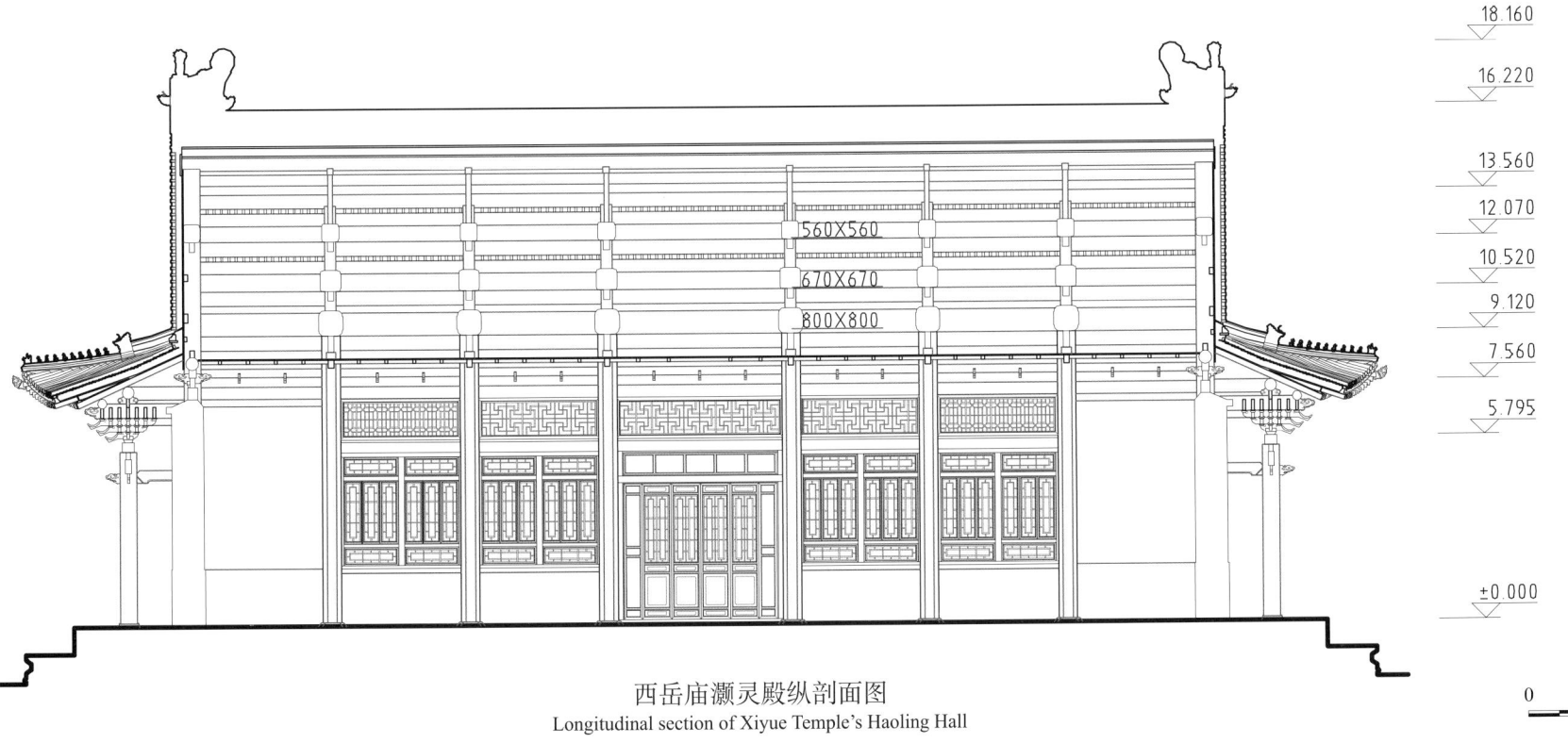

18.160

16.220

13.560

12.070

560X560

10.520

670X670

9.120

800X800

7.560

5.795

±0.000

西岳庙灏灵殿纵剖面图
Longitudinal section of Xiyue Temple's Haoling Hall

0   4   8m

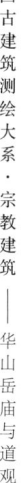

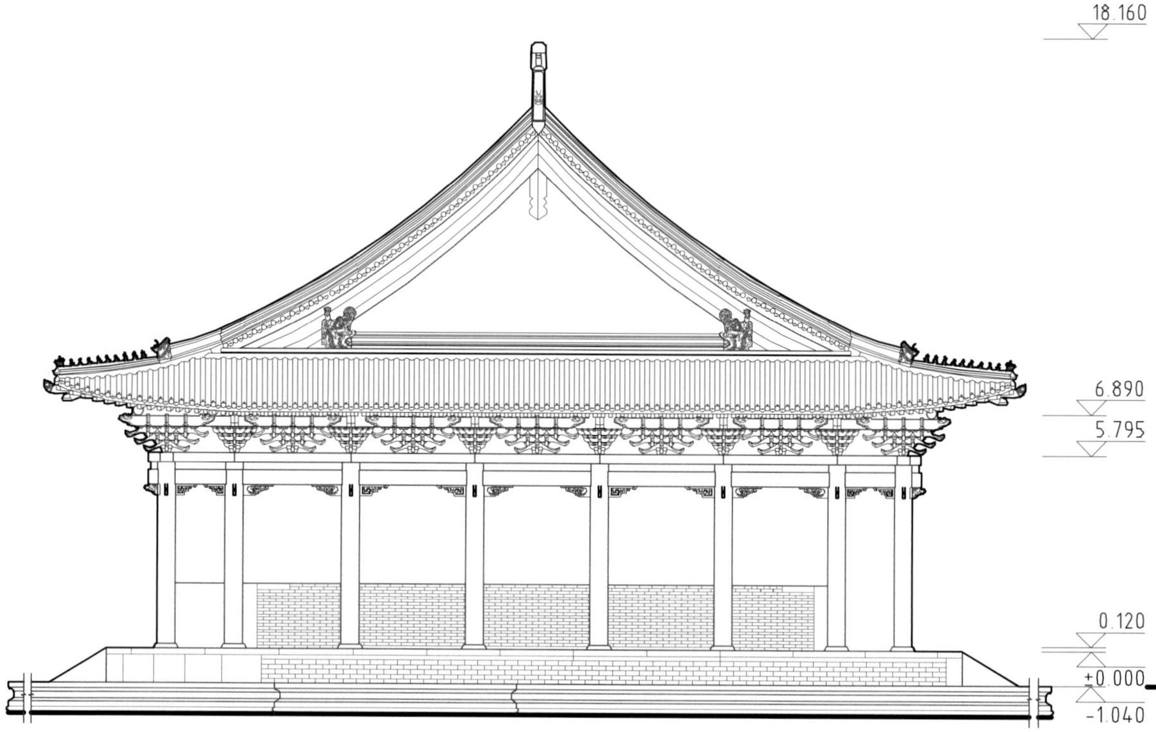

18.160

18.160

16.220

14.720

12.970

11.470

10.220

8.480

6.890

5.795

7.560

5.795

0.120

+0.000

−1.040

0.120

+0.000

−1.040

西岳庙灏灵殿侧立面图
Side elevation of Xiyue Temple's Haoling Hall

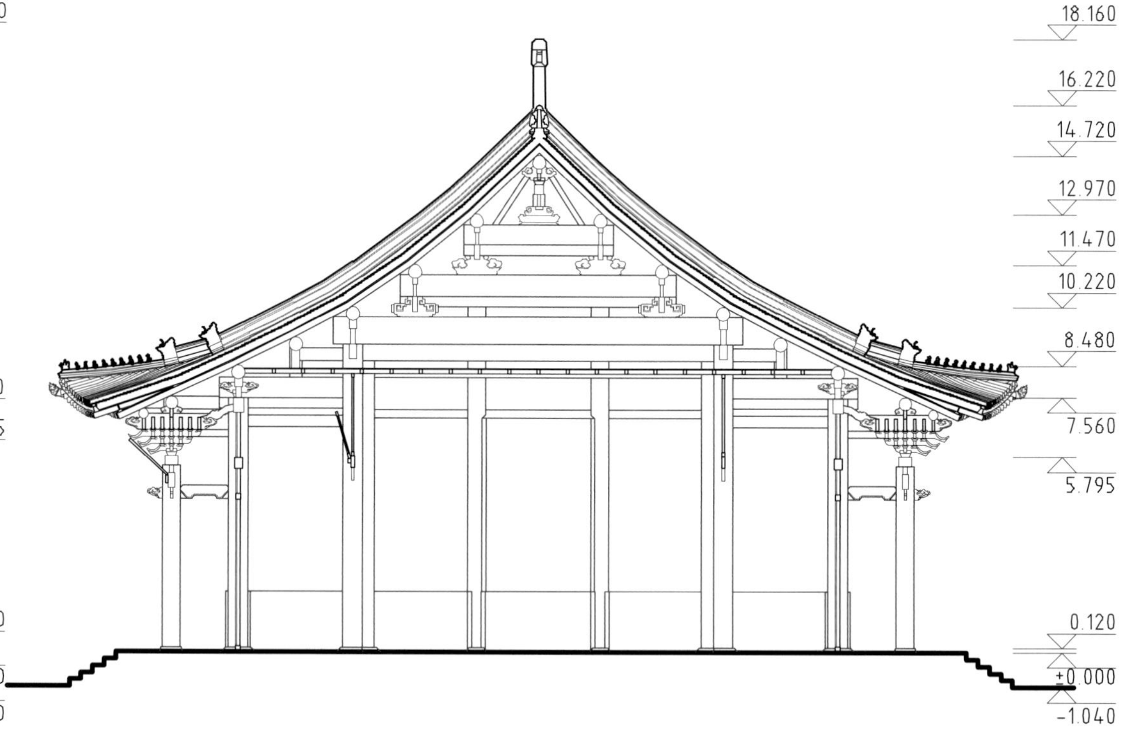

西岳庙灏灵殿横剖面图
Cross-section of Xiyue Temple's Haoling Hall

0　　　4　　　8m

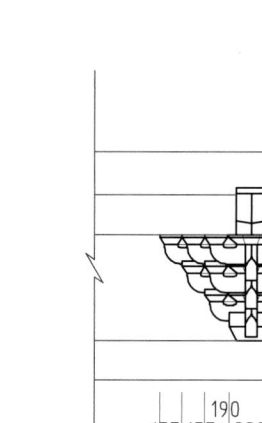
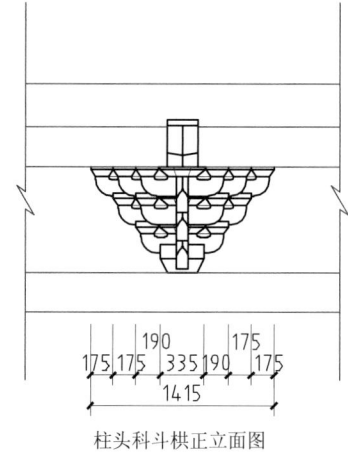

ø380

ø320

柱头科斗栱正立面图

柱头科斗栱侧立面图

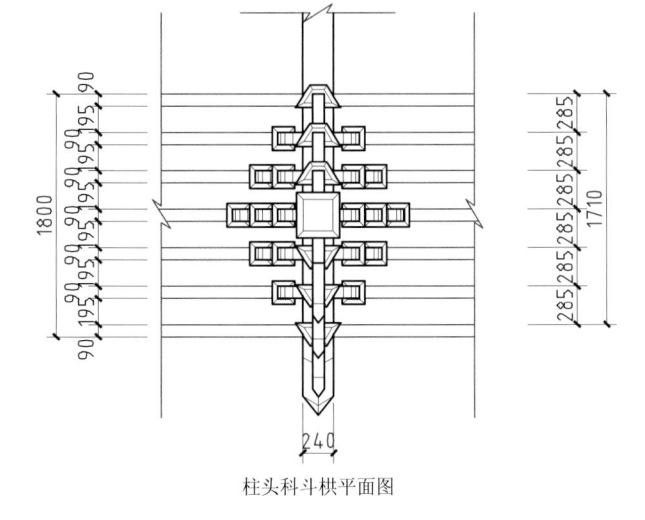

柱头科斗栱平面图

斗栱位置示意

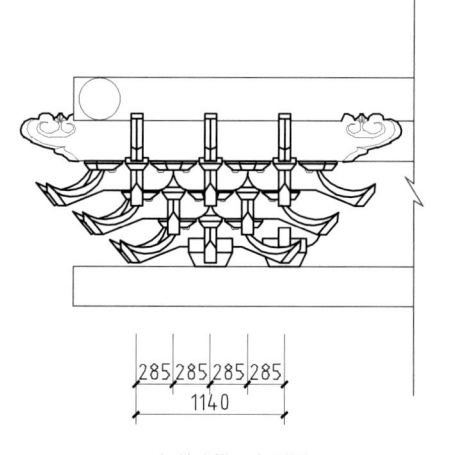
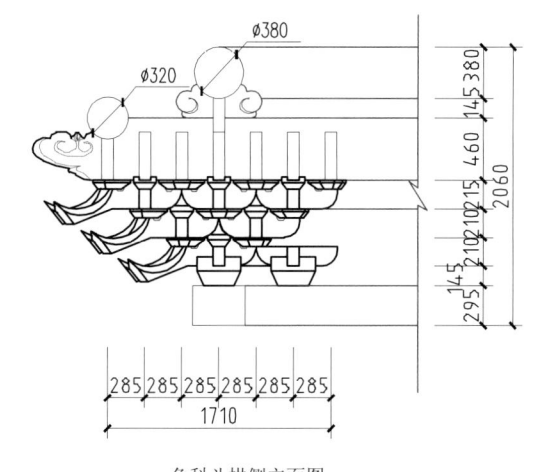

ø380

ø320

角科斗栱正立面图

角科斗栱侧立面图

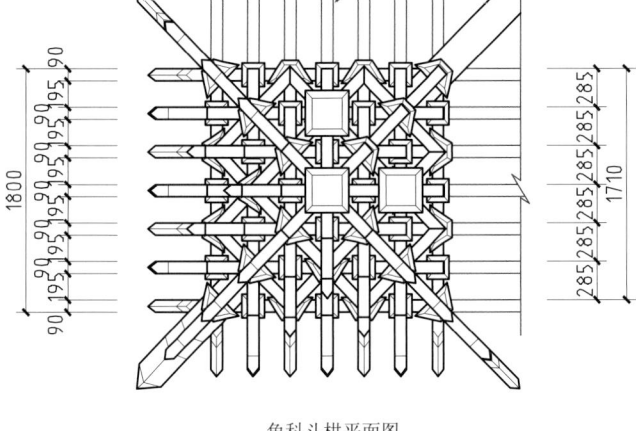

角科斗栱平面图

斗栱位置示意

西岳庙灏灵殿斗栱大样图（一）
Bracket set of Xiyue Temple's Haoling Hall (1)

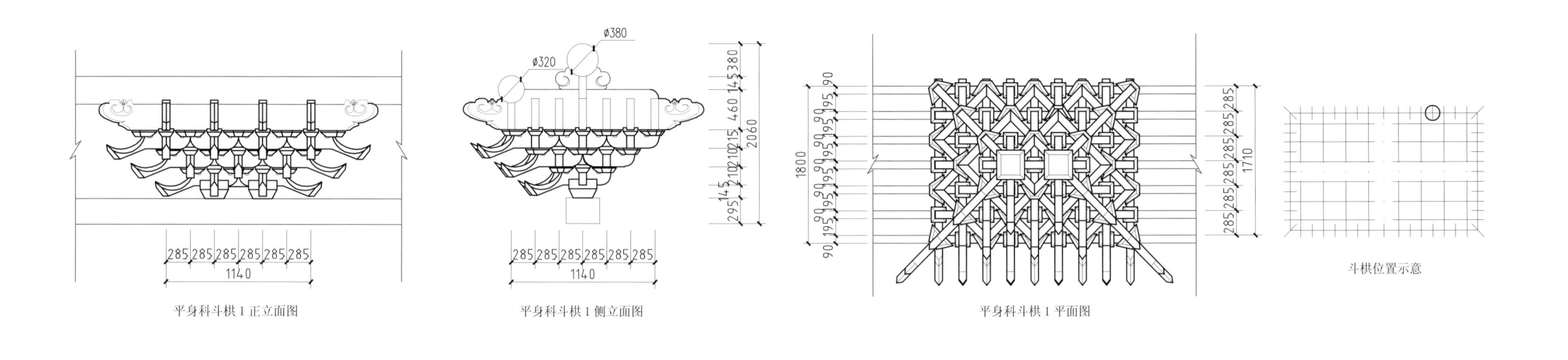

平身科斗栱 1 正立面图　　　　平身科斗栱 1 侧立面图　　　　平身科斗栱 1 平面图　　　　斗栱位置示意

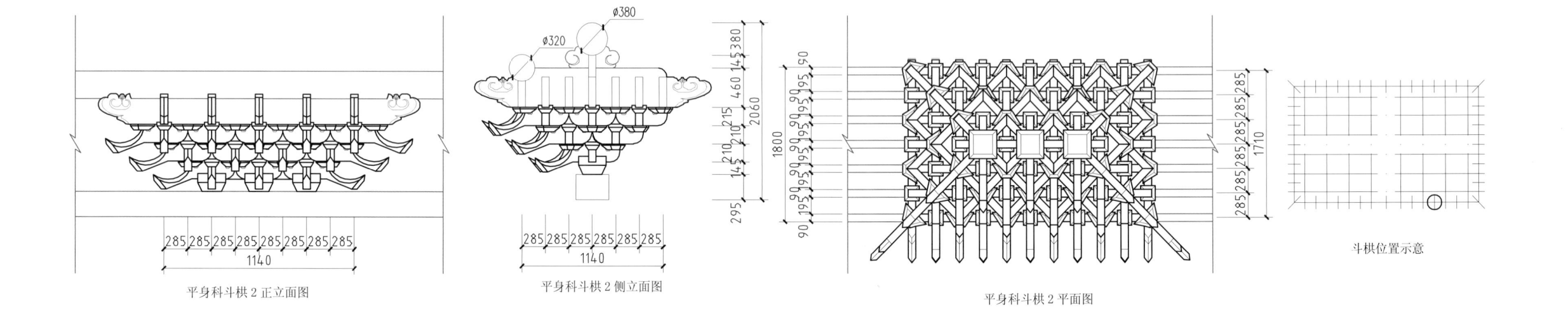

平身科斗栱 2 正立面图　　　　平身科斗栱 2 侧立面图　　　　平身科斗栱 2 平面图　　　　斗栱位置示意

西岳庙灏灵殿斗栱大样图（二）
Bracket set of Xiyue Temple's Haoling Hall (2)

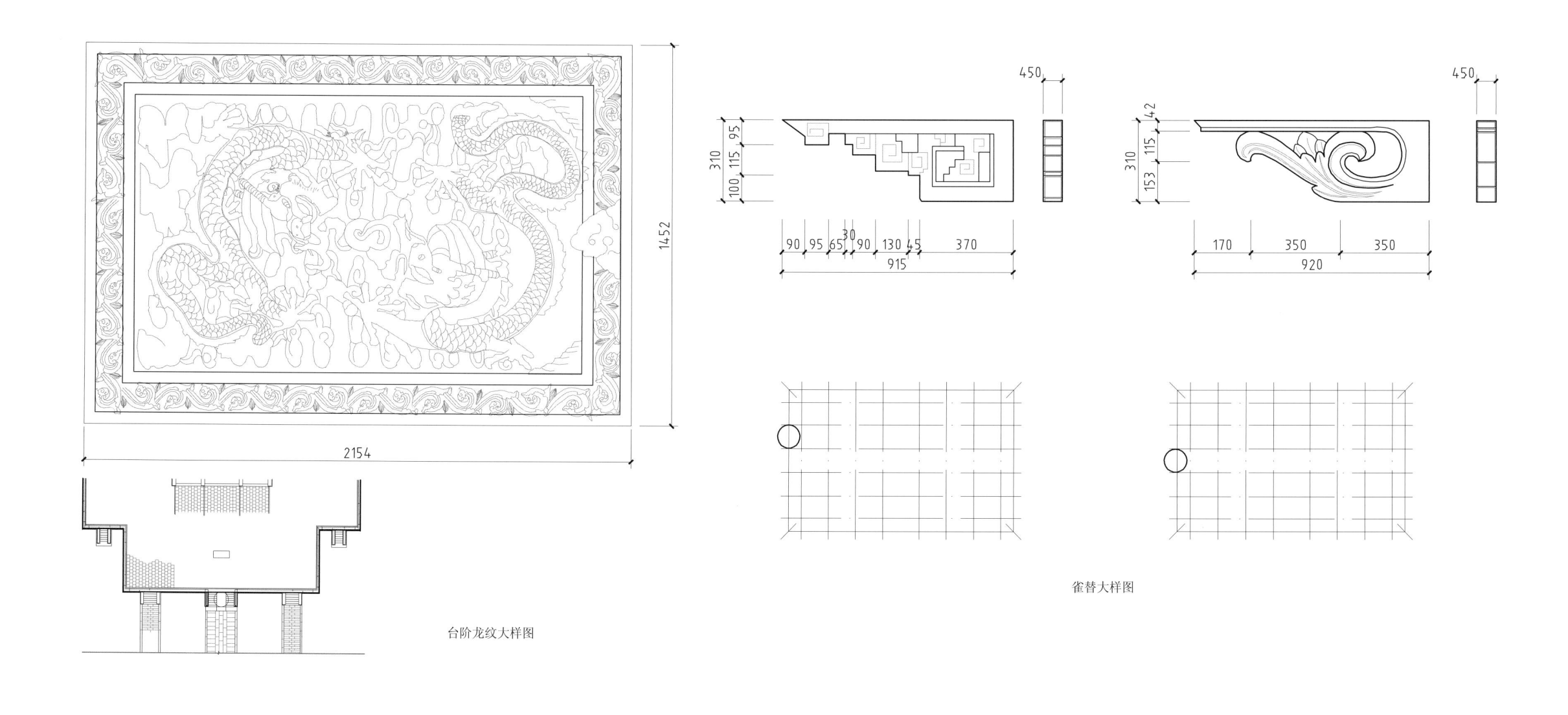

台阶龙纹大样图

雀替大样图

西岳庙灏灵殿构件大样图（一）
Structural component of Xiyue Temple's Haoling Hall (1)

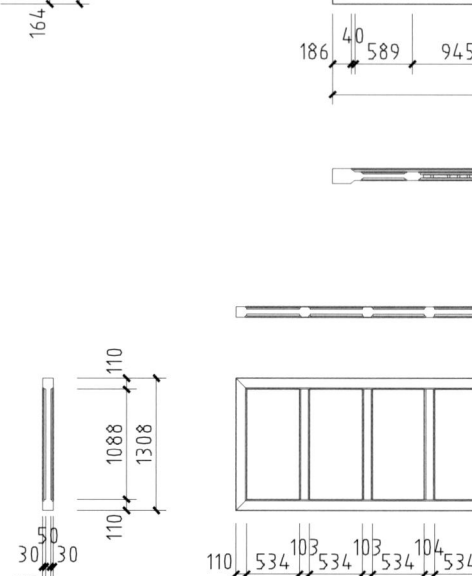

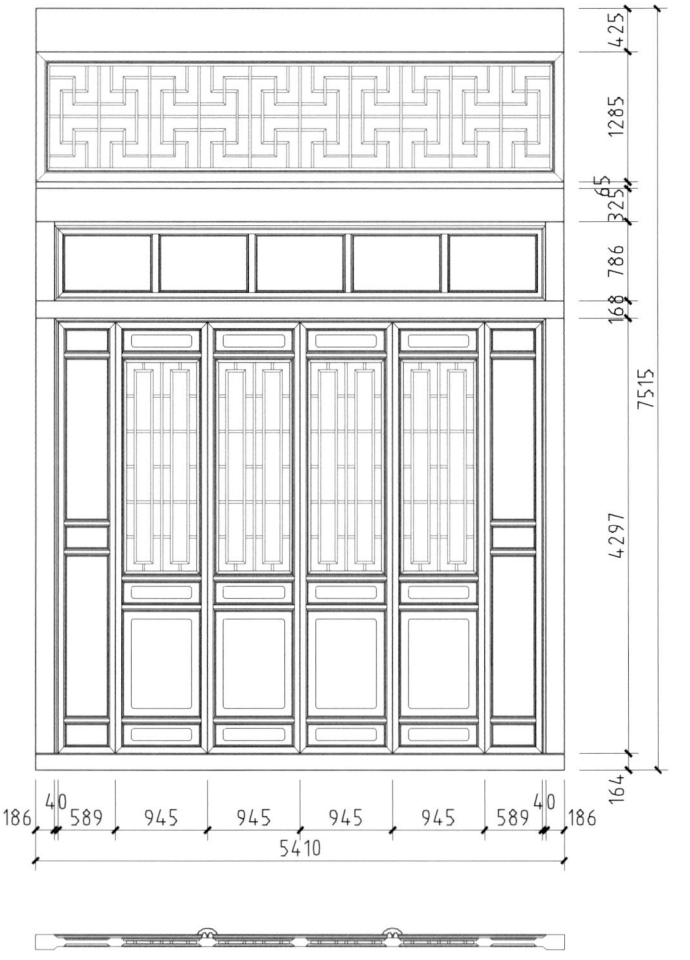

门窗大样一

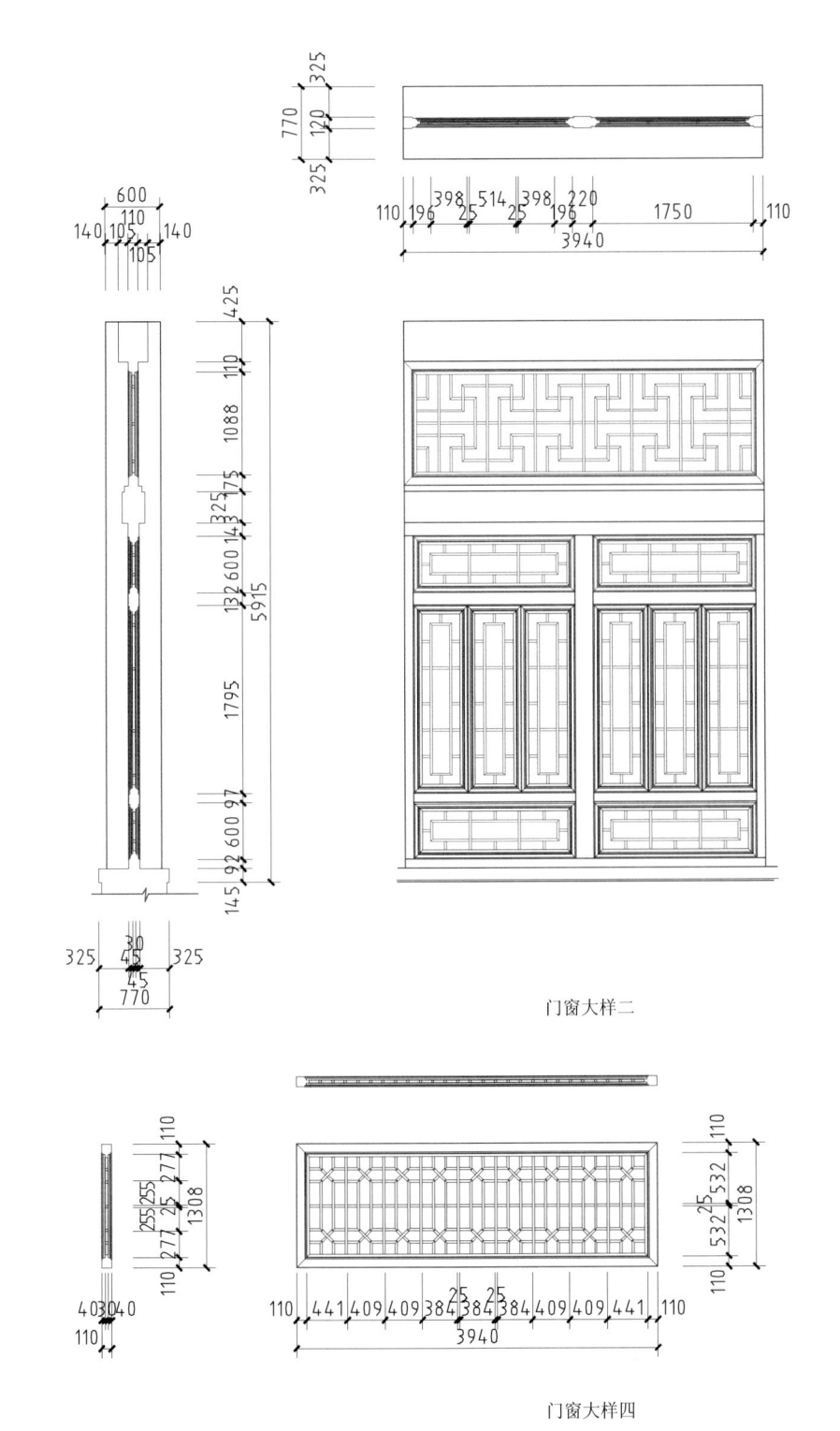

门窗大样二

门窗大样三

门窗大样四

西岳庙灏灵殿构件大样图（二）
Structural component of Xiyue Temple's Haoling Hall (2)

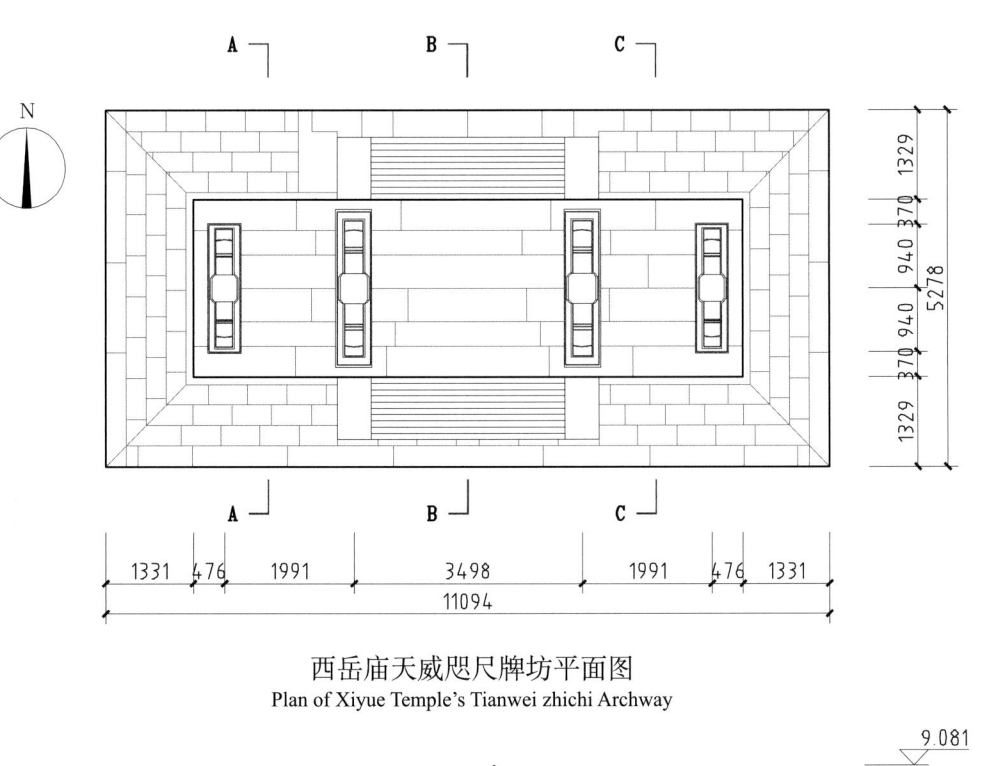

西岳庙天威咫尺牌坊平面图
Plan of Xiyue Temple's Tianwei zhichi Archway

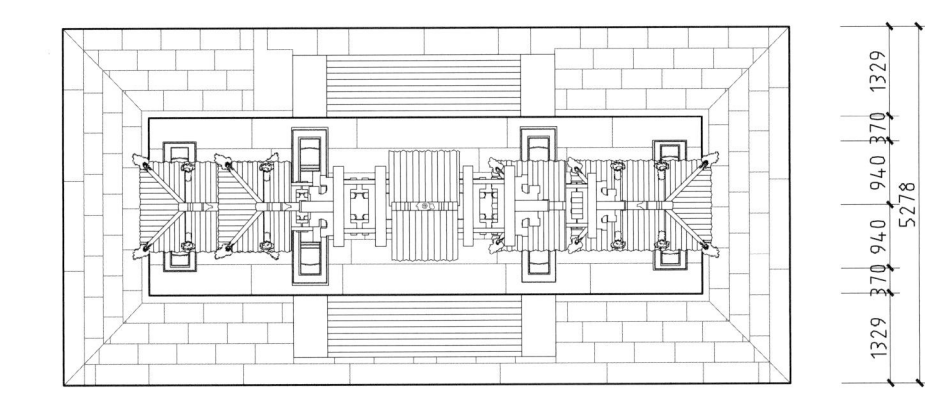
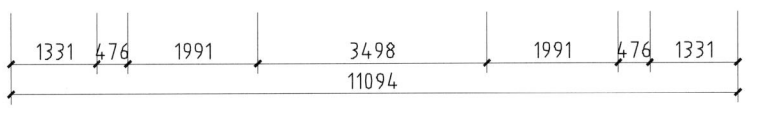
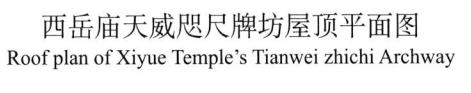

西岳庙天威咫尺牌坊屋顶平面图
Roof plan of Xiyue Temple's Tianwei zhichi Archway

西岳庙天威咫尺牌坊南立面图
South elevation of Xiyue Temple's Tianwei zhichi Archway

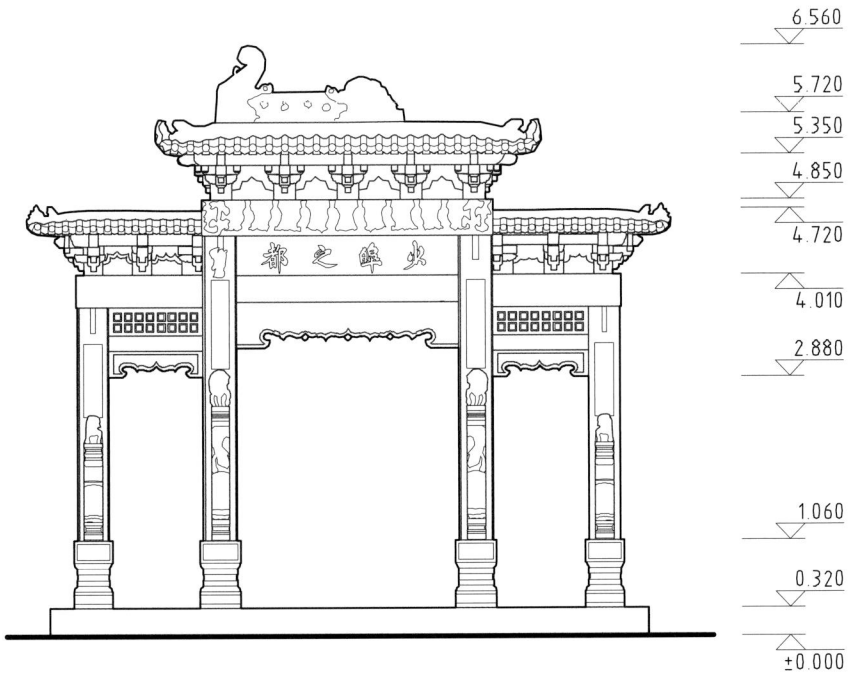

西岳庙天威咫尺牌坊北立面图
North elevation of Xiyue Temple's Tianwei zhichi Archway

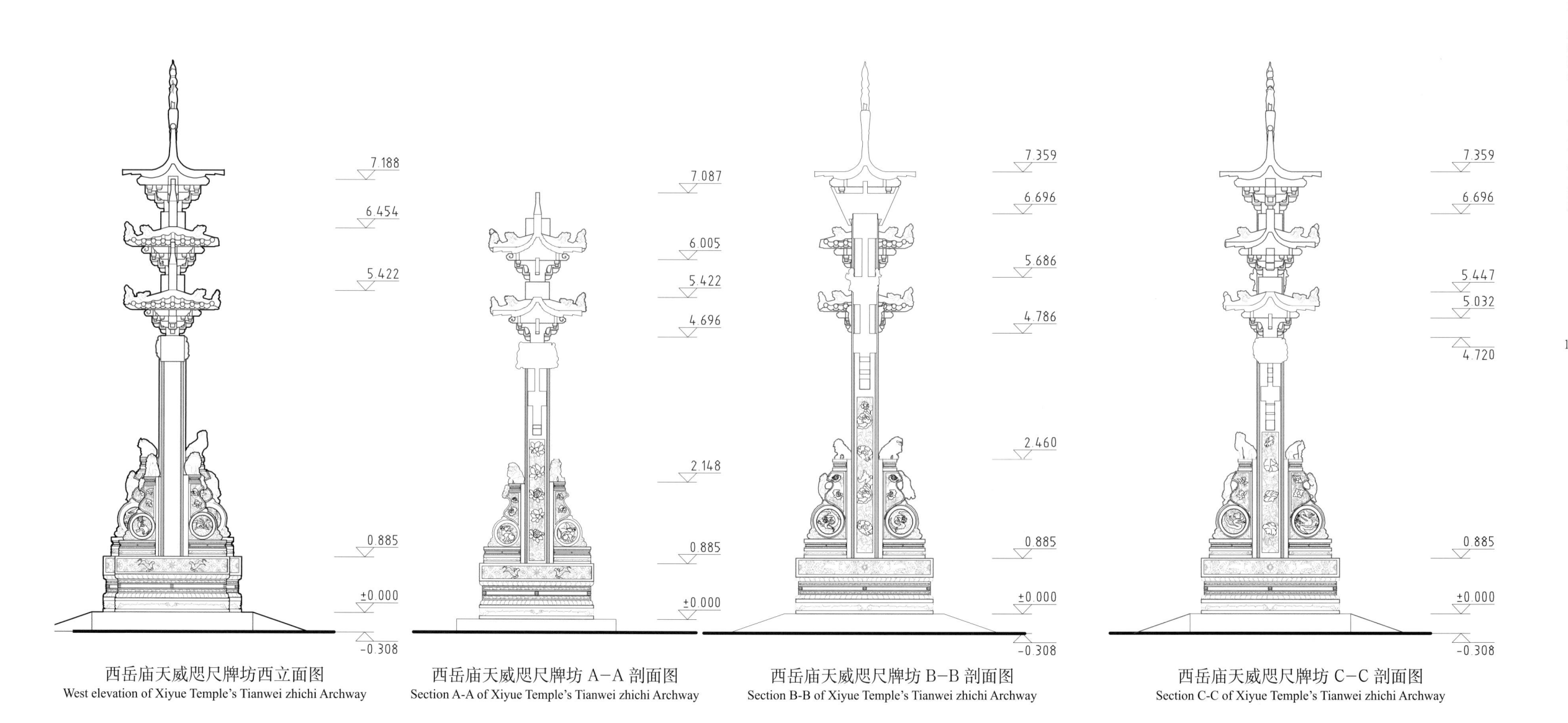

西岳庙天威咫尺牌坊西立面图
West elevation of Xiyue Temple's Tianwei zhichi Archway

西岳庙天威咫尺牌坊 A-A 剖面图
Section A-A of Xiyue Temple's Tianwei zhichi Archway

西岳庙天威咫尺牌坊 B-B 剖面图
Section B-B of Xiyue Temple's Tianwei zhichi Archway

西岳庙天威咫尺牌坊 C-C 剖面图
Section C-C of Xiyue Temple's Tianwei zhichi Archway

西立面石鼓大样图

西侧次间石鼓大样图

主间石鼓大样图

东次间石鼓大样图

南立面西开间大样图

南立面主间大样图

南立面东开间大样图

西岳庙天威咫尺牌坊构件大样图
Structural component of Xiyue Temple's Tianwei zhichi Archway

0   0.4   0.8m

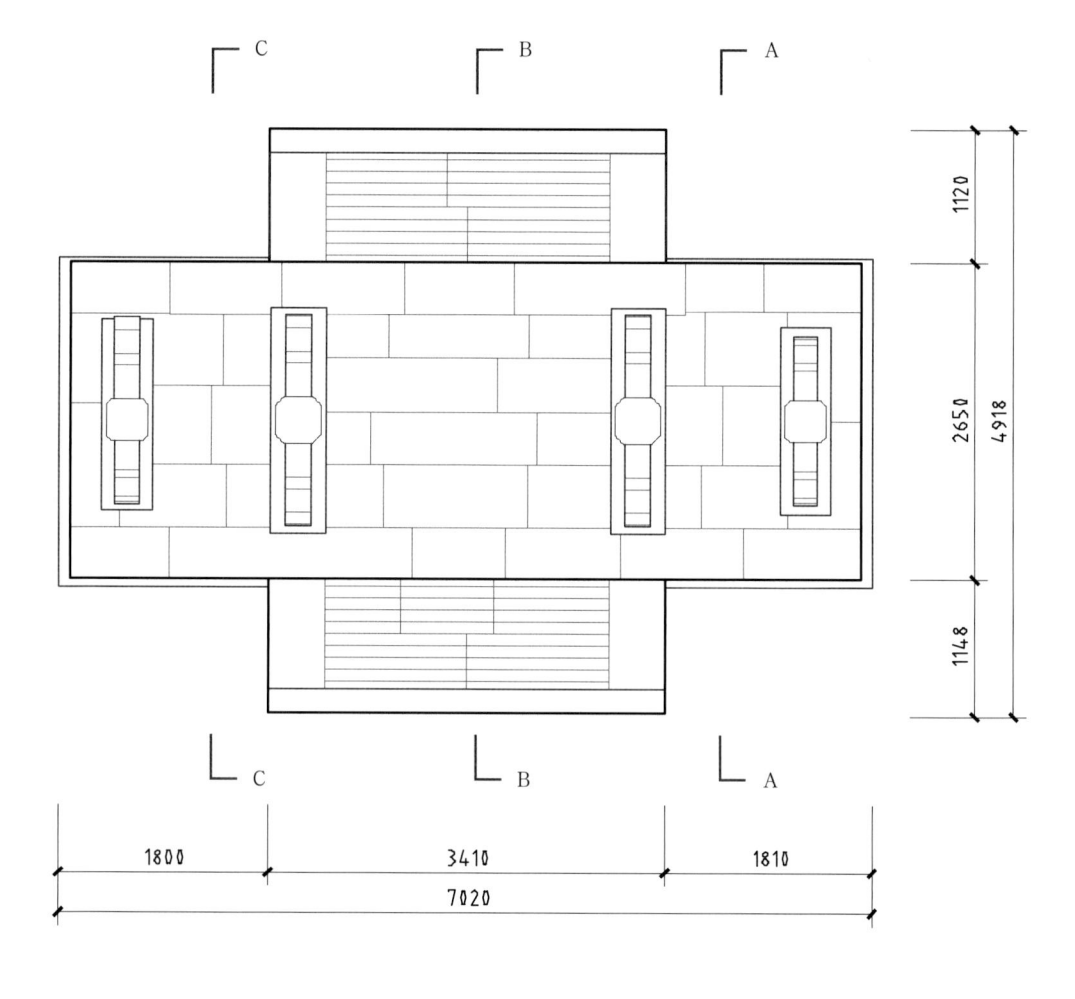

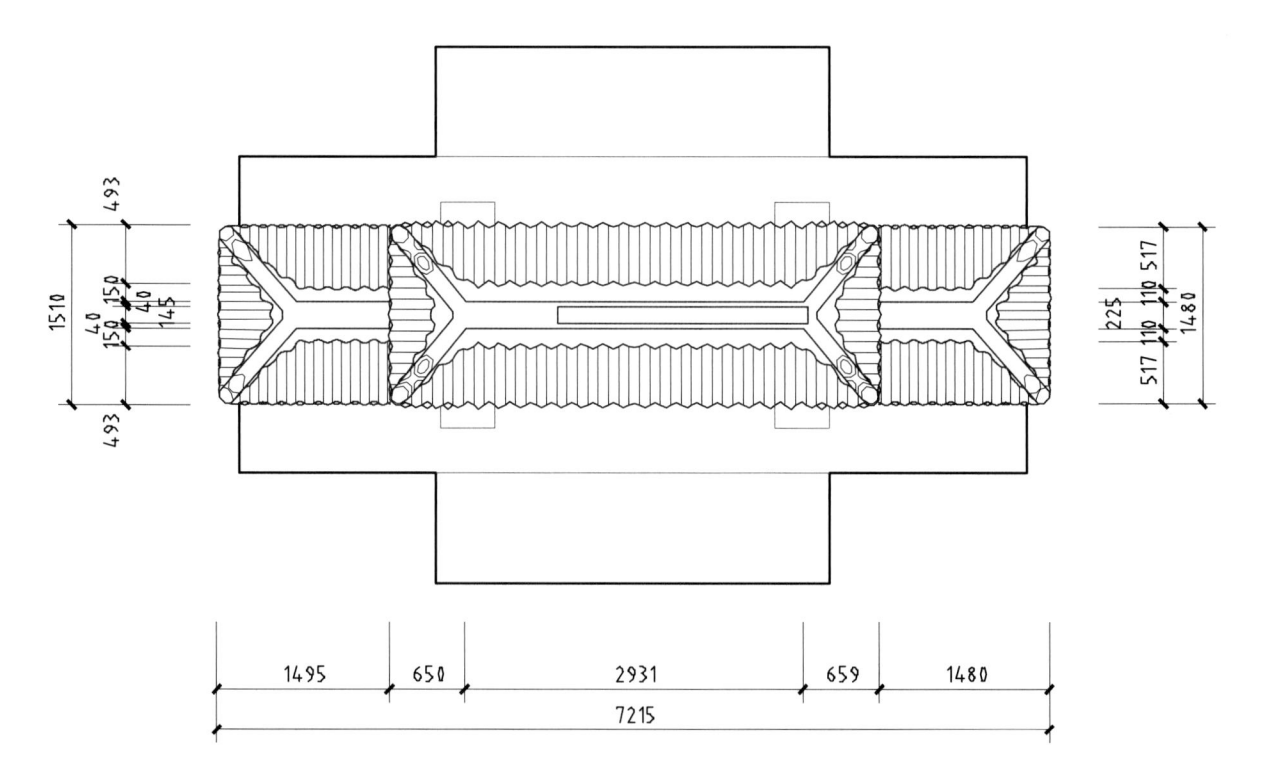

西岳庙少皞之都牌坊平面图
Plan of Xiyue Temple's Shaohao zhidu Archway

西岳庙少皞之都牌坊屋顶平面图
Roof plan of Xiyue Temple's Shaohao zhidu Archway

西岳庙少皞之都牌坊南立面图
South elevation of Xiyue Temple's Shaohao zhidu Archway

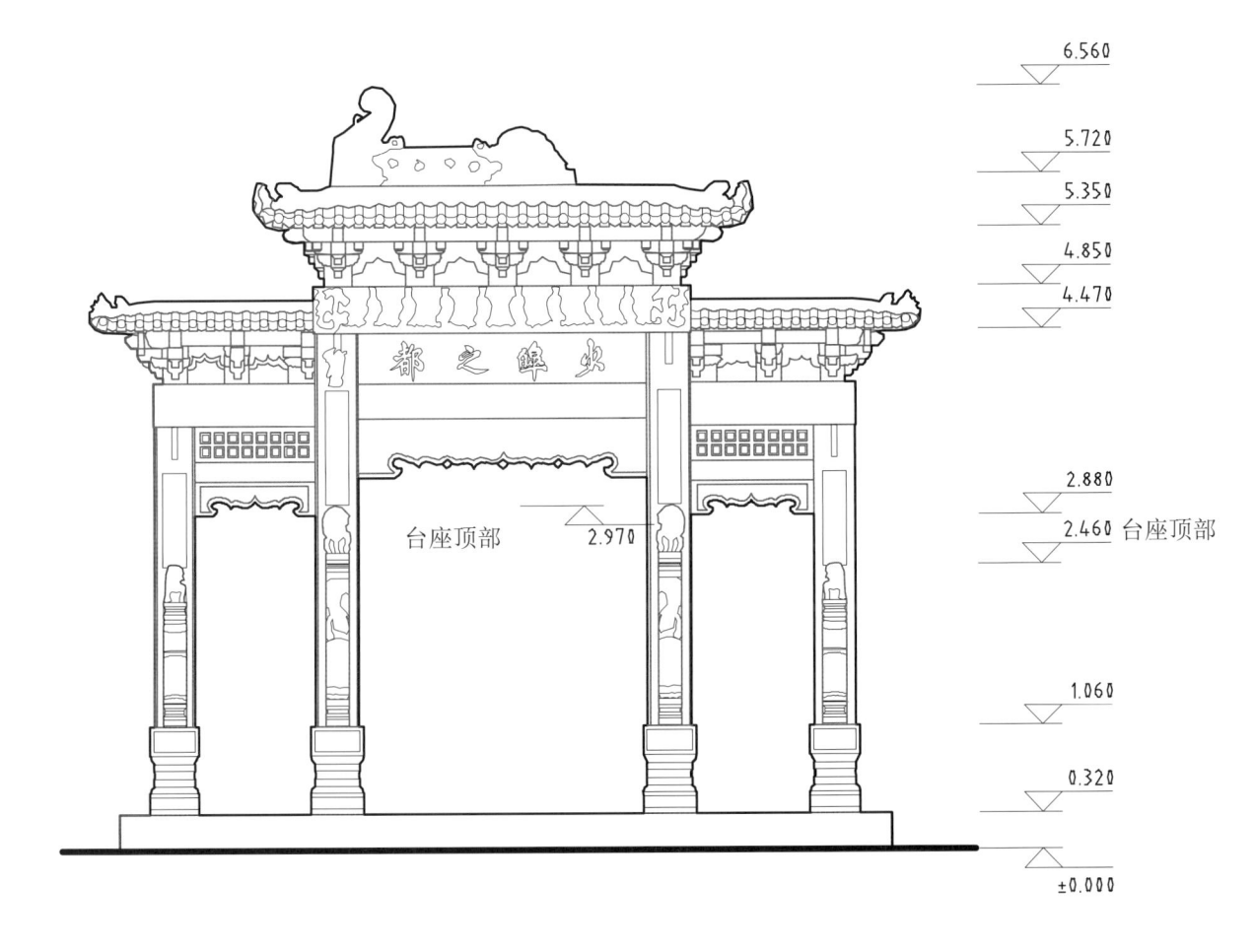

西岳庙少皞之都牌坊北立面图
North elevation of Xiyue Temple's Shaohao zhidu Archway

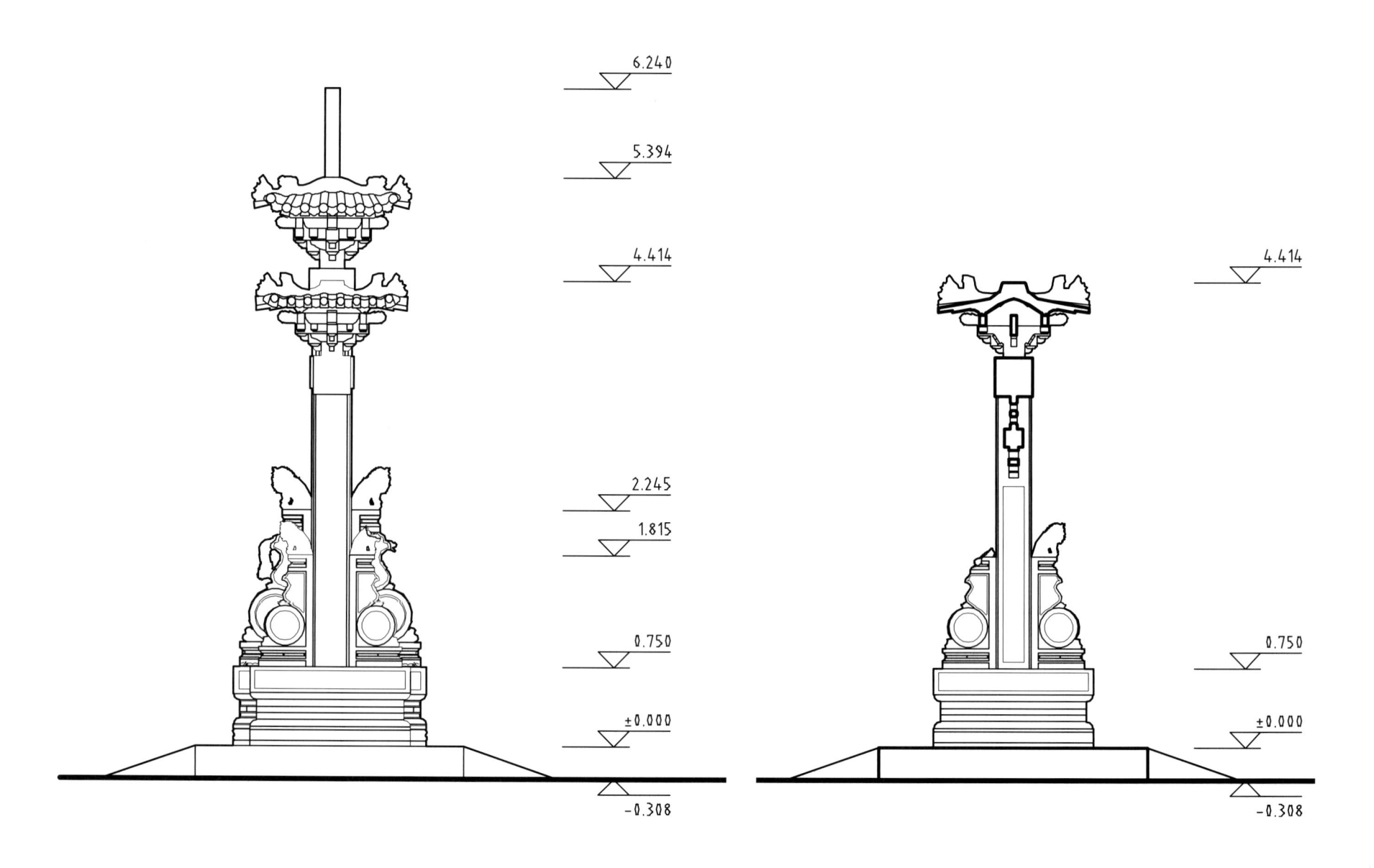

6.240

5.394

4.414

2.245

1.815

0.750

±0.000

-0.308

4.414

0.750

±0.000

-0.308

西岳庙少皞之都牌坊西立面图
West elevation of Xiyue Temple's Shaohao zhidu Archway

西岳庙少皞之都牌坊 A—A 剖面图
Section A-A of Xiyue Temple's Shaohao zhidu Archway

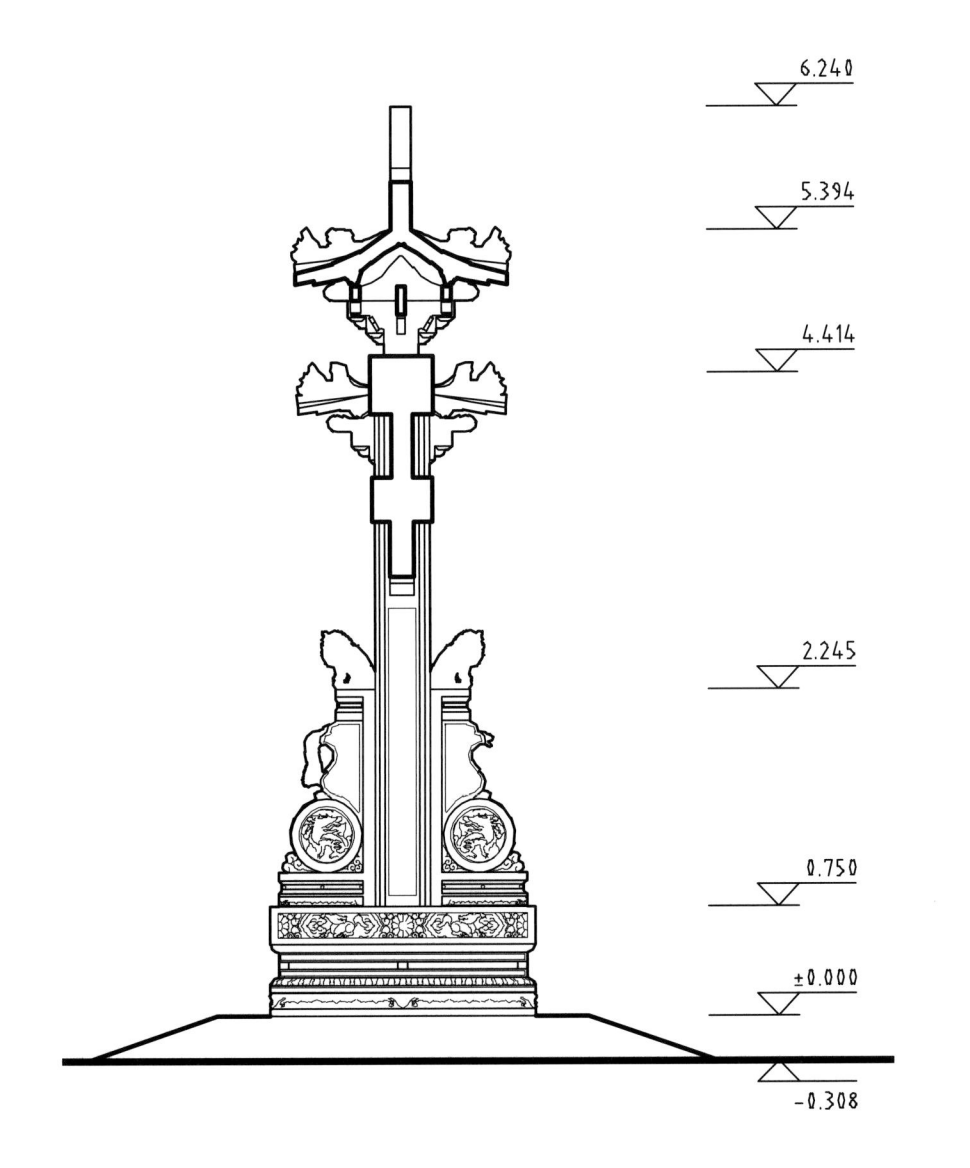

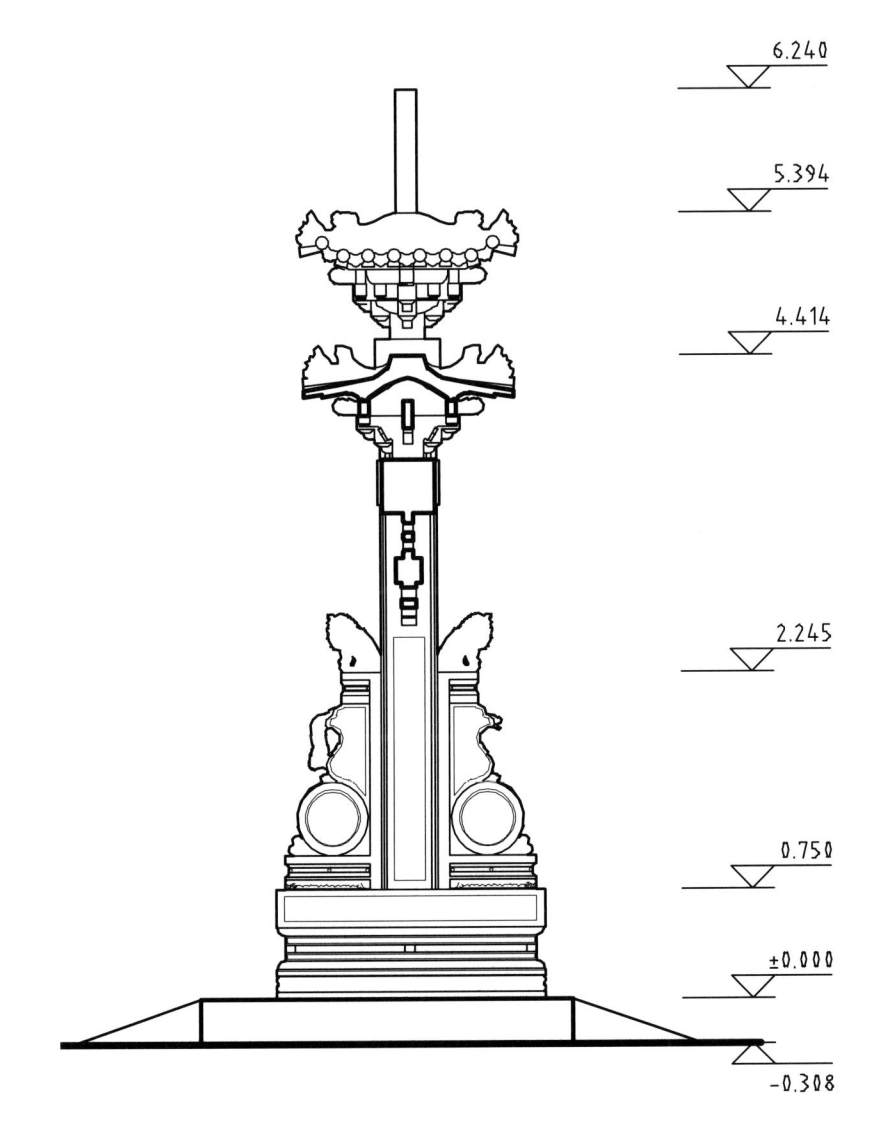

6.240

5.394

4.414

2.245

0.750

±0.000

-0.308

6.240

5.394

4.414

2.245

0.750

±0.000

-0.308

西岳庙少皞之都牌坊 B-B 剖面图
Section B-B of Xiyue Temple's Shaohao zhidu Archway

西岳庙少皞之都牌坊 C-C 剖面图
Section C-C of Xiyue Temple's Shaohao zhidu Archway

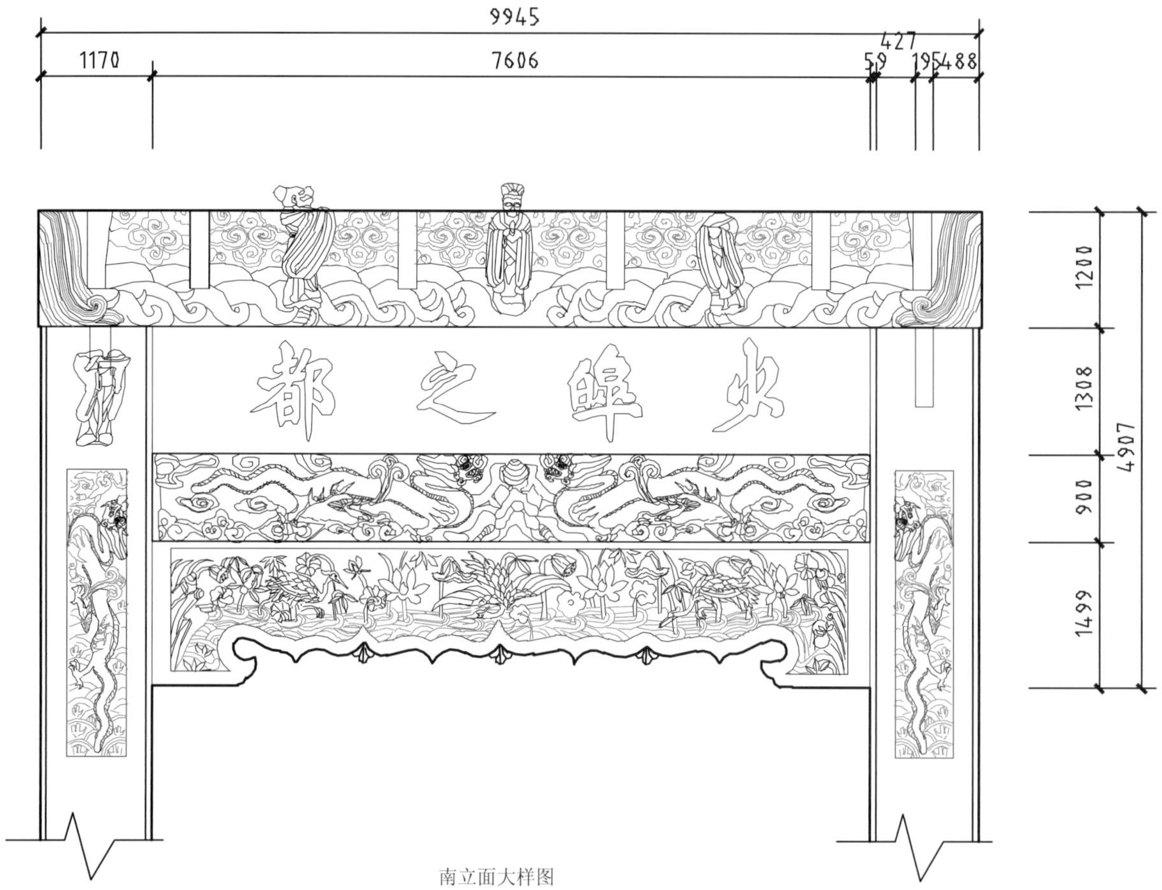

9945

1170　7606　427　59　195 488

1200

1308

4907

900

1499

南立面大样图

9945

1170　7606　428　59　195 488

1200

1308

4946

900

1538

北立面大样图

主开间石鼓大样图

西岳庙少皞之都牌坊构件大样图
Structural component of Xiyue Temple's Shaohao zhidu Archway

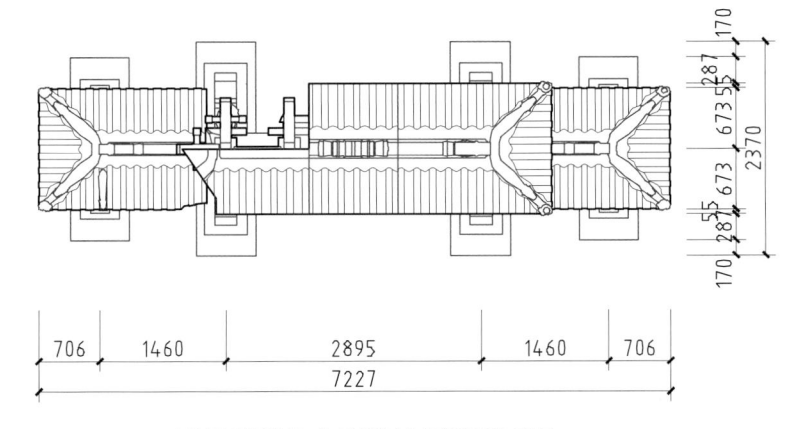

西岳庙蓐收之府牌坊平面图
Plan of Xiyue Temple's Rushou zhifu Archway

西岳庙蓐收之府牌坊屋顶平面图
Roof plan of Xiyue Temple's Rushou zhifu Archway

西岳庙蓐收之府牌坊南立面图
South elevation of Xiyue Temple's Rushou zhifu Stone Archway

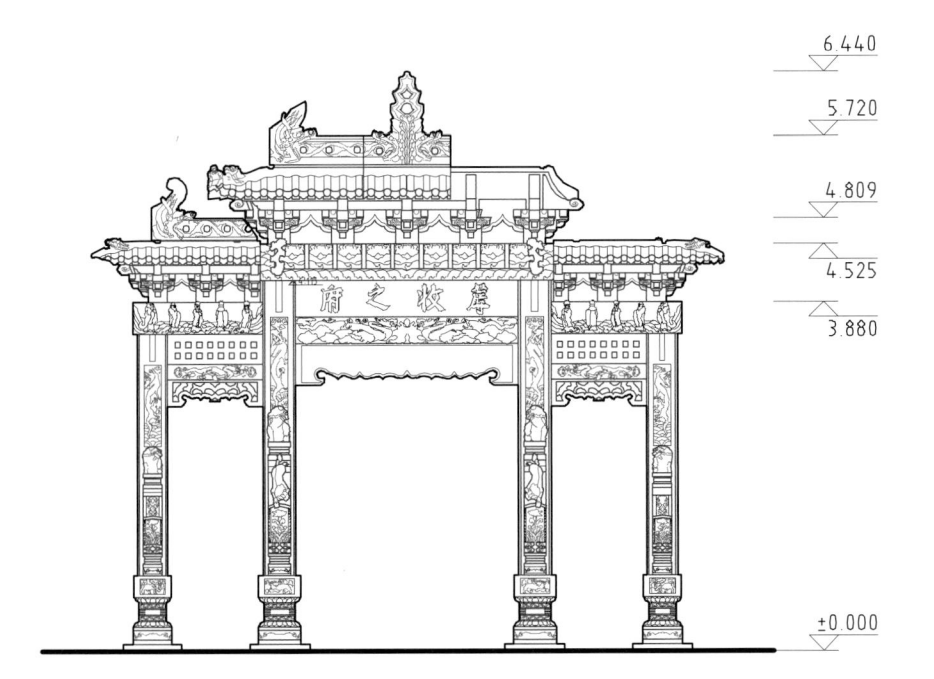

西岳庙蓐收之府牌坊北立面图
North elevation of Xiyue Temple's Rushou zhifu Stone Archway

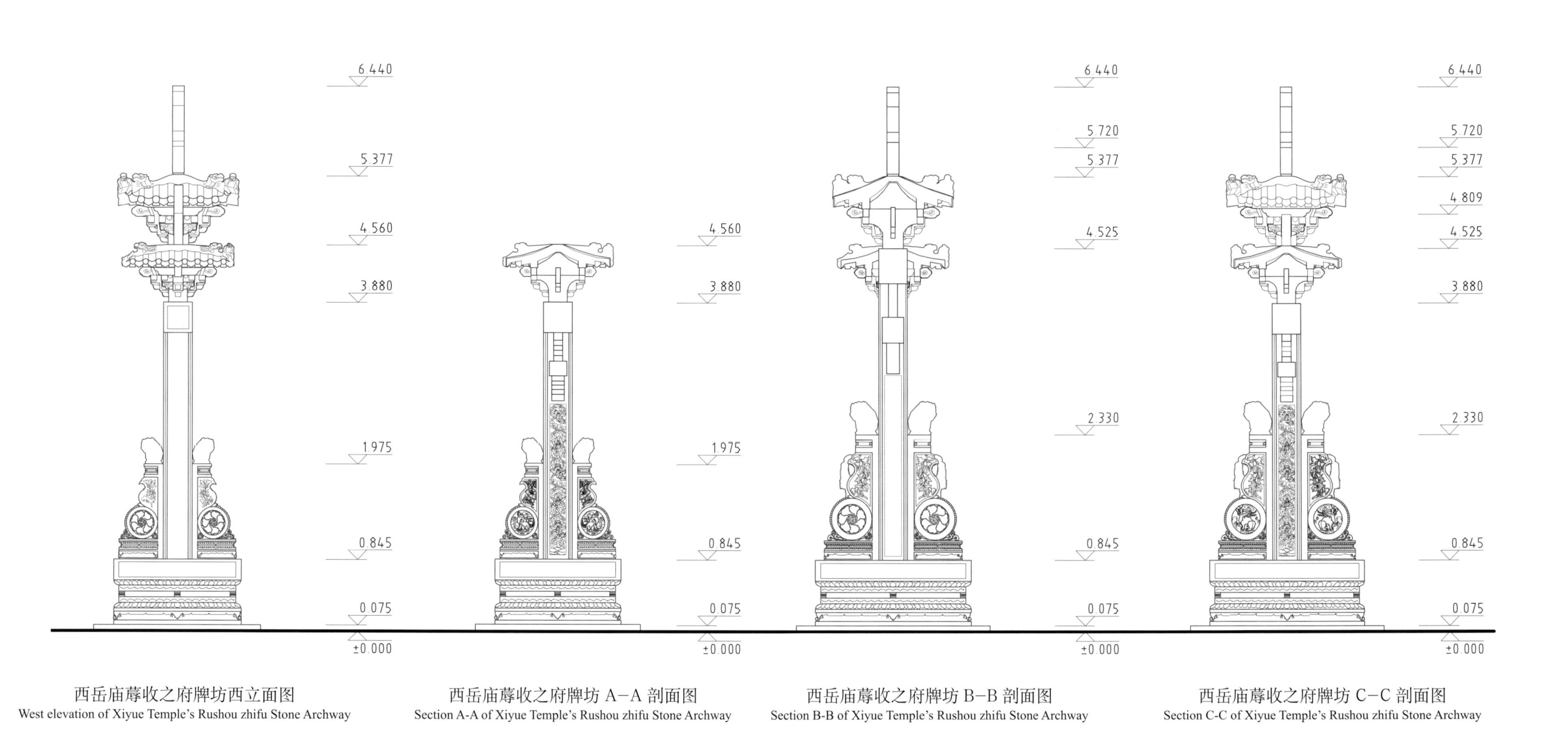

西岳庙蓐收之府牌坊西立面图
West elevation of Xiyue Temple's Rushou zhifu Stone Archway

西岳庙蓐收之府牌坊 A-A 剖面图
Section A-A of Xiyue Temple's Rushou zhifu Stone Archway

西岳庙蓐收之府牌坊 B-B 剖面图
Section B-B of Xiyue Temple's Rushou zhifu Stone Archway

西岳庙蓐收之府牌坊 C-C 剖面图
Section C-C of Xiyue Temple's Rushou zhifu Stone Archway

抱石鼓大样图

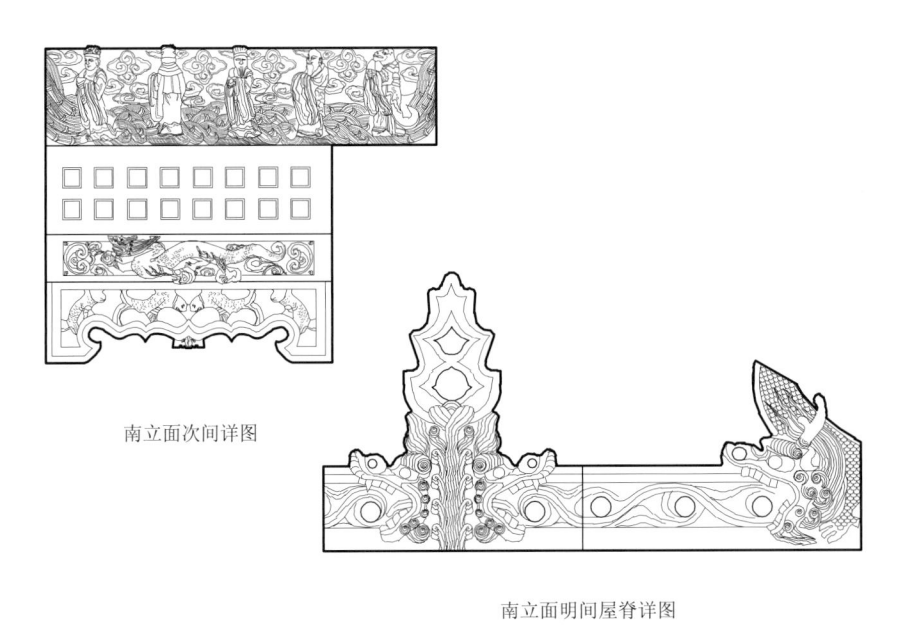

南立面次间详图

南立面明间屋脊详图

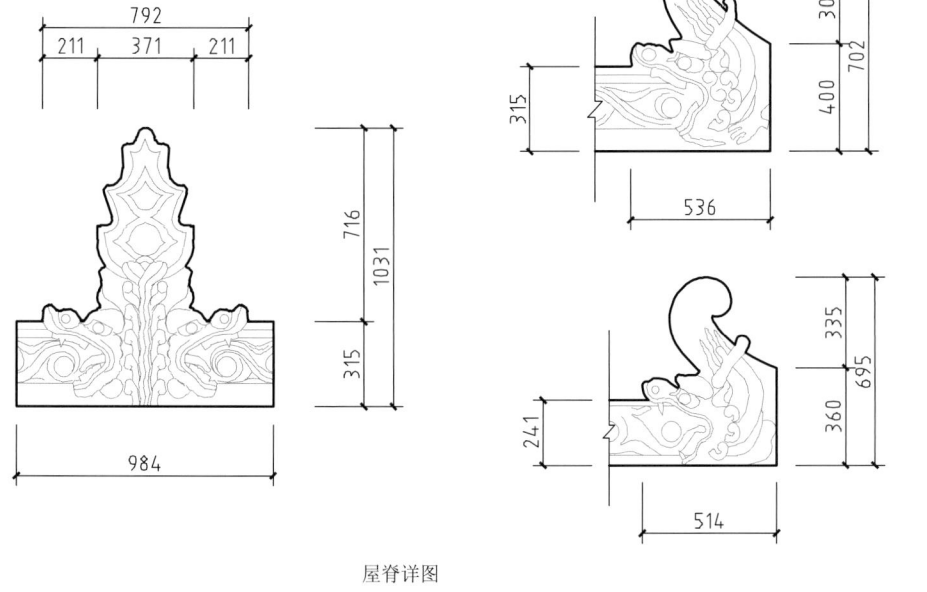

屋脊详图

南立面明间详图

西岳庙蓐收之府牌坊构件大样图
Structural component of Xiyue Temple's Rushou zhifu Stone Archway

0   0.4   0.8m

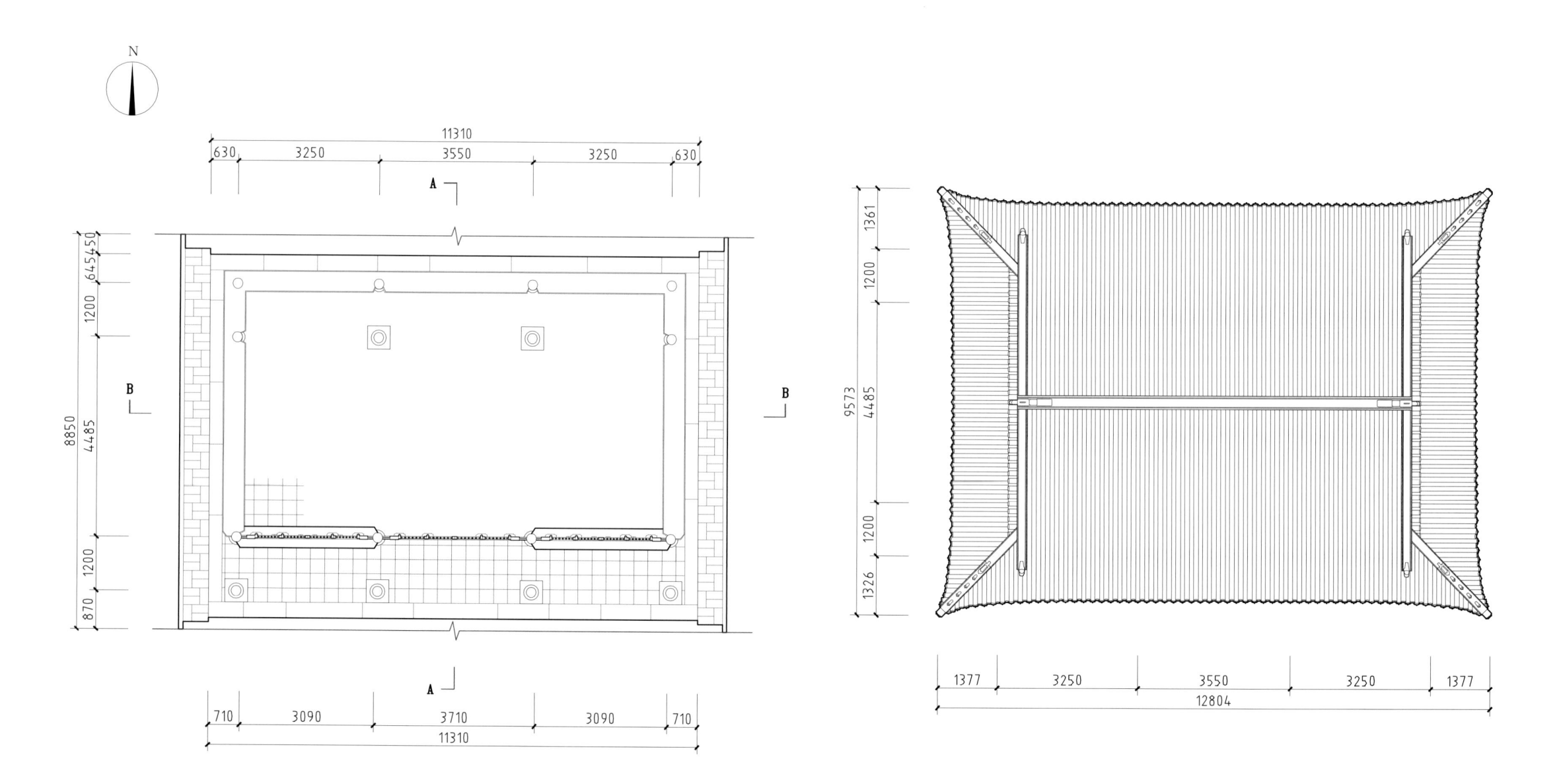

西岳庙游岳牌坊平面图
Plan of Xiyue Temple's Youyue Archway

西岳庙游岳牌坊屋顶平面图
Roof plan of Xiyue Temple's Youyue Archway

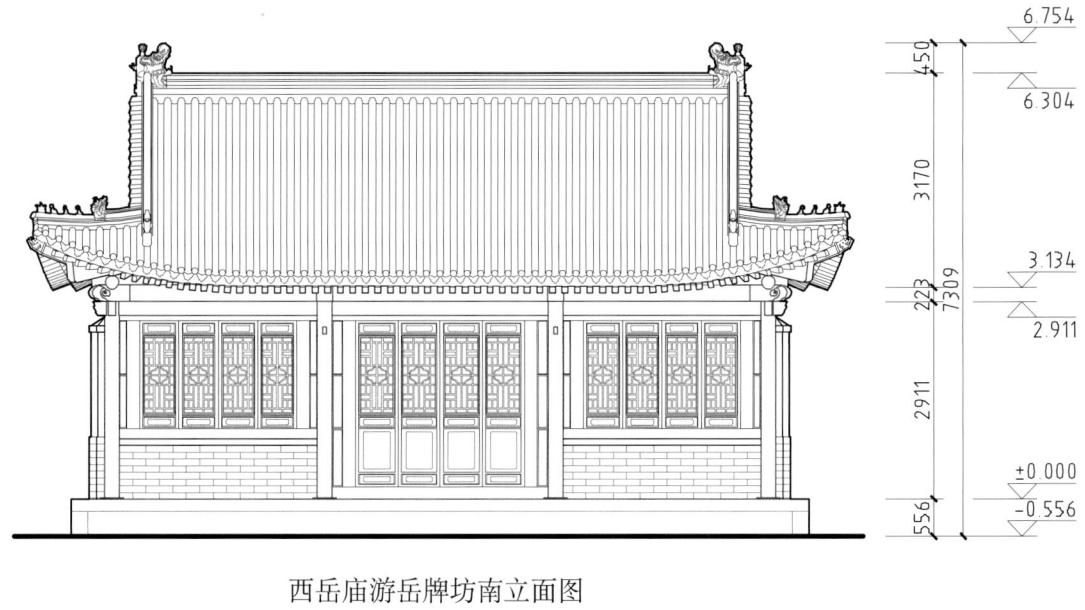

西岳庙游岳牌坊南立面图
South elevation of Xiyue Temple's Youyue Archway

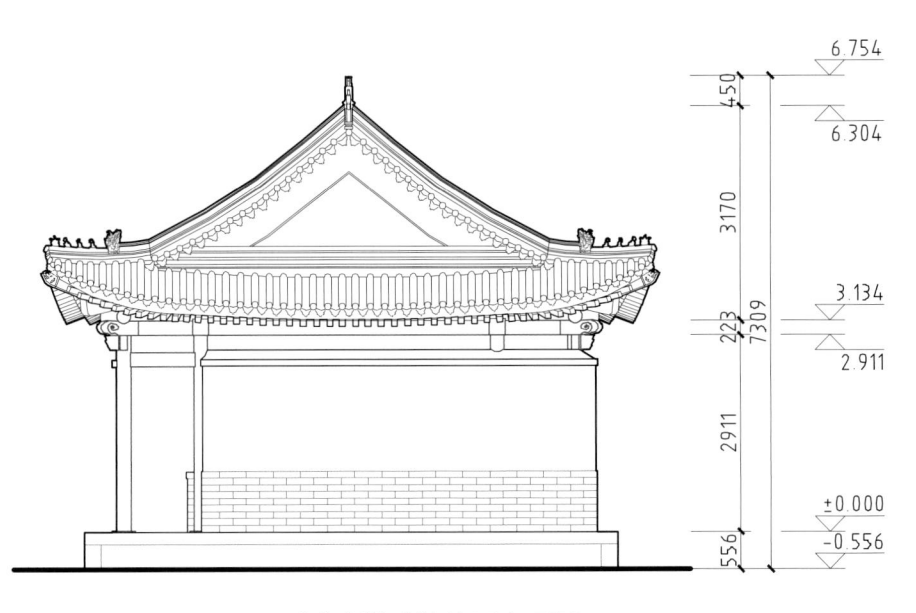

西岳庙游岳牌坊西立面图
West elevation of Xiyue Temple's Youyue Archway

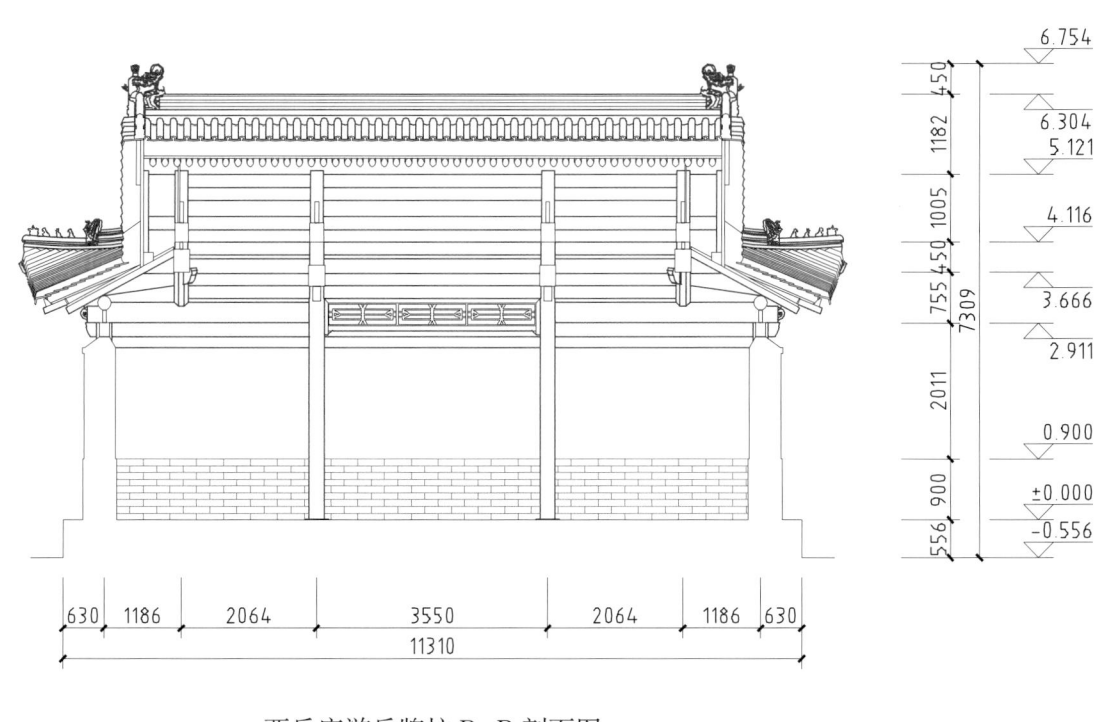

西岳庙游岳牌坊 B−B 剖面图
Section B-B of Xiyue Temple's Youyue Archway

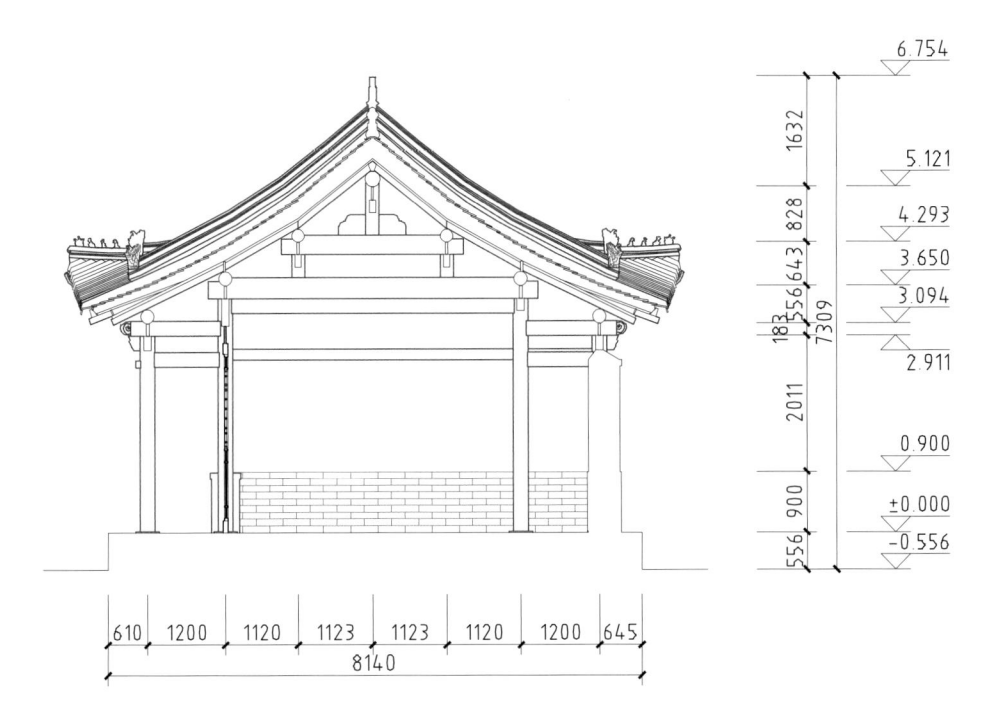

西岳庙游岳牌坊 A−A 剖面图
Section A-A of Xiyue Temple's Youyue Archway

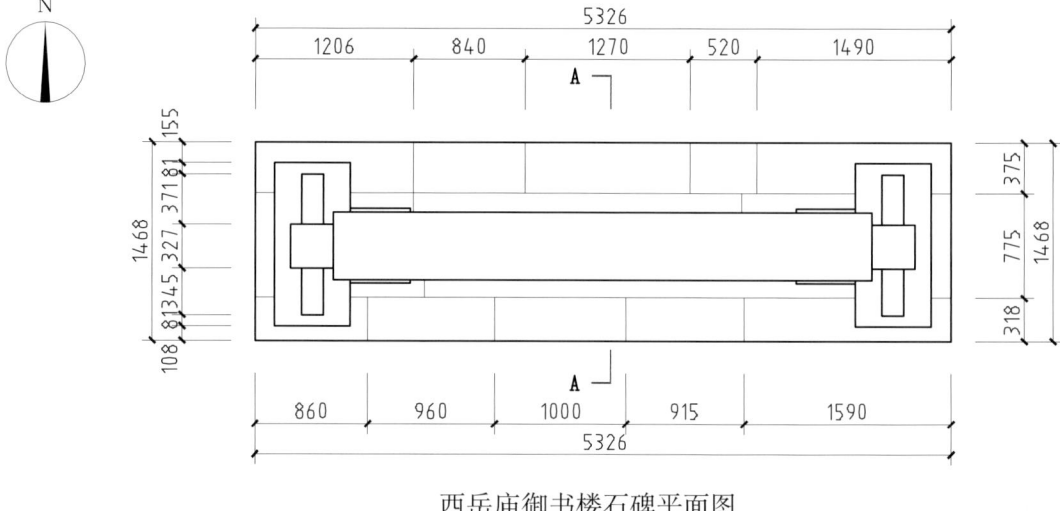

西岳庙御书楼石碑平面图
Plan of Xiyue Temple's Yushulou's stone tablet

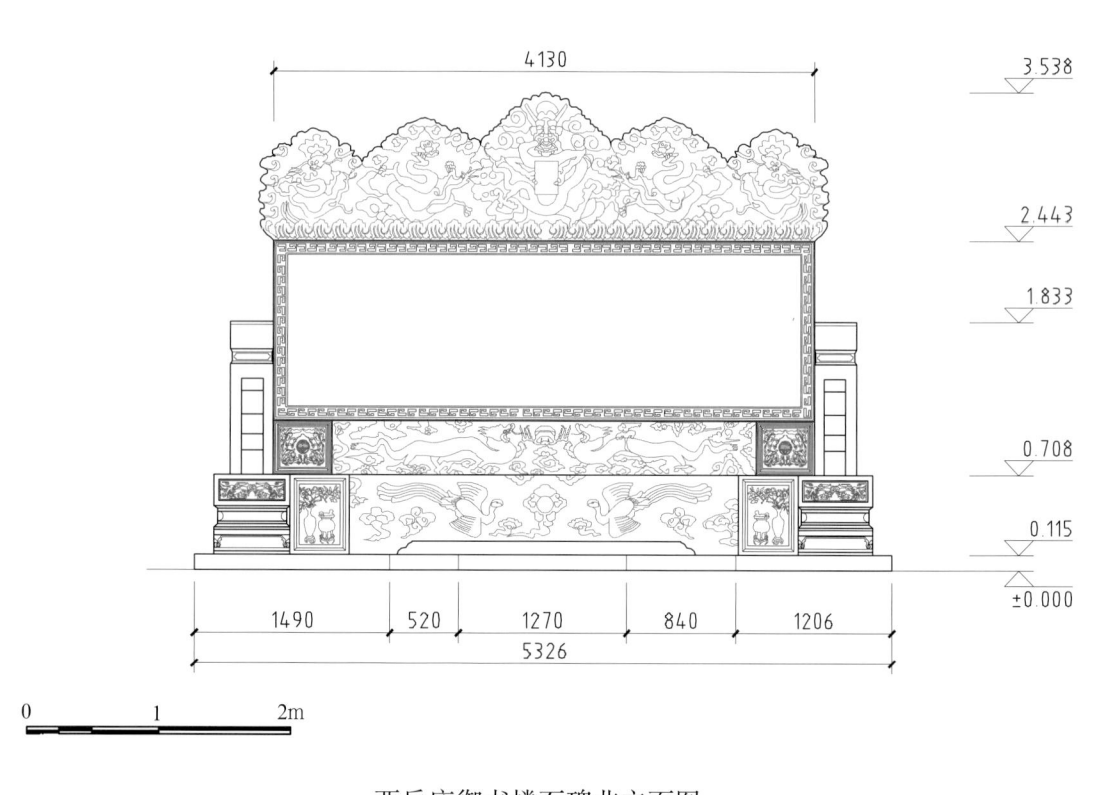

西岳庙御书楼石碑北立面图
North elevation of Xiyue Temple's Yushulou's stone tablet

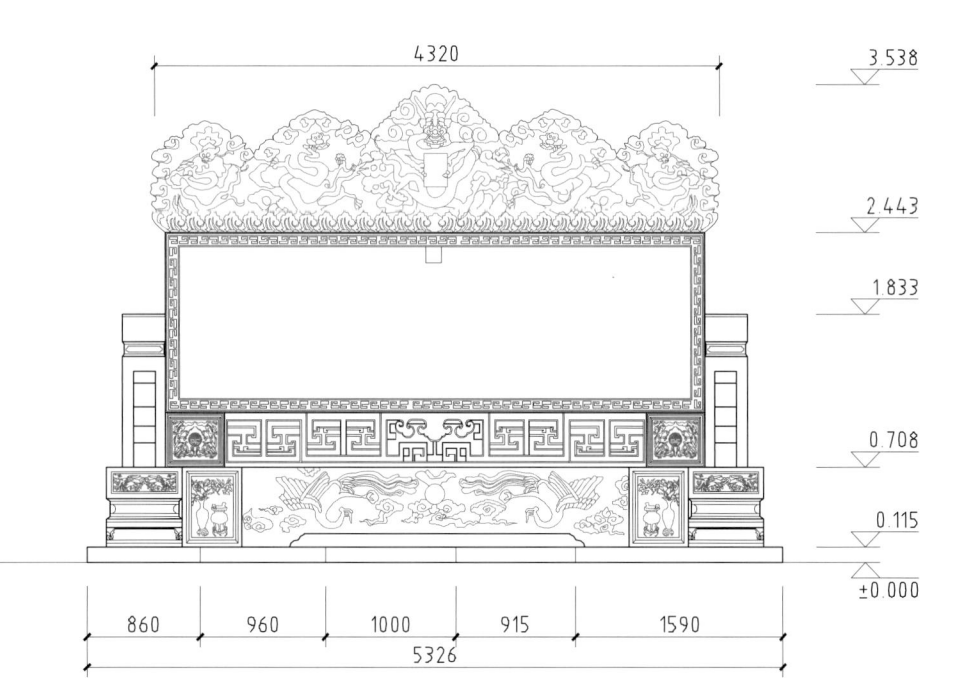

西岳庙御书楼石碑南立面图
South elevation of Xiyue Temple's Yushulou's stone tablet

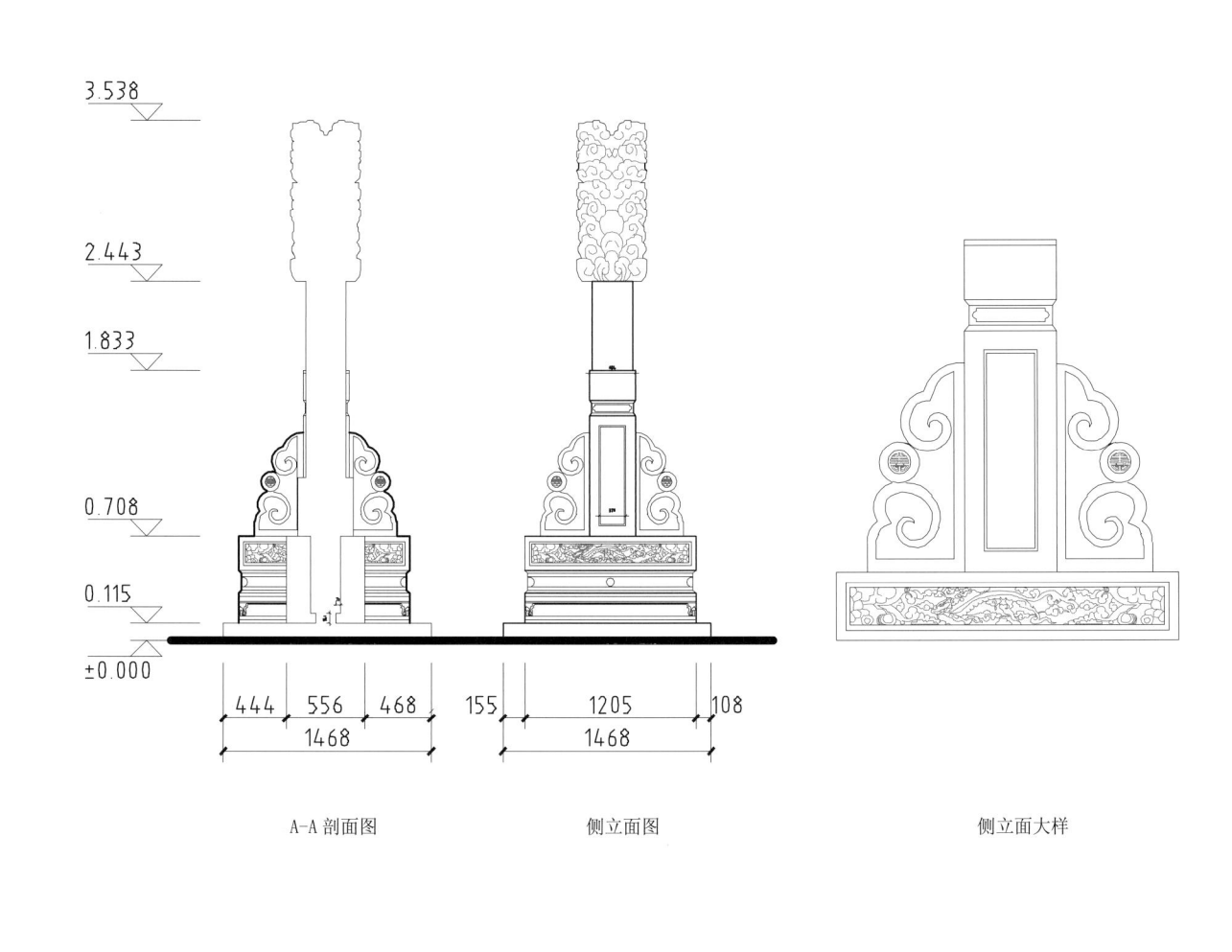

A-A 剖面图　　　　　　　側立面图　　　　　　　側立面大样

正立面大样

0　　　0.8　　　1.6m

西岳庙御书楼石碑构件大样图
Structural component of Xiyue Temple's Yushulou's stone tablet

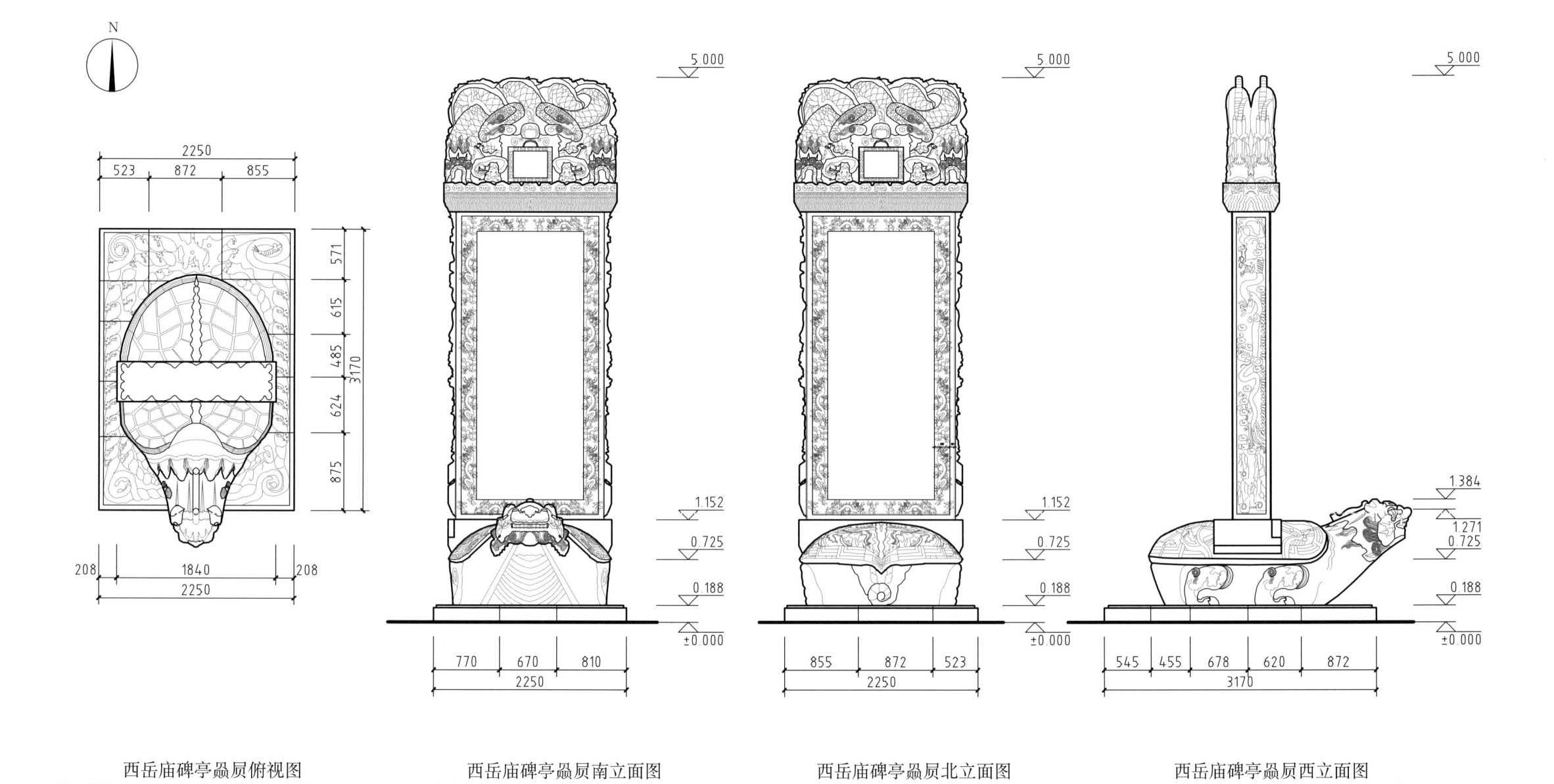

西岳庙碑亭赑屃俯视图
Bixi of Xiyue Temple's stele pavilion as seen from above

西岳庙碑亭赑屃南立面图
South elevation of *bixi* of Xiyue Temple's stele pavilion

西岳庙碑亭赑屃北立面图
North elevation of *bixi* of Xiyue Temple's stele pavilion

西岳庙碑亭赑屃西立面图
West elevation of *bixi* of Xiyue Temple's stele pavilion

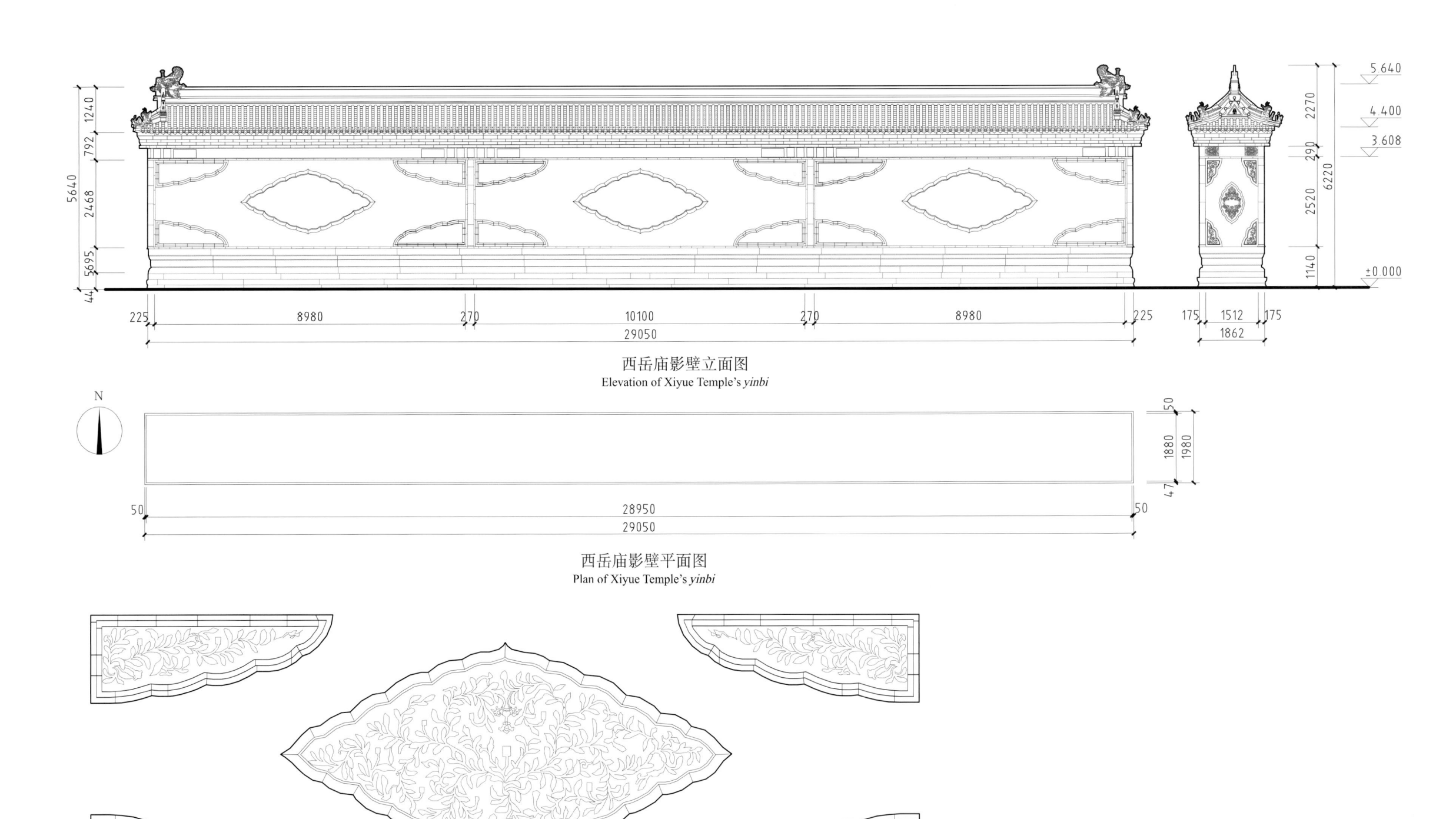

西岳庙影壁立面图
Elevation of Xiyue Temple's *yinbi*

西岳庙影壁平面图
Plan of Xiyue Temple's *yinbi*

西岳庙影壁正立面大样图
Front elevation of Xiyue Temple's *yinbi*

0   0.5   1m

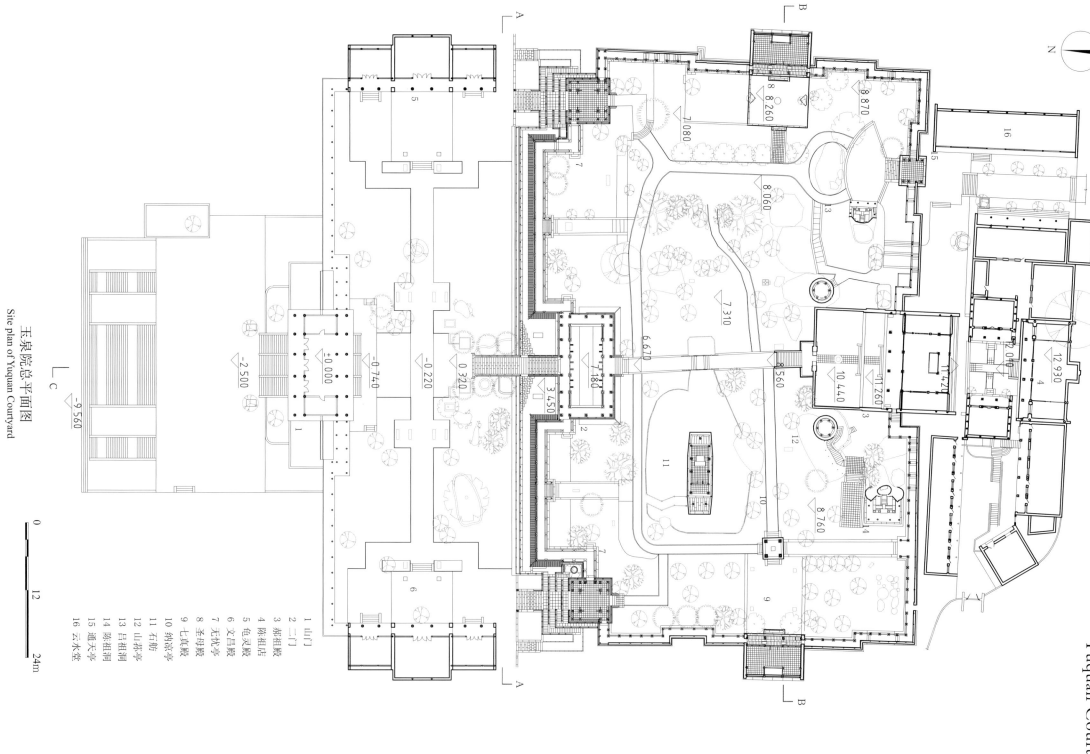

玉泉院
Yuquan Courtyard

玉泉院总平面图
Site plan of Yuquan Courtyard

1 山门
2 二门
3 郝祖殿
4 陈祖店
5 色灵殿
6 文昌殿
7 无忧亭
8 圣母殿
9 七真殿
10 纳凉亭
11 石舫
12 山荪亭
13 吕祖殿
14 陈抟洞
15 通天亭
16 云水堂

0 12 24m

溯源与思辨——渭南建筑·华山道教建筑甲编

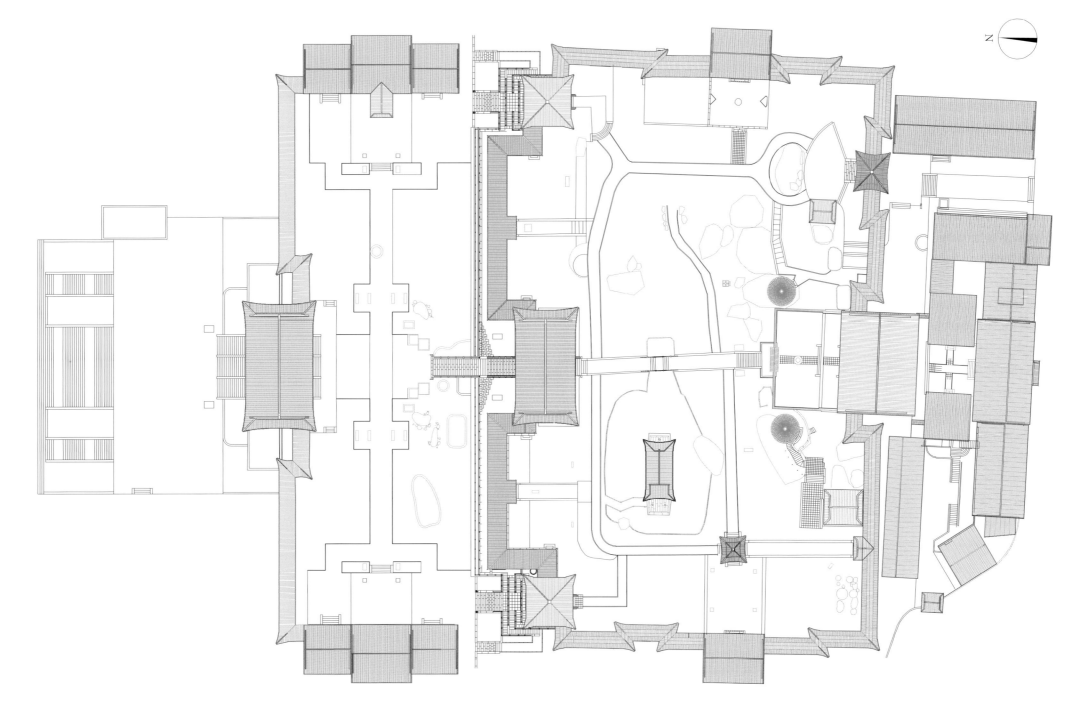

玉泉院屋顶总平面图
Site roof plan of Yuquan Courtyard

0   12   24m

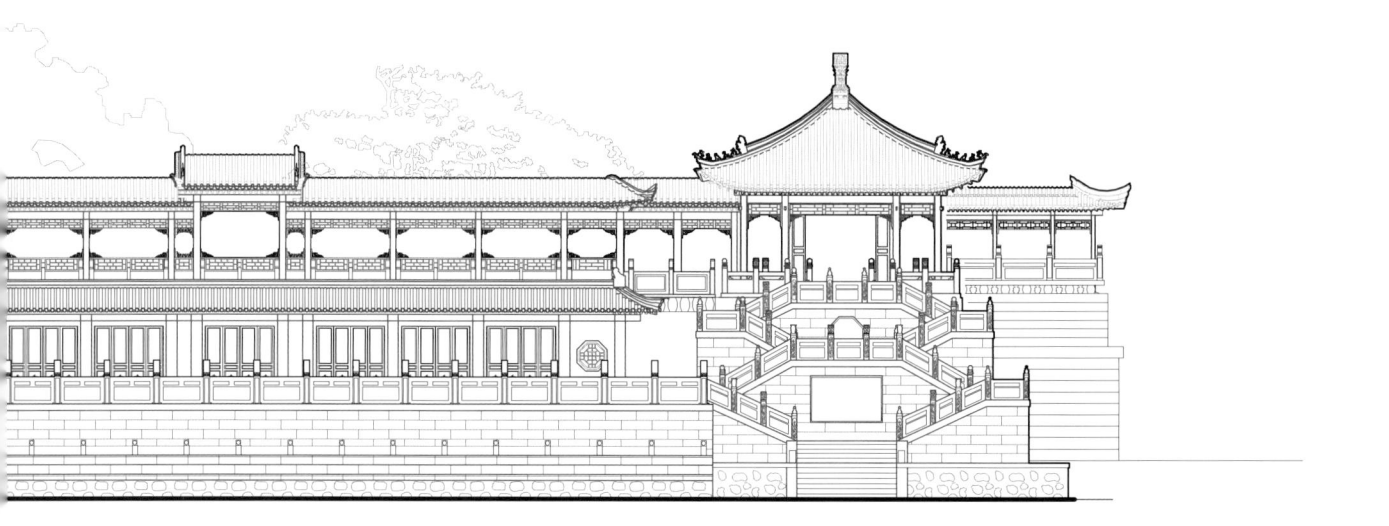

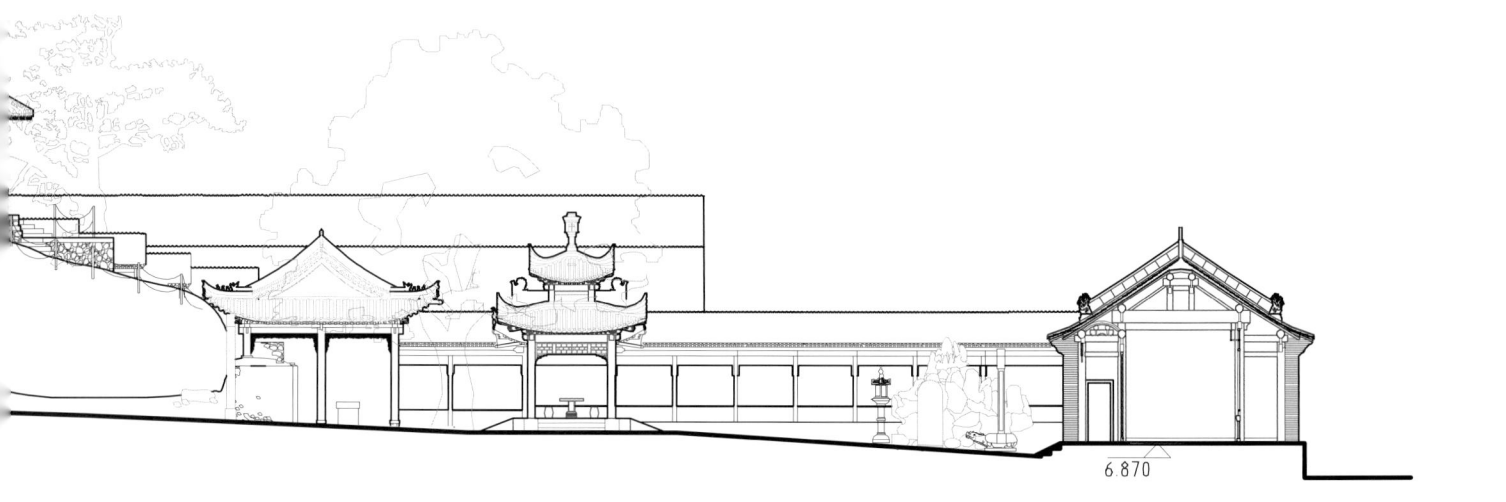

6.870

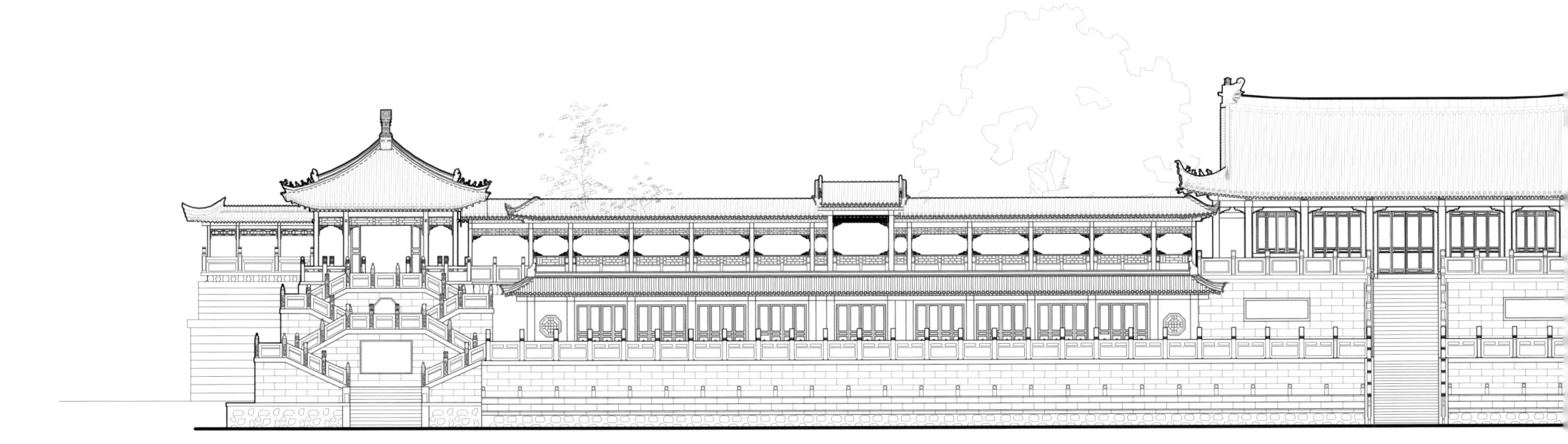

玉泉院 A—A 剖面图
Section A-A of Yuquan Courtyard

11.780

7 820

8 560

0   4   8m

玉泉院 B—B 剖面图
Section B-B of Yuquan Courtyard

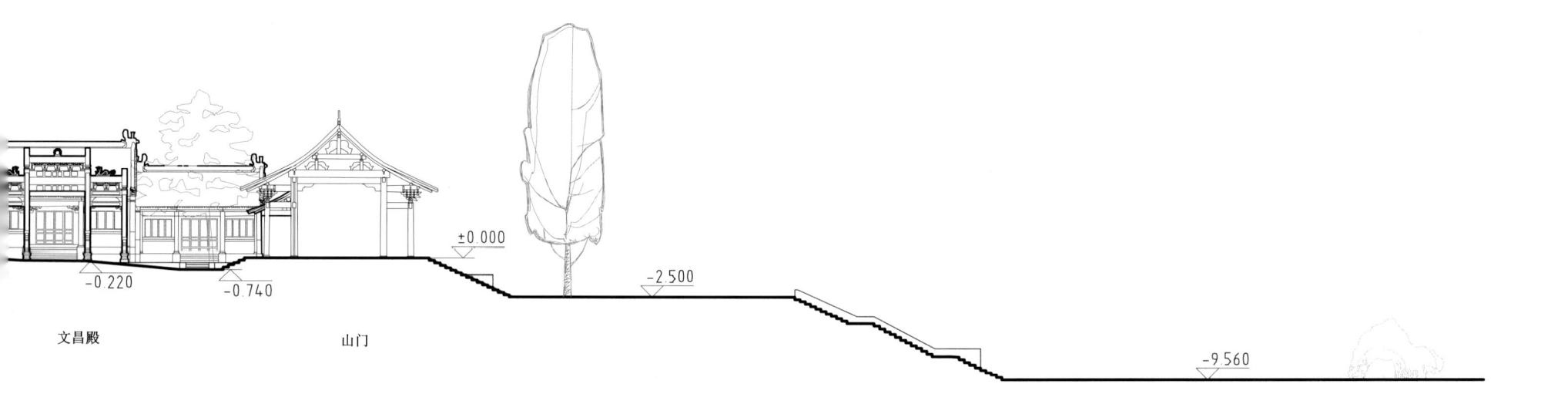

±0.000

−0.220

−0.740

−2.500

−9.560

文昌殿

山门

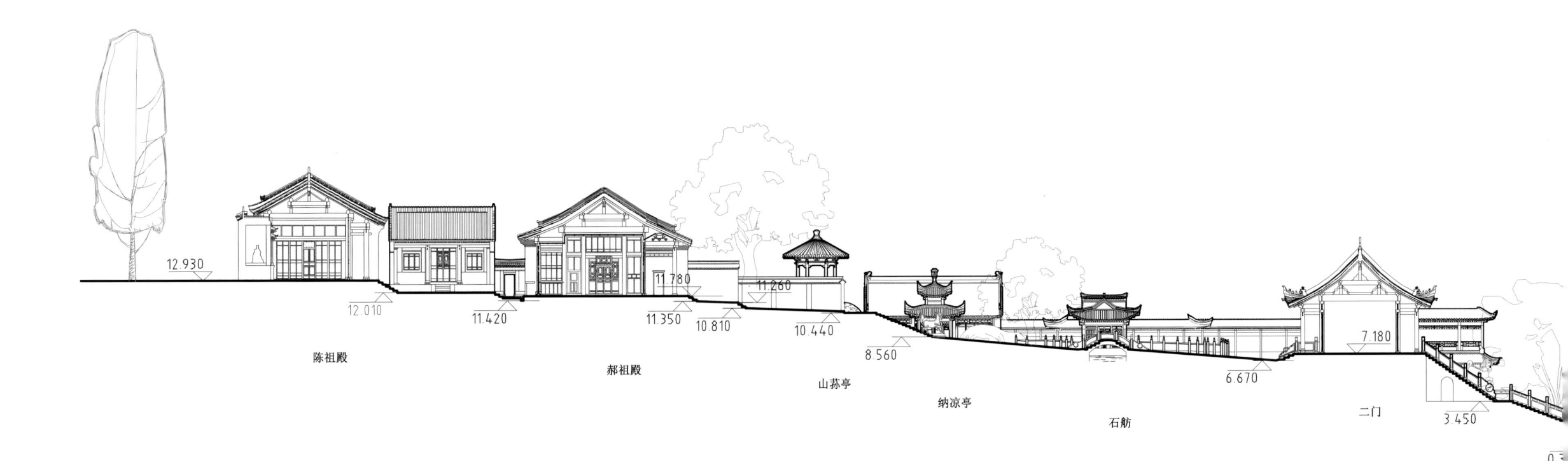

12.930

12.010

11.780

11.420

11.350 10.810

11.260

10.440

8.560

7.180

6.670

3.450

3.450

0.3

陈祖殿

郝祖殿

山荪亭

纳凉亭

石舫

二门

0 5 10m

玉泉院 C-C 剖面图
Section C-C of Yuquan Courtyard

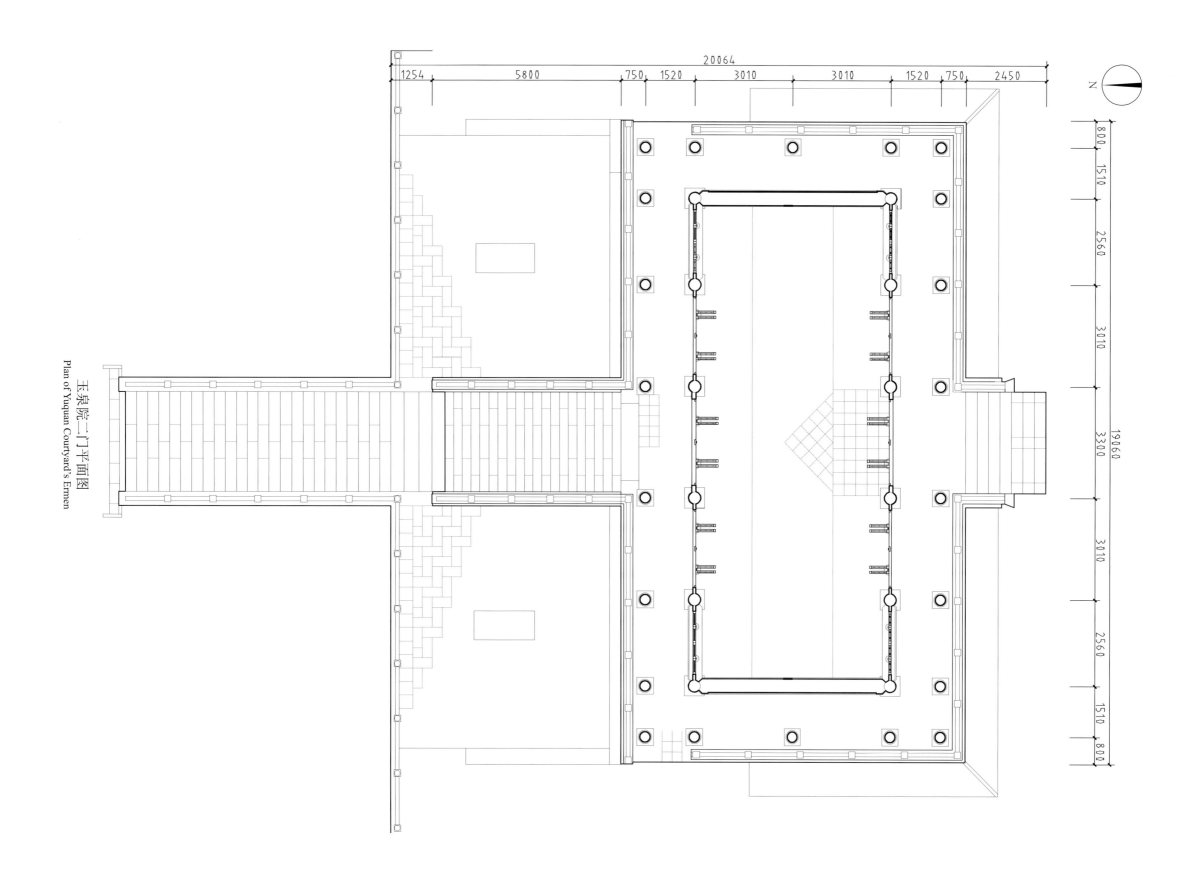

玉泉院一门平面图
Plan of Yuquan Courtyard's Ermen

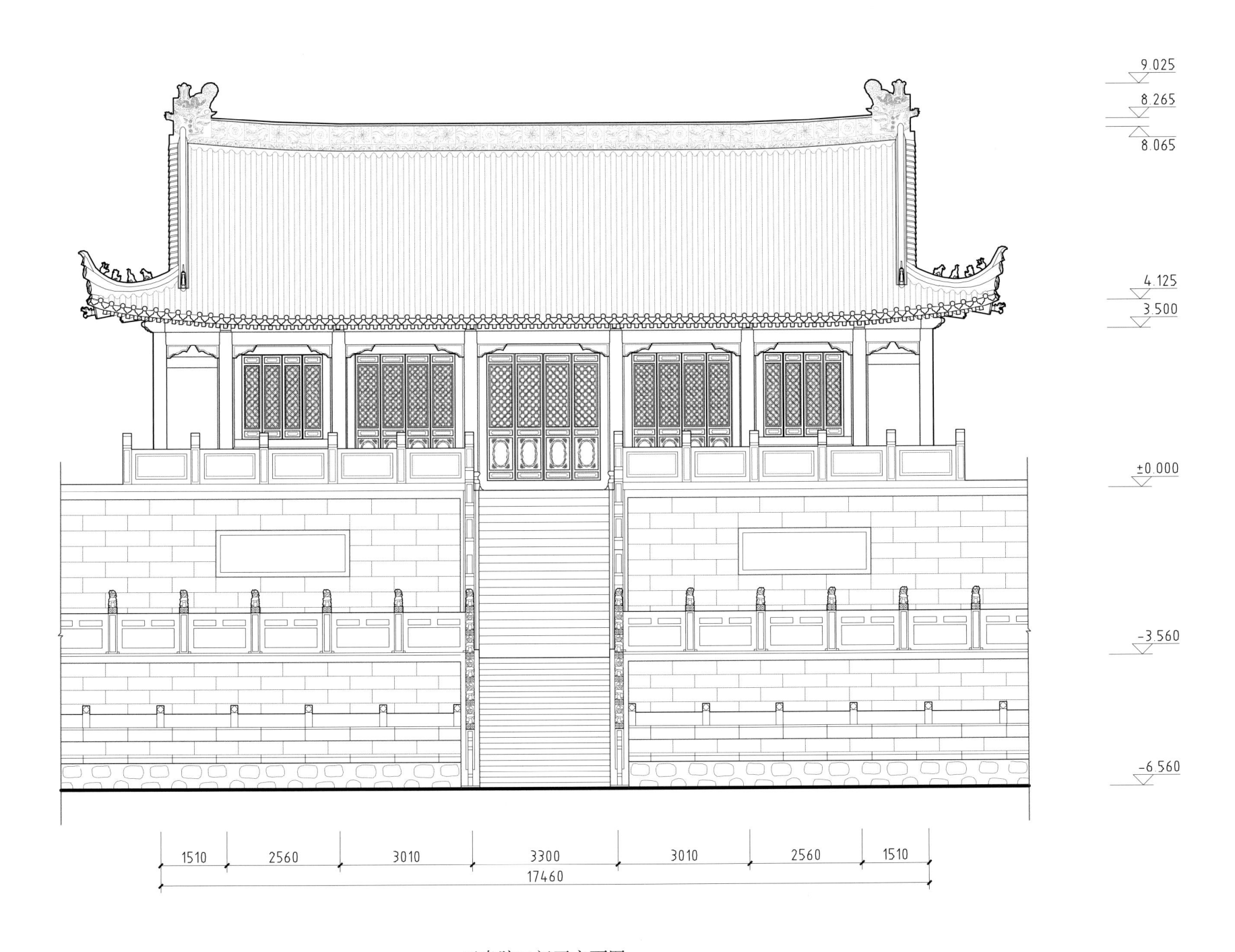

9.025
8.265
8.065
4.125
3.500
±0.000
-3.560
-6.560

1510 2560 3010 3300 3010 2560 1510
17460

玉泉院二门正立面图
Front elevation of Yuquan Courtyard's Ermen

0   2   4m

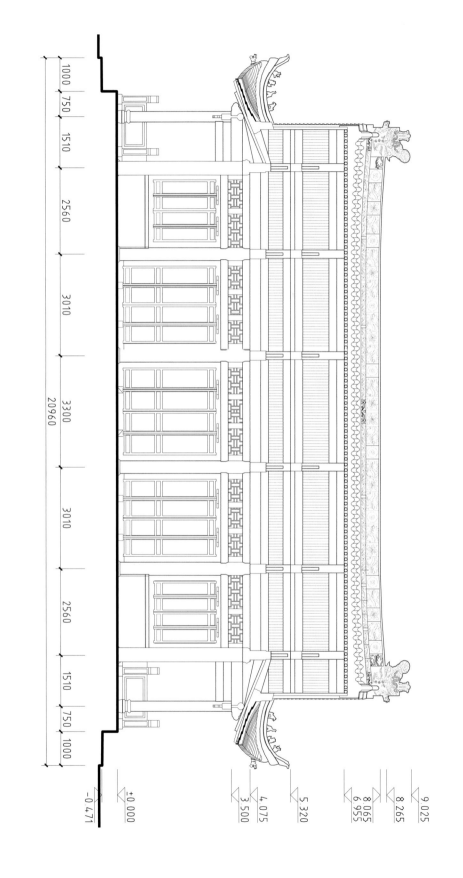

玉泉院二门纵剖面图
Longitudinal section of Yuquan Courtyard's Ermen

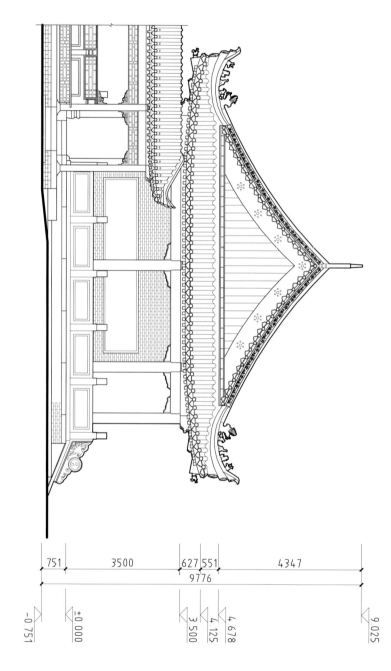

玉泉院二门侧立面图
Side elevation of Yuquan Courtyard's Ermen

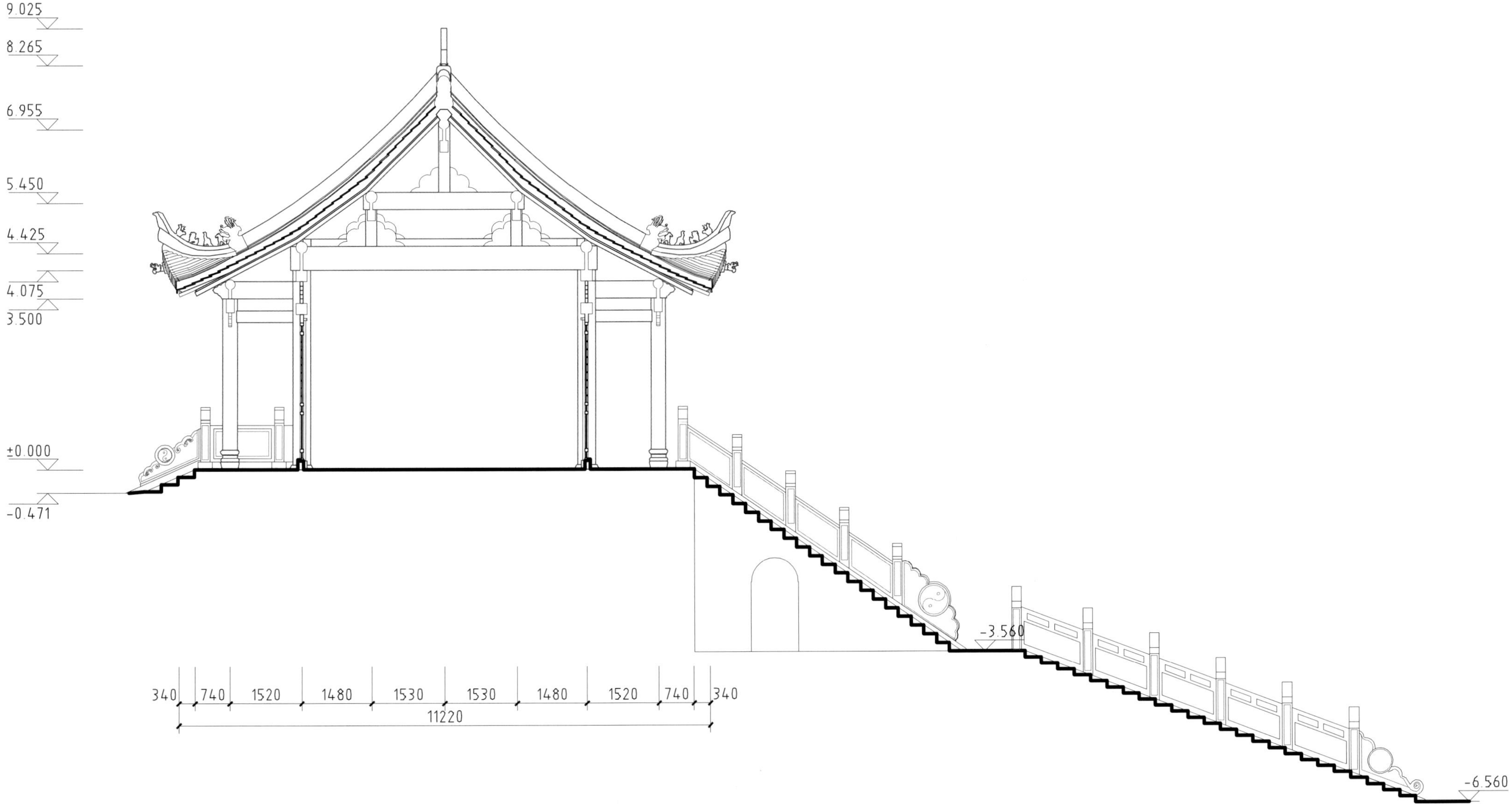

9.025

8.265

6.955

5.450

4.425

4.075

3.500

±0.000

-0.471

-3.560

-6.560

340 740 1520 1480 1530 1530 1480 1520 740 340

11220

玉泉院二门横剖面图

Cross-section of Yuquan Courtyard's Ermen

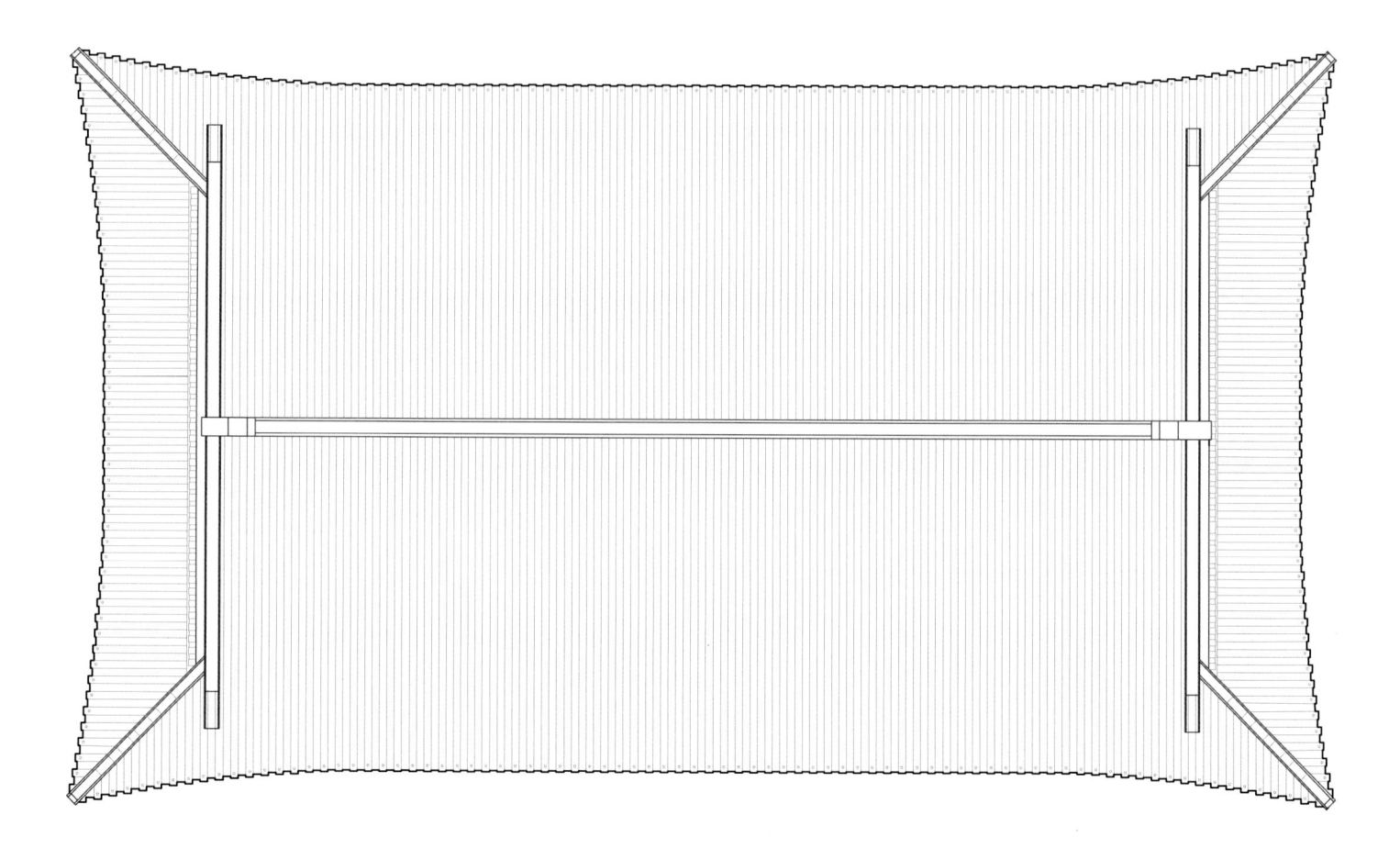

玉泉院二门屋顶平面图
Roof plan of Yuquan Courtyard's Ermen

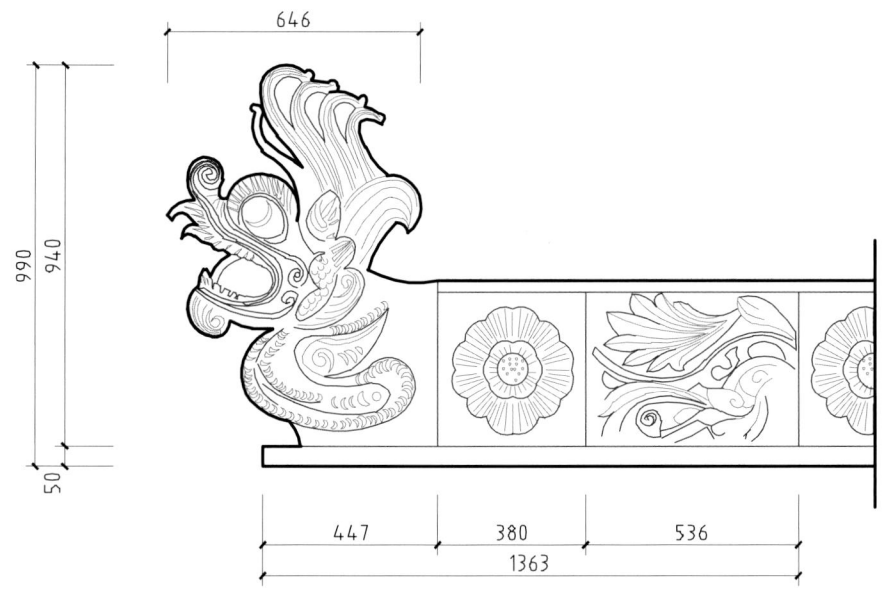

正脊吻兽大样

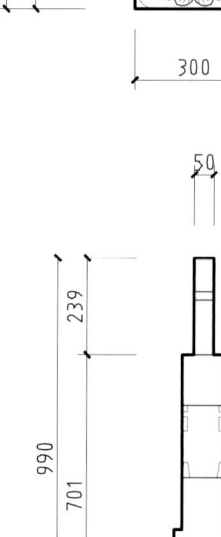

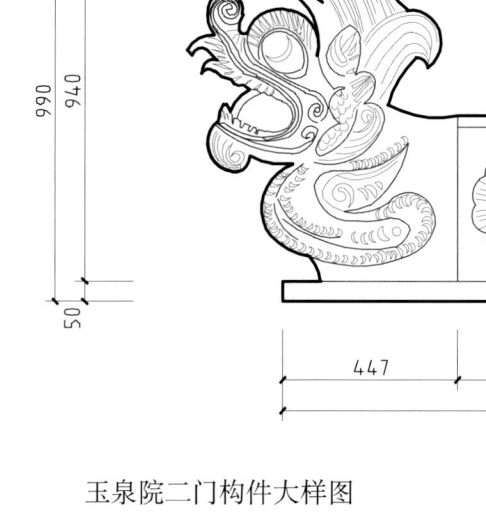

垂脊吻兽大样

玉泉院二门构件大样图
Structural component of Yuquan Courtyard's Ermen

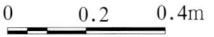

0   0.2   0.4m

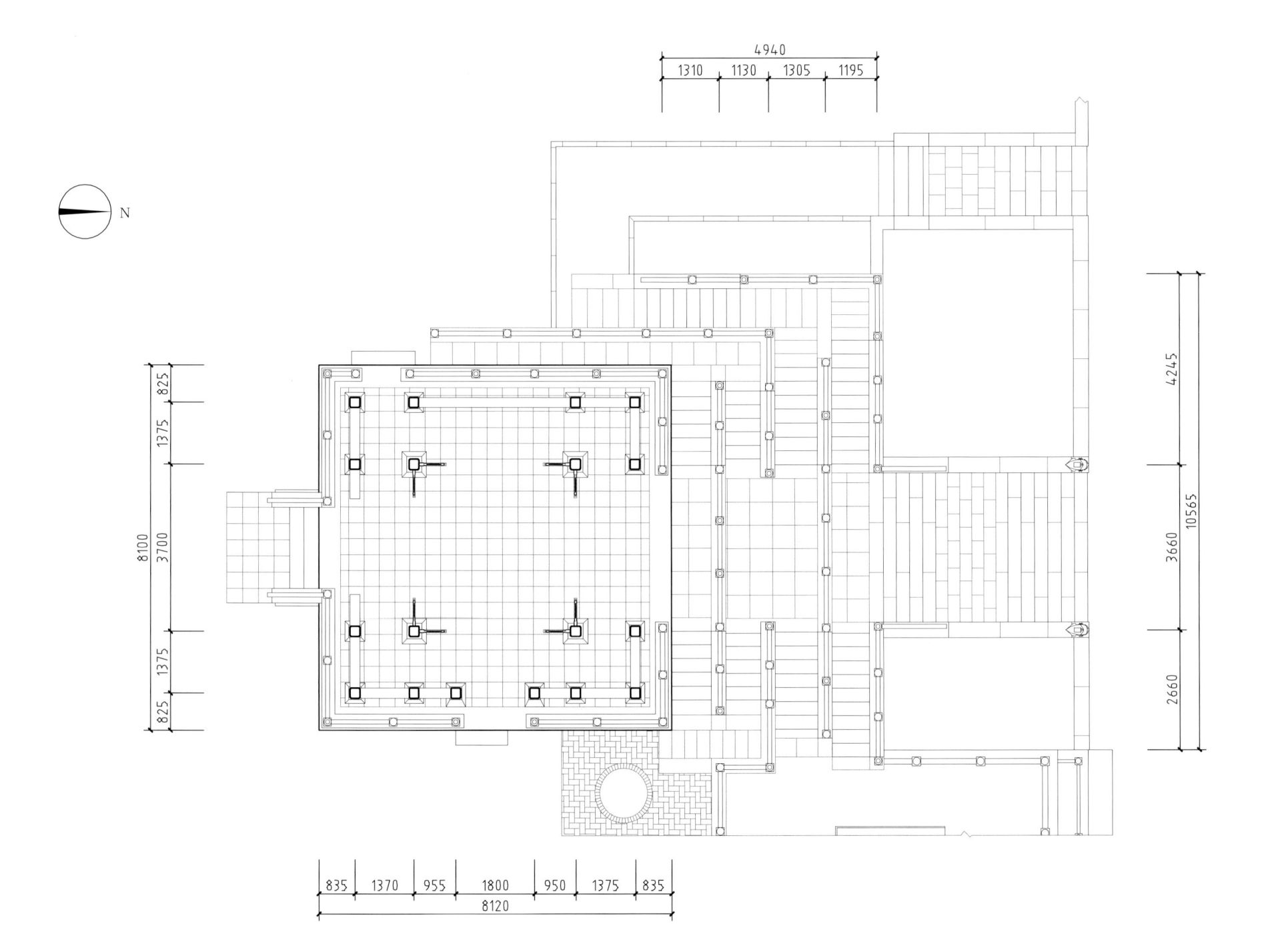

N

4940
1310 1130 1305 1195

825
1375
8100
3700
1375
825

4245
10565
3660
2660

835 1370 955 1800 950 1375 835
8120

玉泉院无忧亭平面图
Plan of Yuquan Courtyard's Wuyou Pavilion

0   2   4m

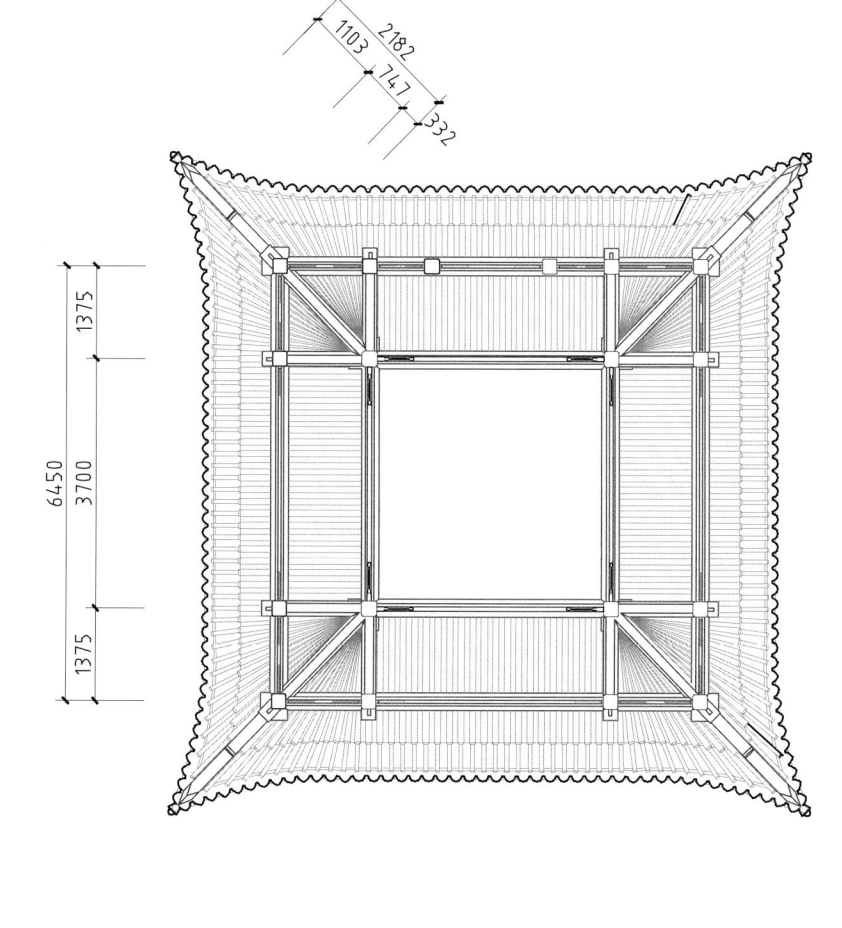

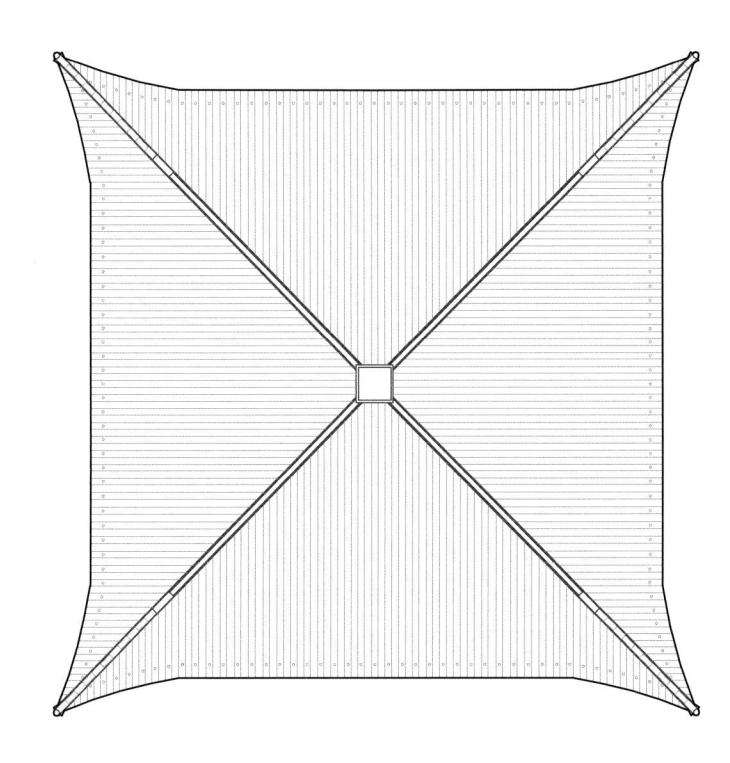

玉泉院无忧亭梁架仰视平面图
Plan of framework of Yuquan Courtyard's Wuyou Pavilion as seen from below

玉泉院无忧亭屋顶平面图
Roof plan of Yuquan Courtyard's Wuyou Pavilion

0   2   4m

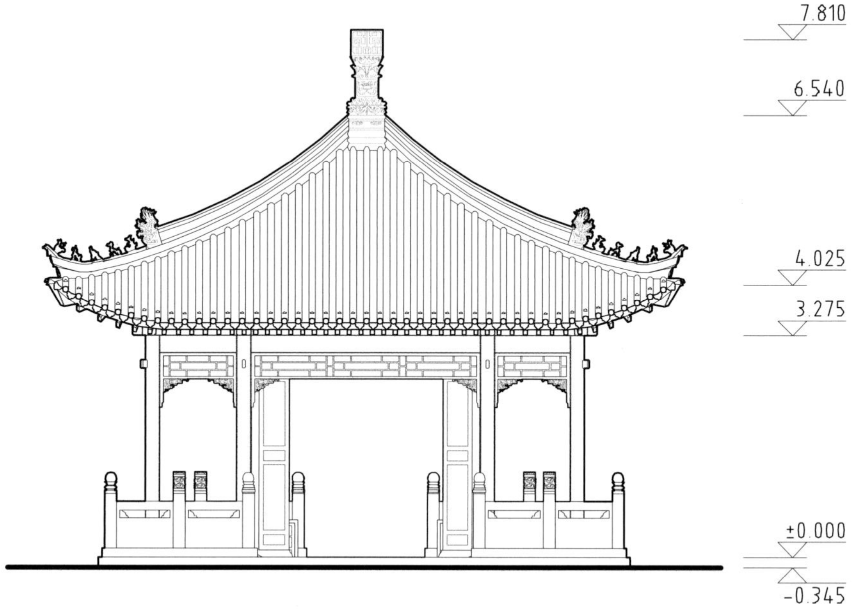

7.810

6.540

7.810

4.025

3.275

4.080

3.455

3.275

+0.000

±0.000

-0.345

-0.345

玉泉院无忧亭北立面图

North elevation of Yuquan Courtyard's Wuyou Pavilion

±0.000

-0.155

-1.080

-1.980

-3.410

-4.565

-5.265

-6.280

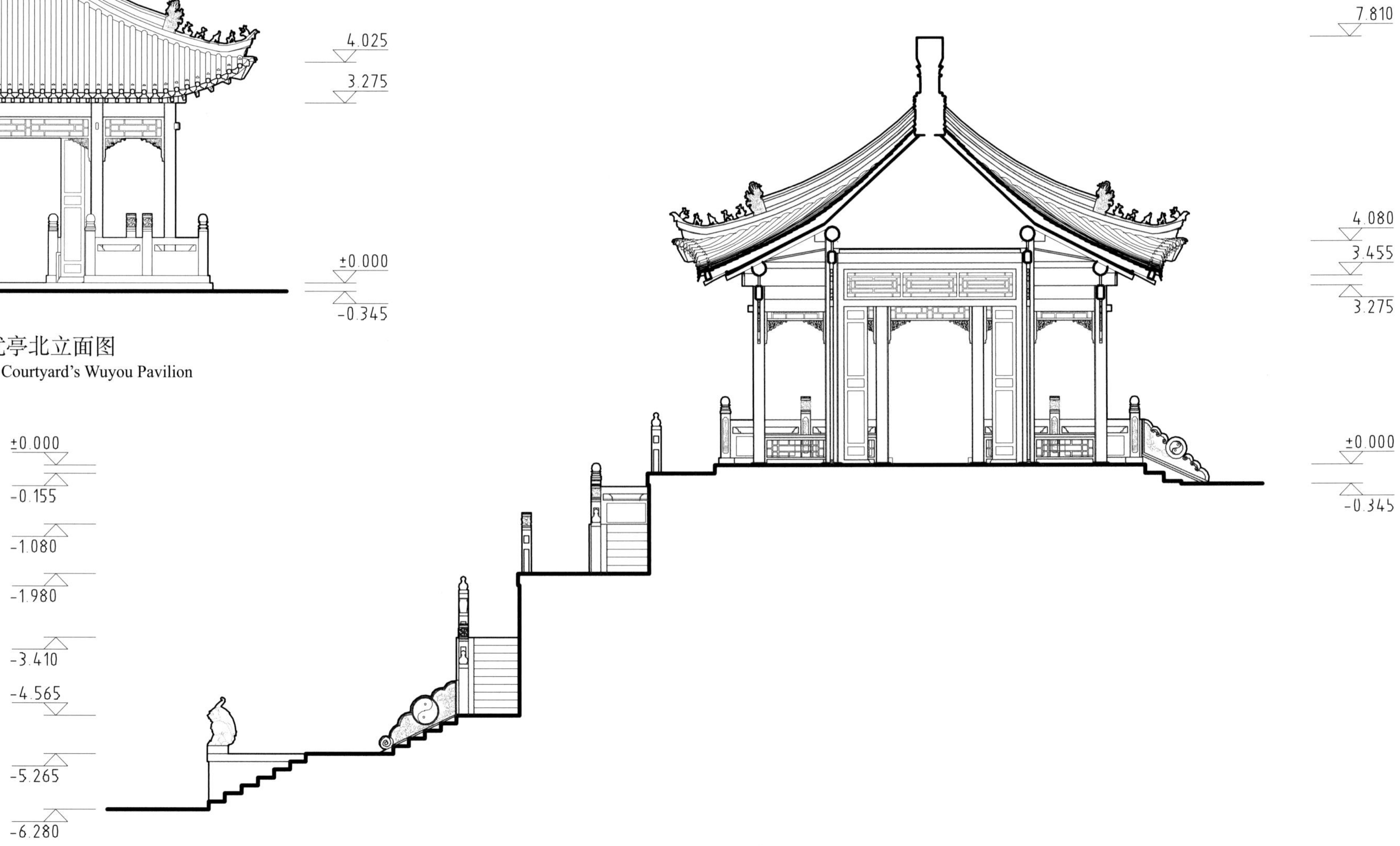

玉泉院无忧亭剖面图

Section of Yuquan Courtyard's Wuyou Pavilion

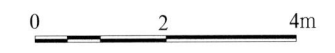

0   2   4m

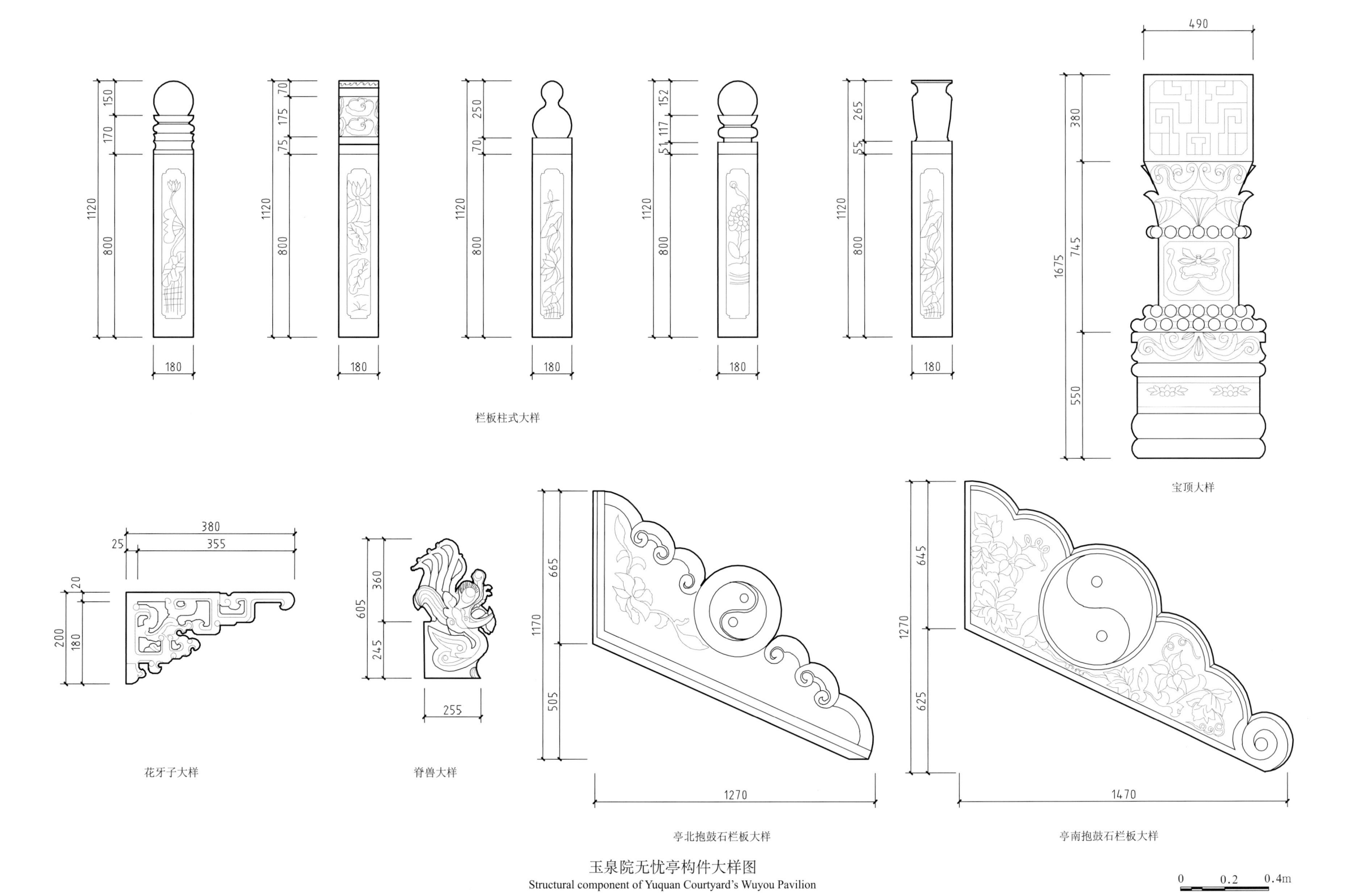

栏板柱式大样

宝顶大样

花牙子大样

脊兽大样

亭北抱鼓石栏板大样

亭南抱鼓石栏板大样

玉泉院无忧亭构件大样图
Structural component of Yuquan Courtyard's Wuyou Pavilion

0   0.2   0.4m

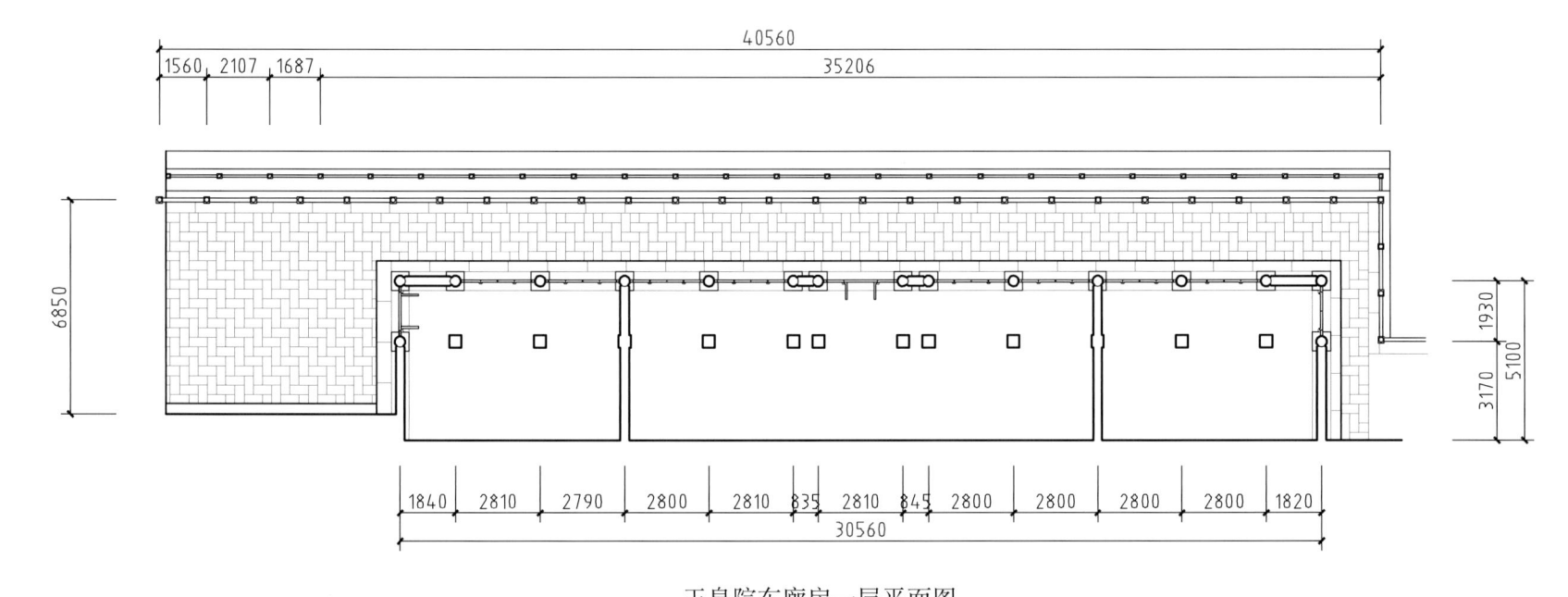

玉泉院东廊房一层平面图
Plan of first floor of Yuquan Courtyard's east *langfang*

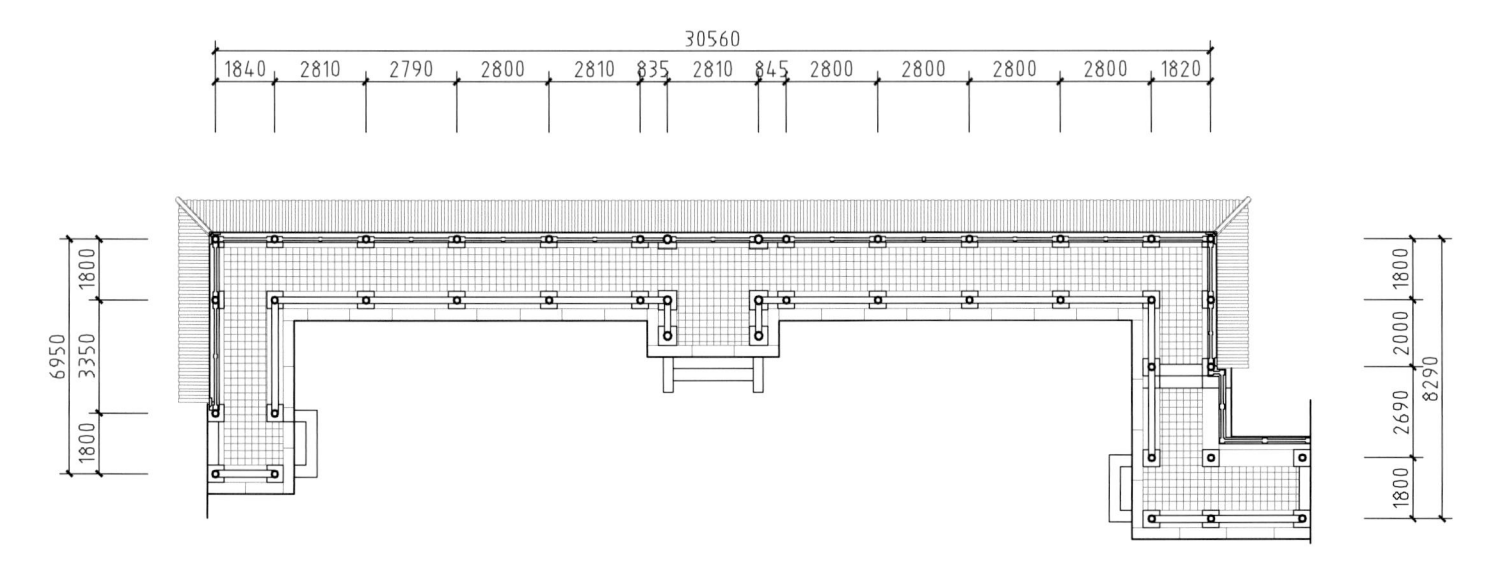

玉泉院东廊房二层平面图
Plan of second floor of Yuquan Courtyard's east *langfang*

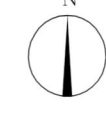

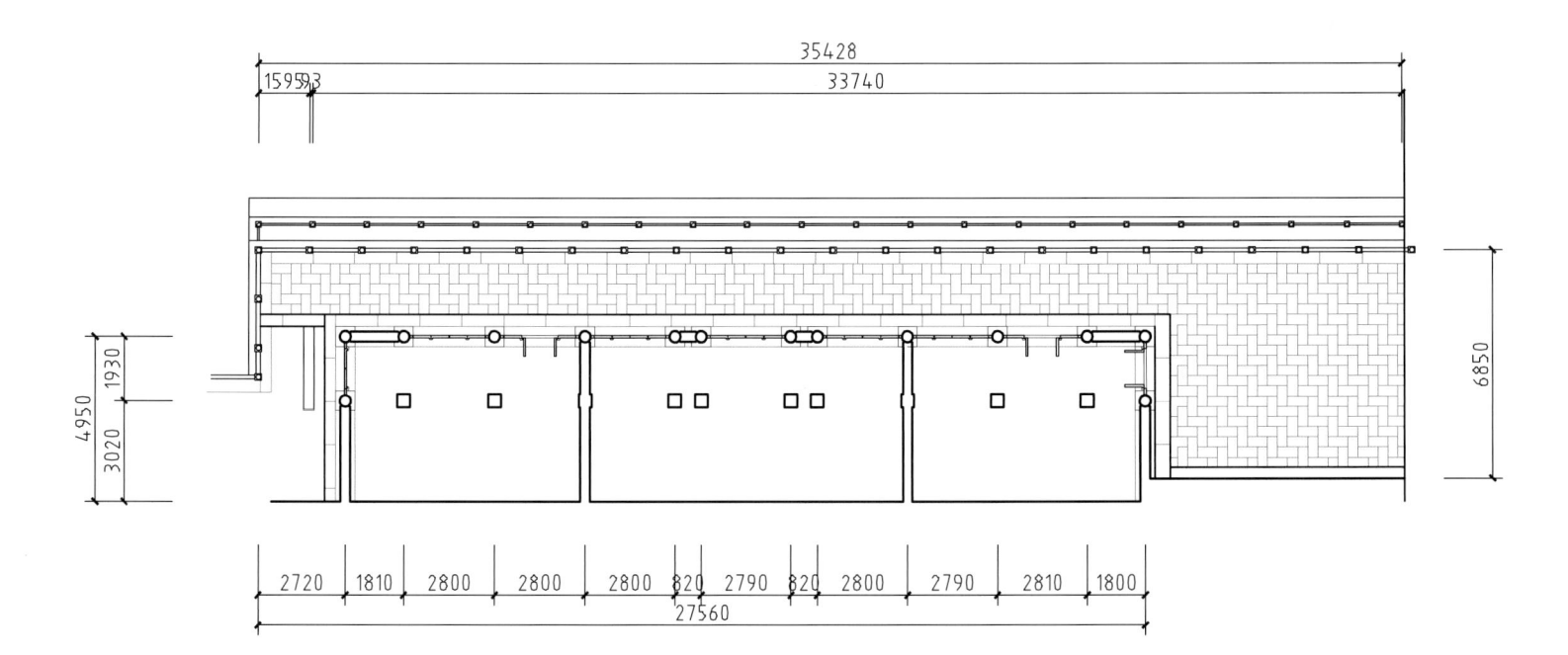

玉泉院西廊房一层平面图

Plan of first floor of Yuquan Courtyard's west *langfang*

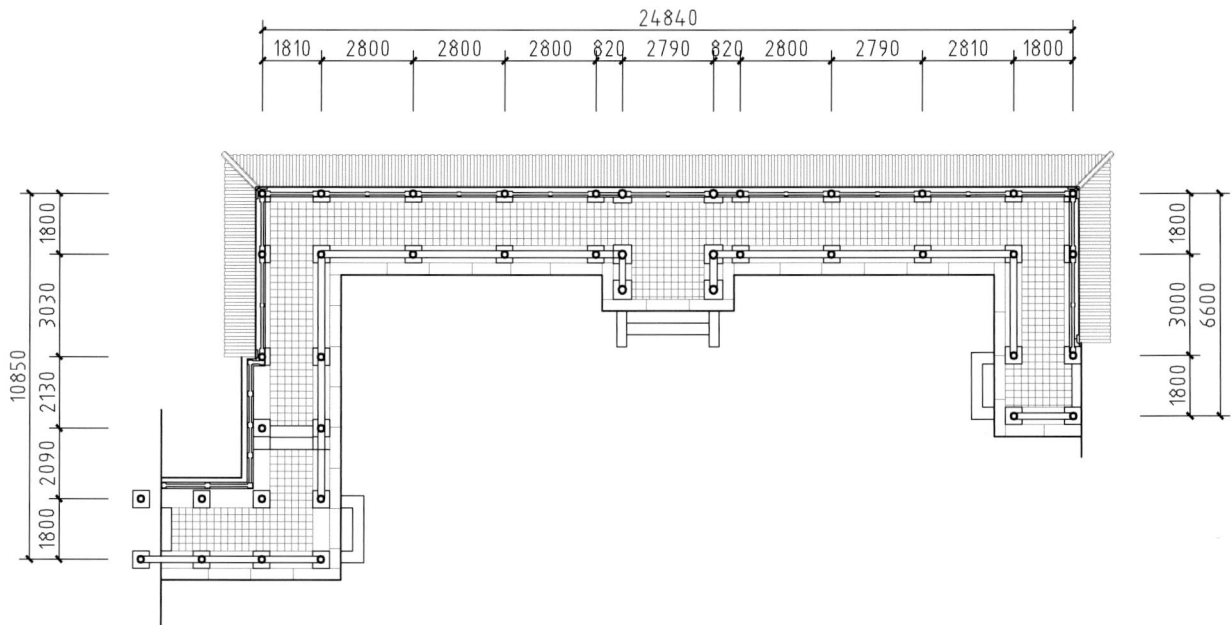

玉泉院西廊房二层平面图

Plan of second floor of Yuquan Courtyard's west *langfang*

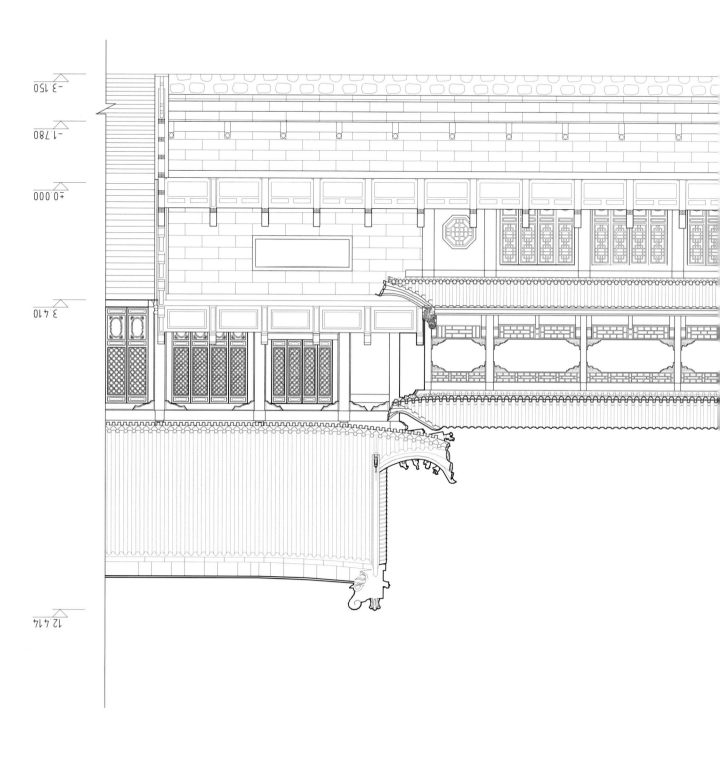

-3 150

-1 780

±0 000

3 410

12 474

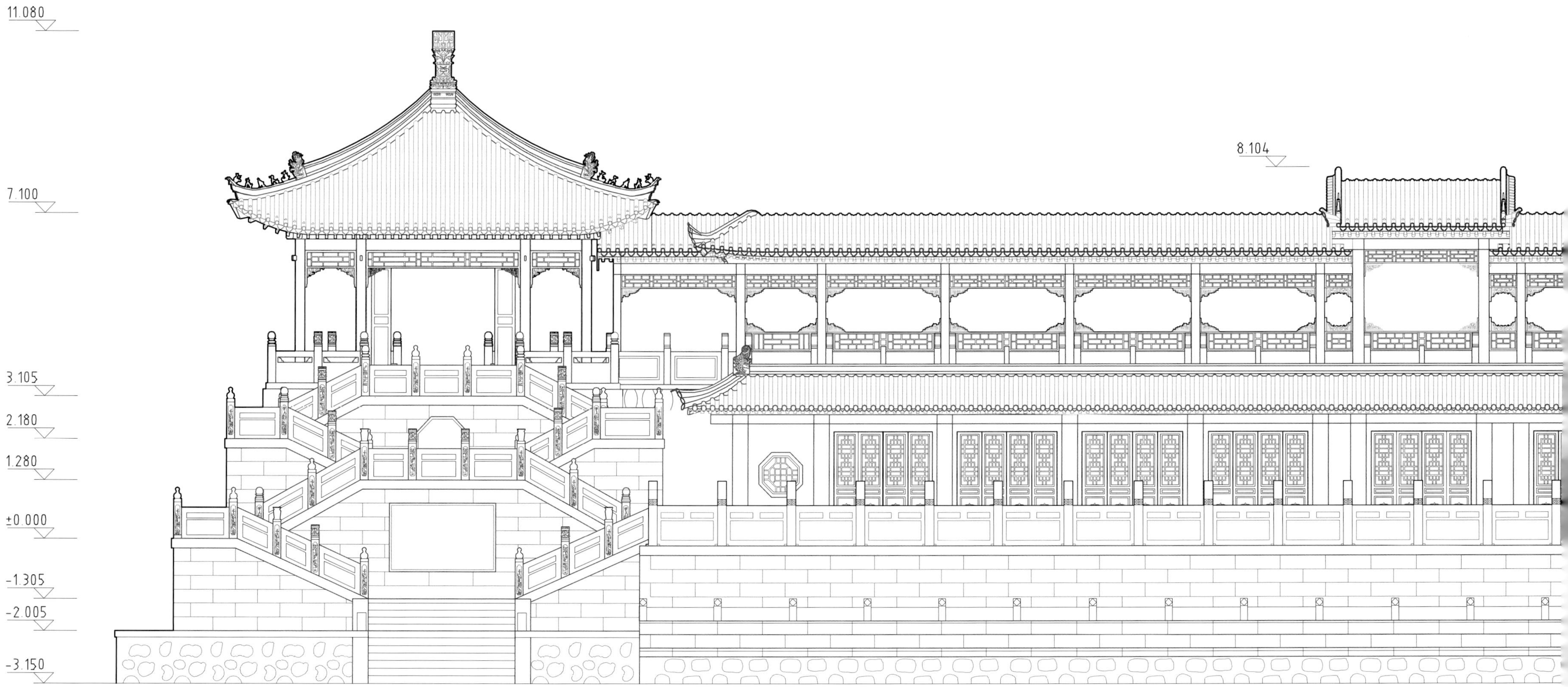

11.080

8.104

7.100

3.105

2.180

1.280

±0.000

-1.305

-2.005

-3.150

玉泉院东廊房正立面图
Front elevation of Yuquan Courtyard's east *langfang*

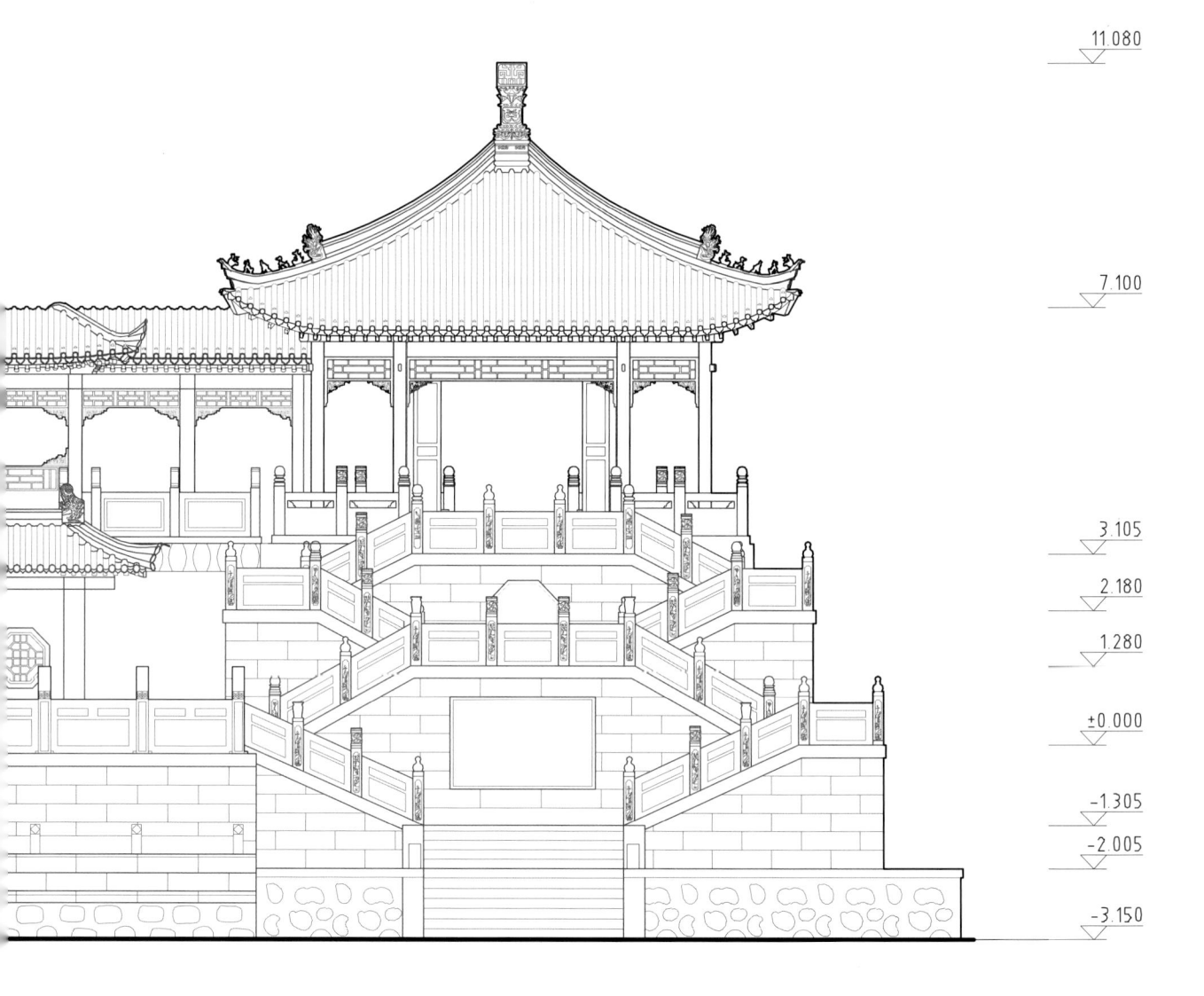

11.080

7.100

3.105

2.180

1.280

±0.000

-1.305

-2.005

-3.150

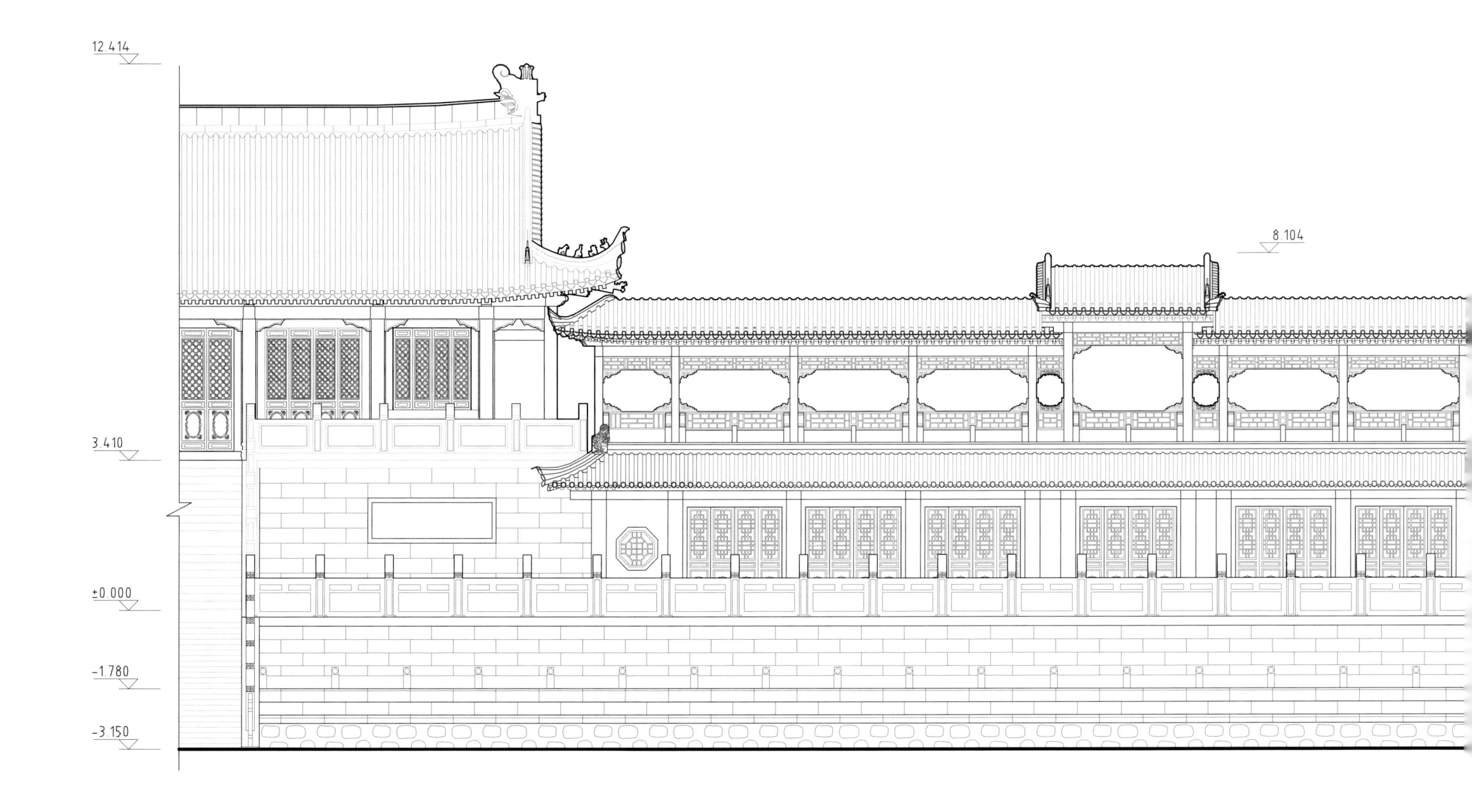

12.414

8.104

3.410

±0.000

-1.780

-3.150

玉泉院西廊房正立面图
Front elevation of Yuquan Courtyard's west *langfang*

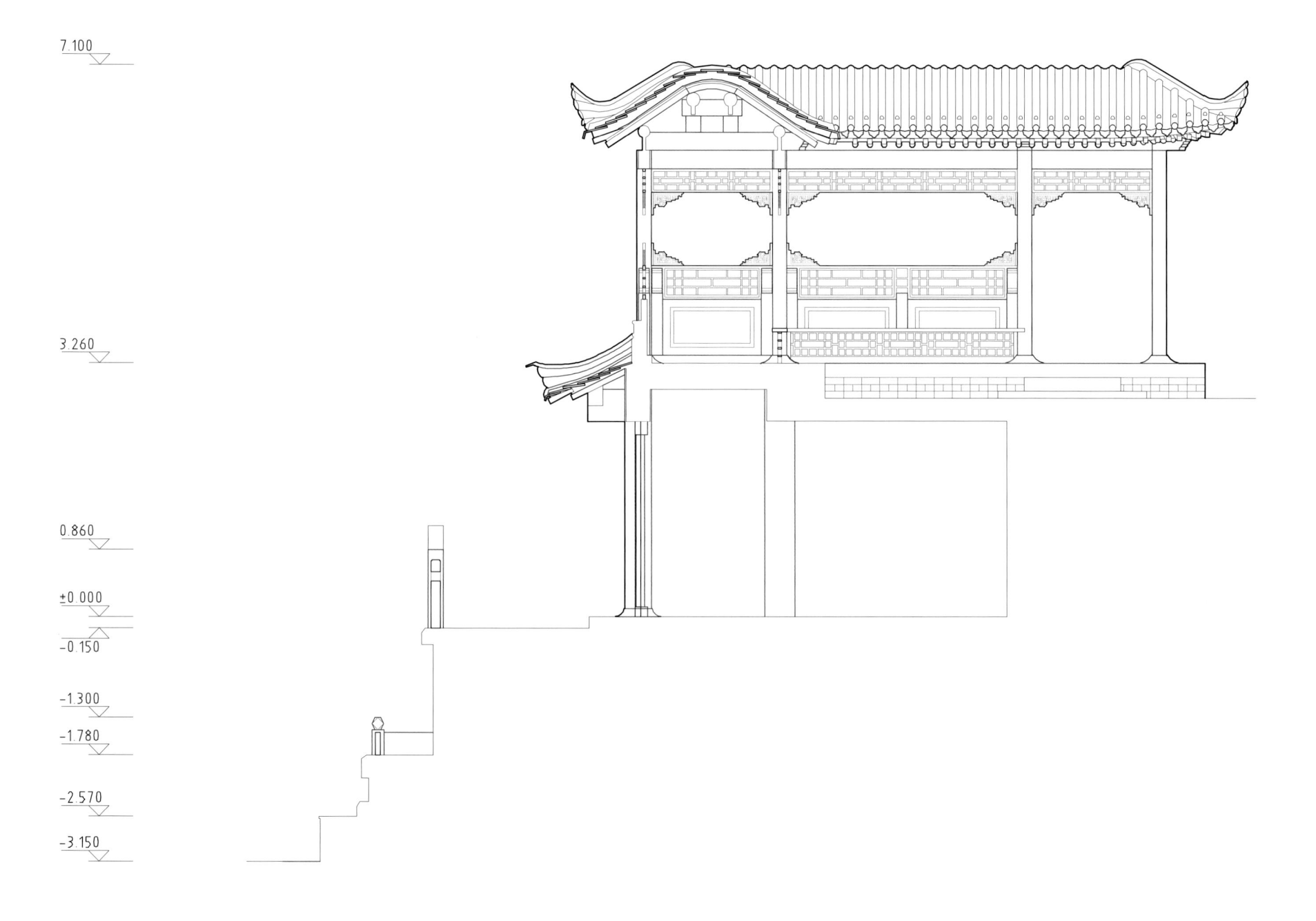

7.100

3.260

0.860

±0.000

-0.150

-1.300

-1.780

-2.570

-3.150

玉泉院廊房剖面图
Section of Yuquan Courtyard's *langfangs*

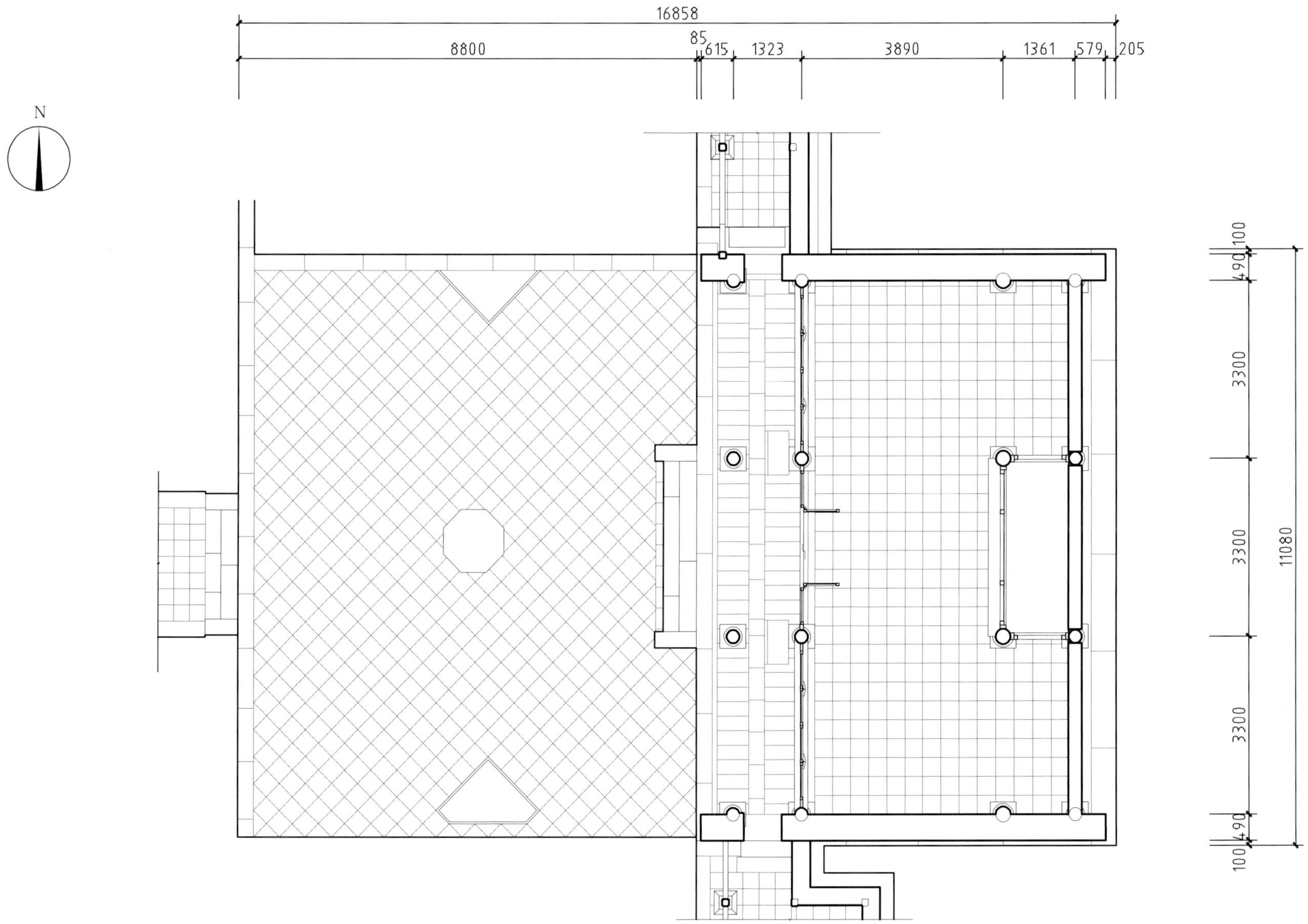

16858

8800　　85　615　1323　　3890　　1361　579　205

490 100

3300

3300　11080

3300

100 490

N

玉泉院圣母殿平面图
Plan of Yuquan Courtyard's Shengmu Hall

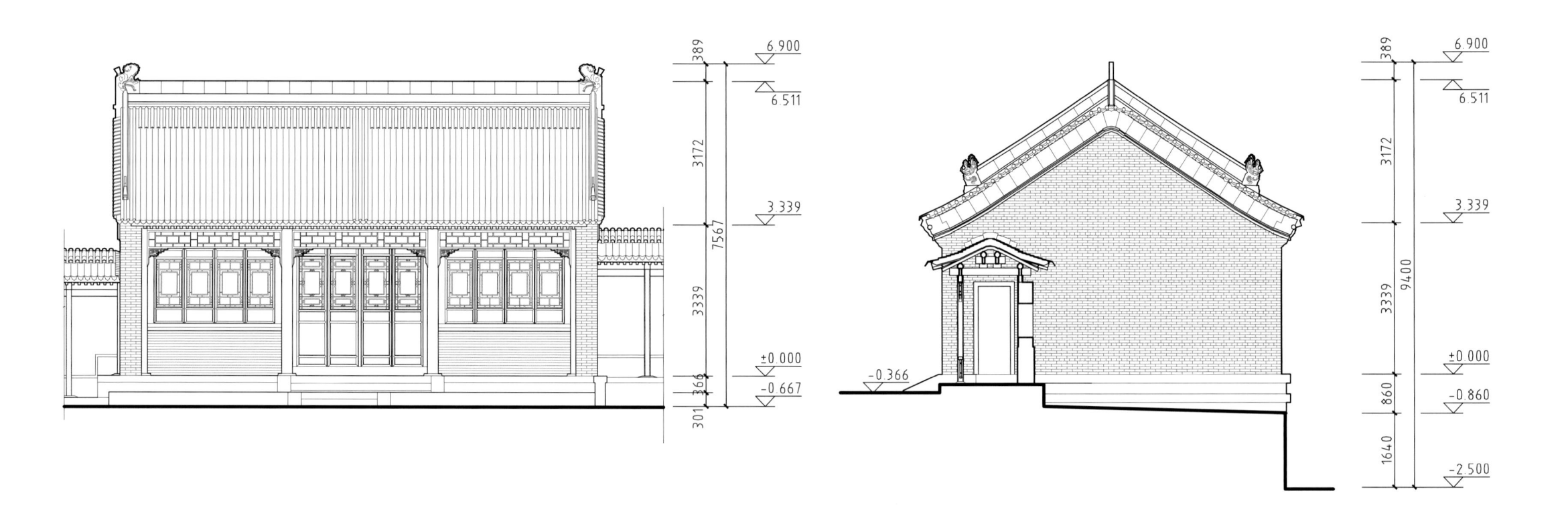

玉泉院圣母殿正立面图
Front elevation of Yuquan Courtyard's Shengmu Hall

玉泉院圣母殿侧立面图
Side elevation of Yuquan Courtyard's Shengmu Hall

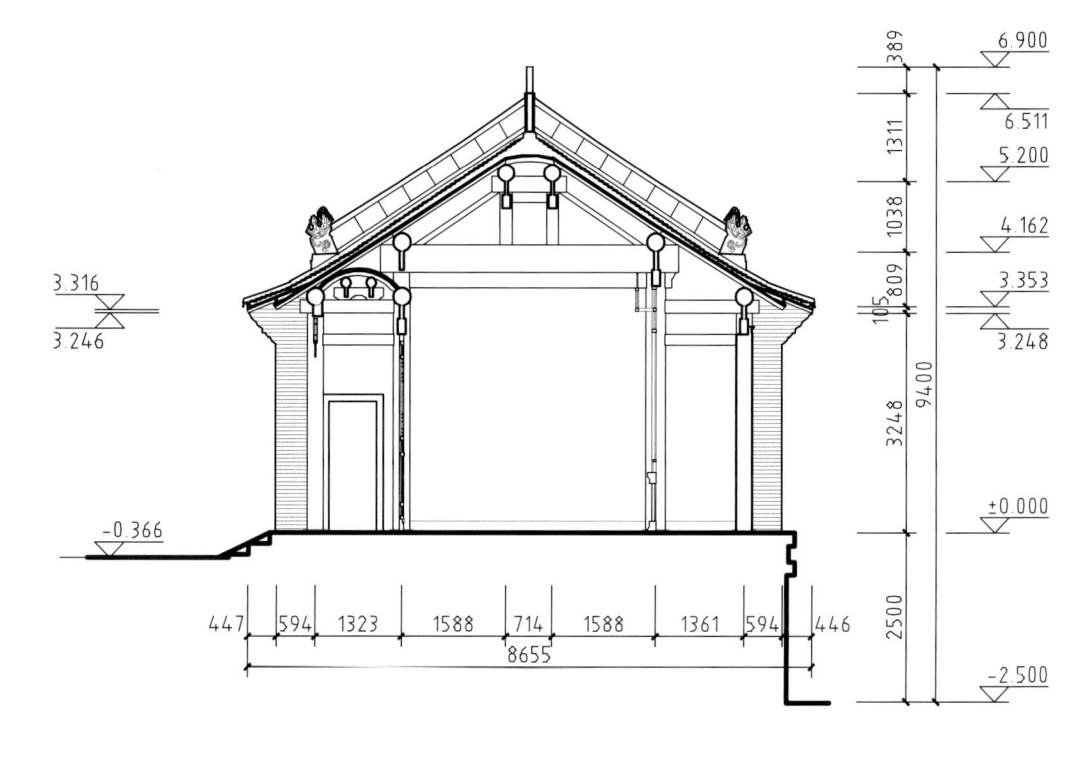

玉泉院圣母殿横剖面图
Cross-section of Yuquan Courtyard's Shengmu Hall

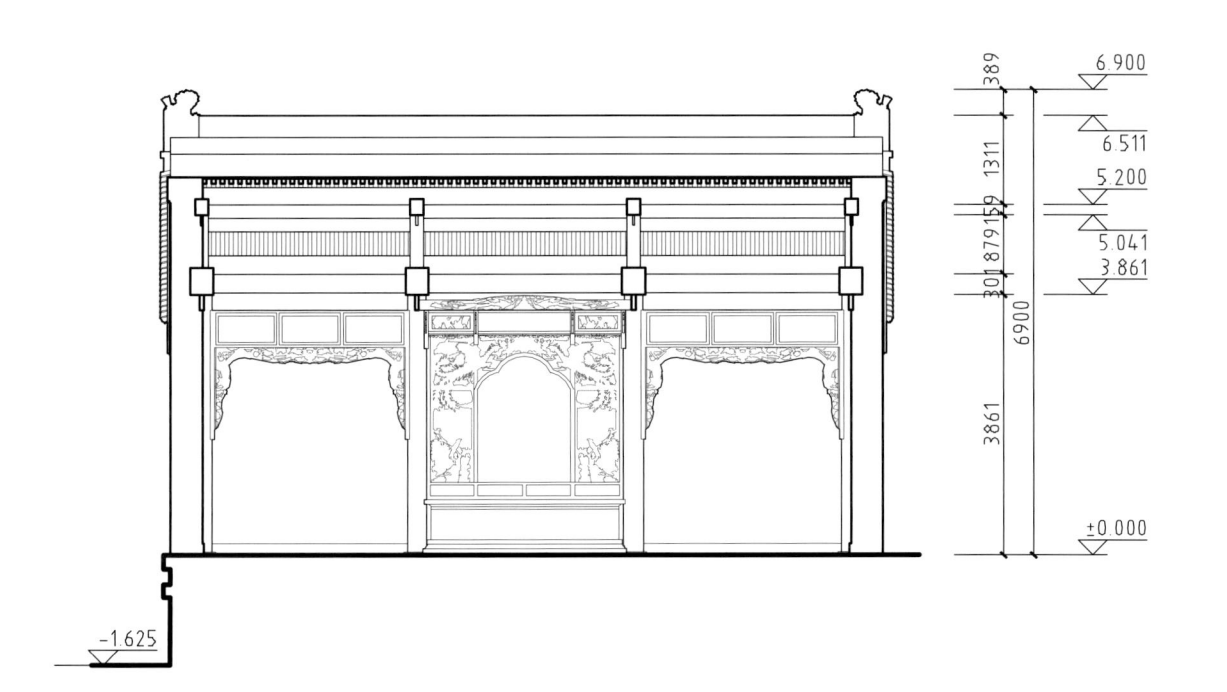

玉泉院圣母殿纵剖面图
Longitudinal section of Yuquan Courtyard's Shengmu Hall

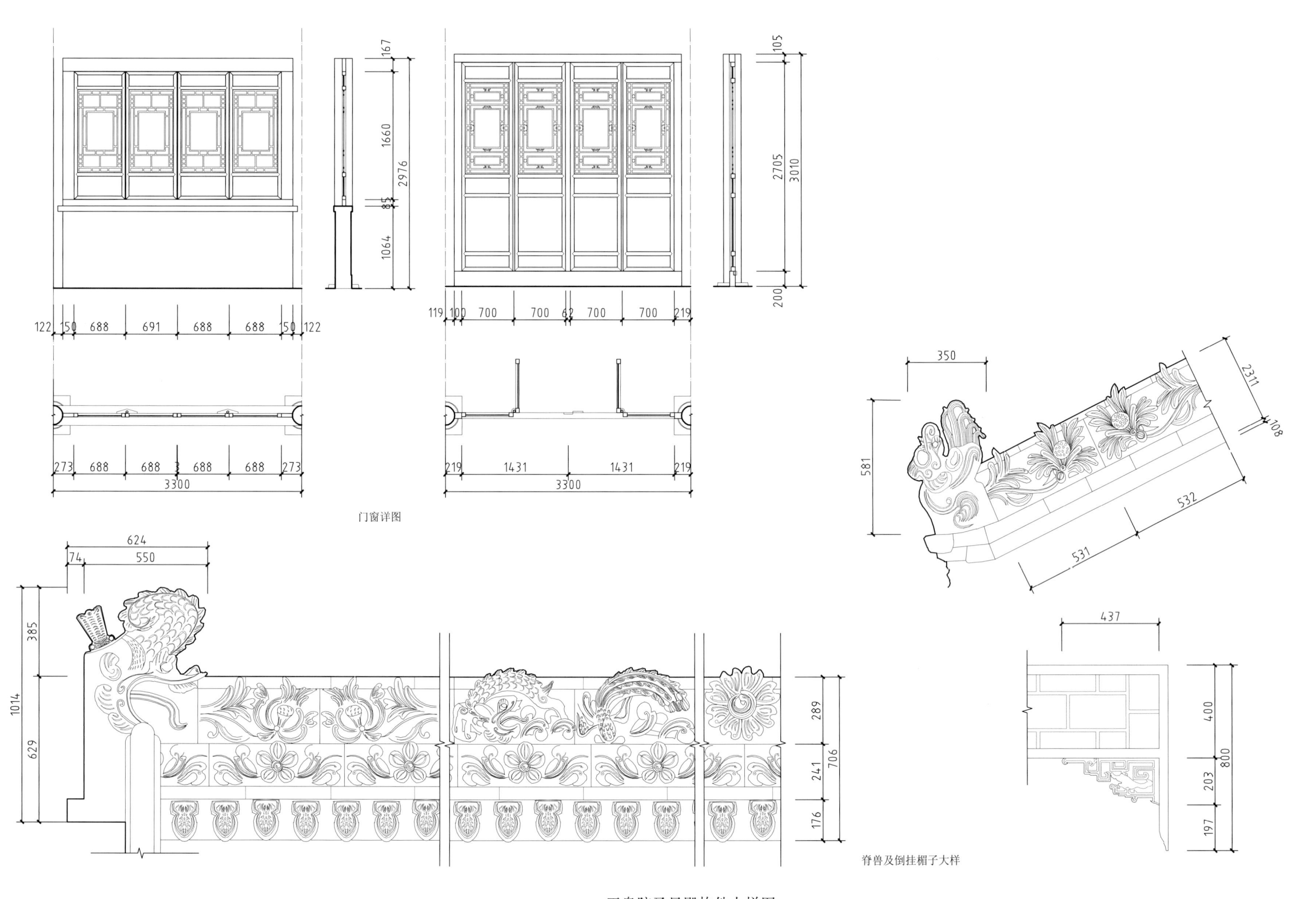

门窗详图

脊兽及倒挂楣子大样

玉泉院圣母殿构件大样图
Structural component of Yuquan Courtyard's Shengmu Hall

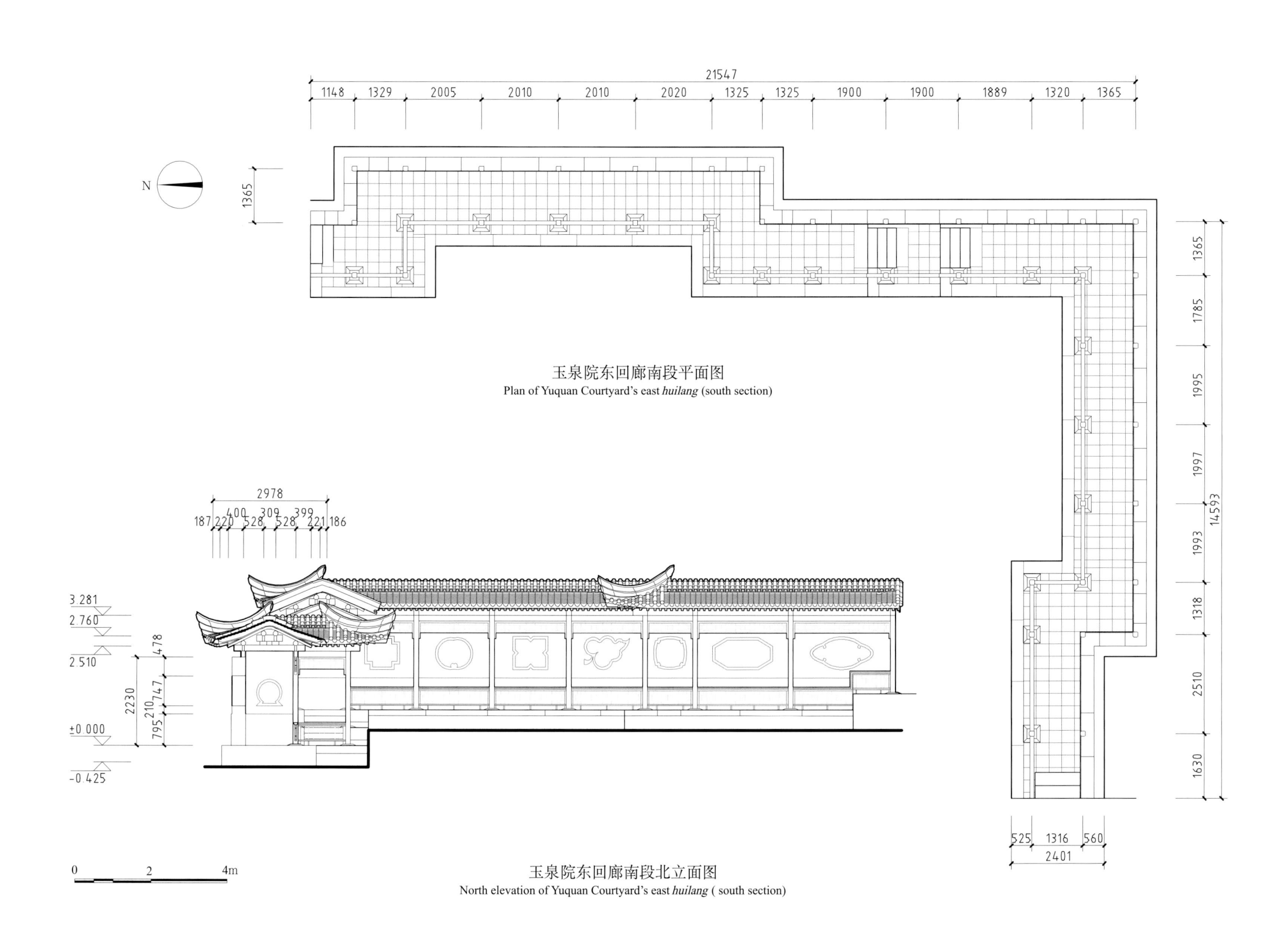

玉泉院东回廊南段平面图
Plan of Yuquan Courtyard's east *huilang* (south section)

玉泉院东回廊南段北立面图
North elevation of Yuquan Courtyard's east *huilang* (south section)

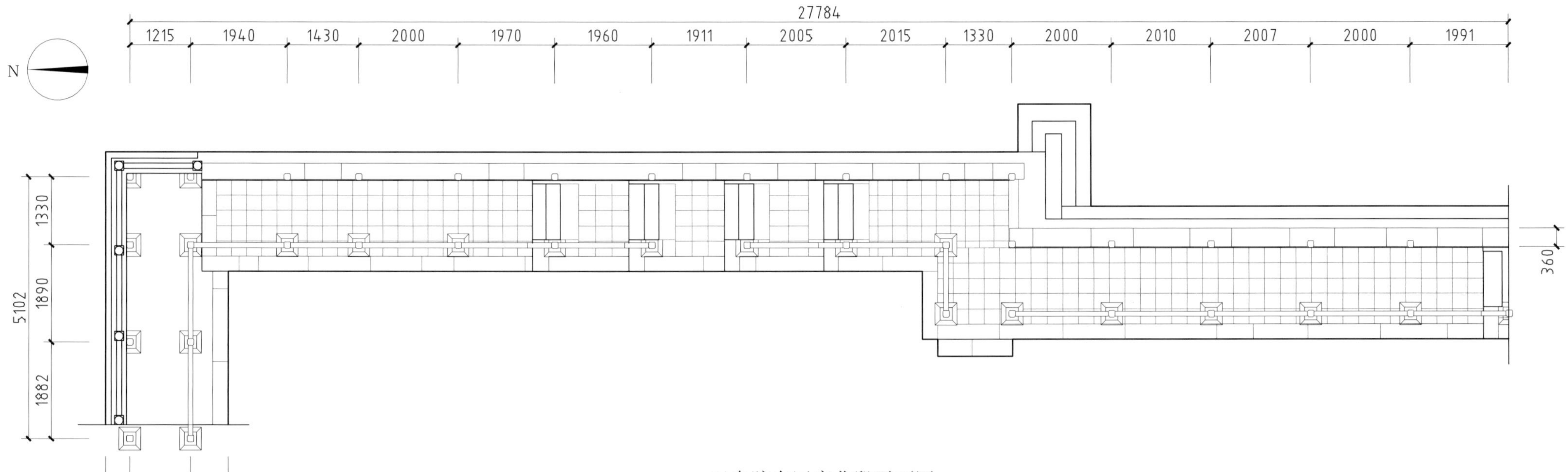

27784

1215　1940　1430　2000　1970　1960　1911　2005　2015　1330　2000　2010　2007　2000　1991

N

1330
5102
1890
1882

360

480　1215　755
2450

玉泉院东回廊北段平面图
Plan of Yuquan Courtyard's east *huilang* (north section)

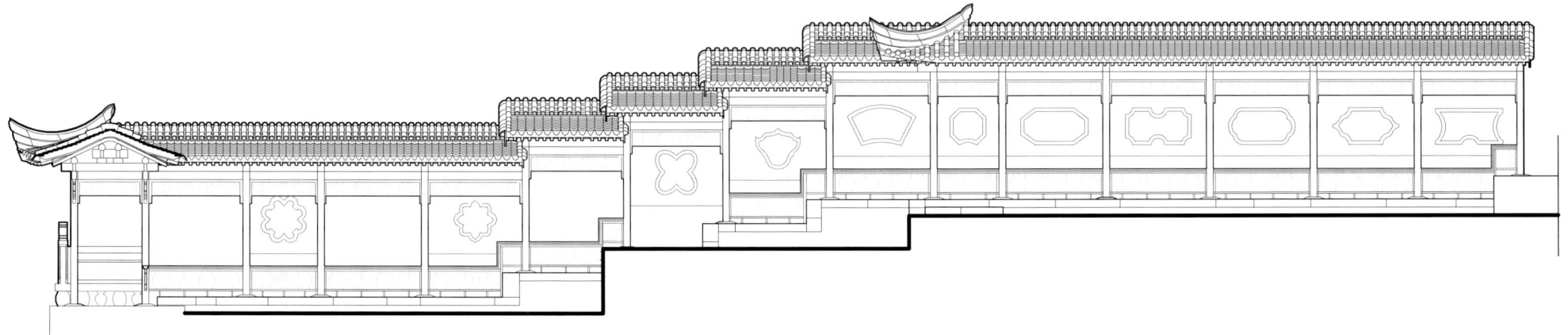

玉泉院东回廊北段西立面图
West elevation of Yuquan Courtyard's east *huilang* (north section)

0　2　4m

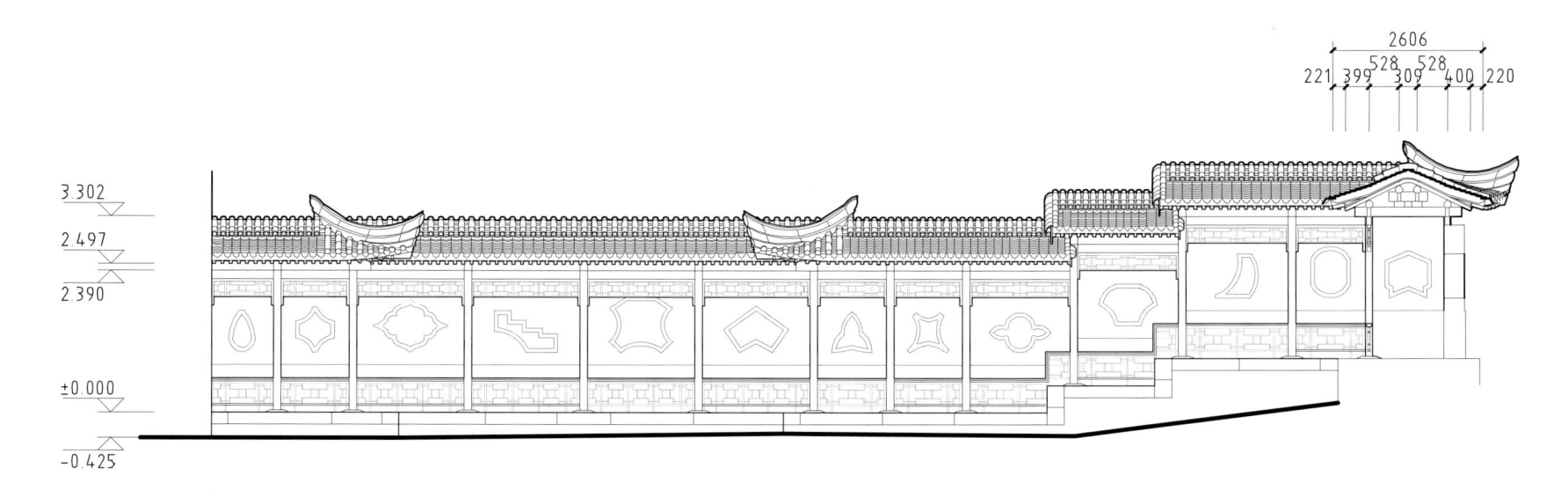

3.302

2.497

2.390

±0.000

-0.425

玉泉院东回廊南段西立面图
West elevation of Yuquan Courtyard's east *huilang* (south section)

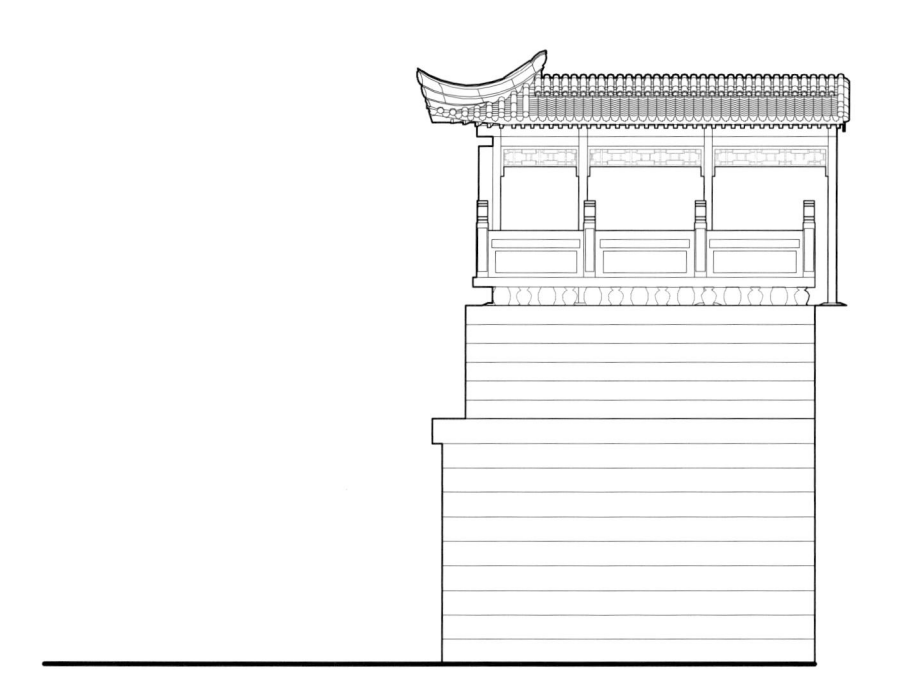

玉泉院东回廊北段北立面图
North elevation of Yuquan Courtyard's east *huilang* (north section)

0   2   4m

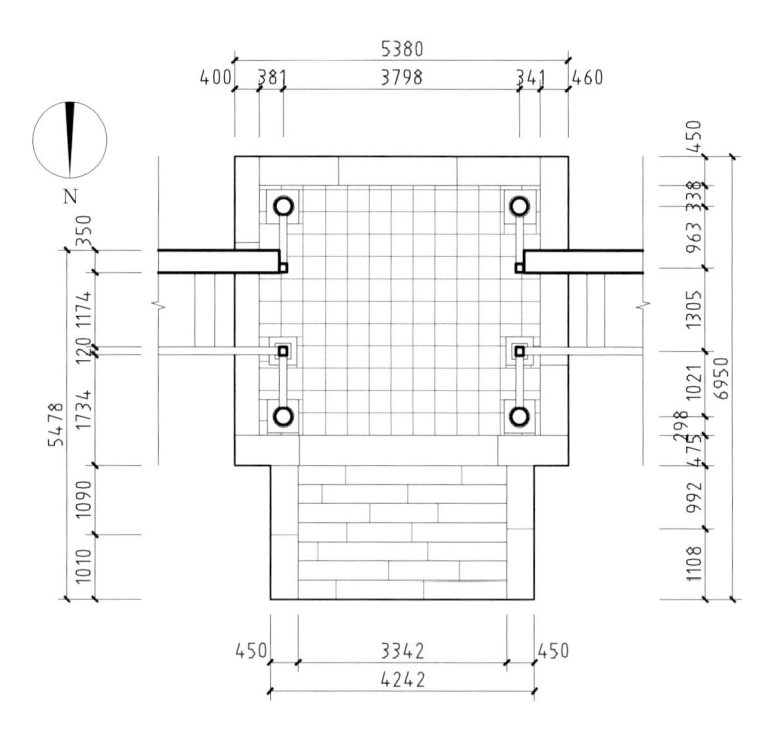

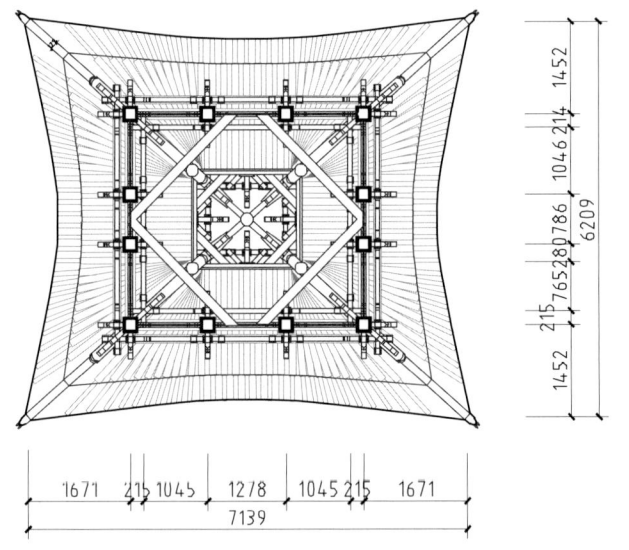

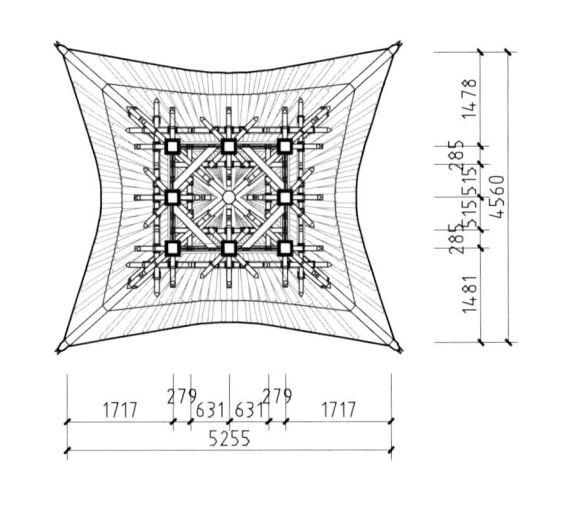

玉泉院通天亭平面图
Plan of Yuquan Courtyard's Tongtian Pavilion

玉泉院通天亭一层梁架仰视平面图
Plan of ground floor framework of Yuquan Courtyard's Tongtian Pavilion as seen from below

玉泉院通天亭二层梁架仰视平面图
Plan of second floor framework of Yuquan Courtyard's Tongtian Pavilion as seen from below

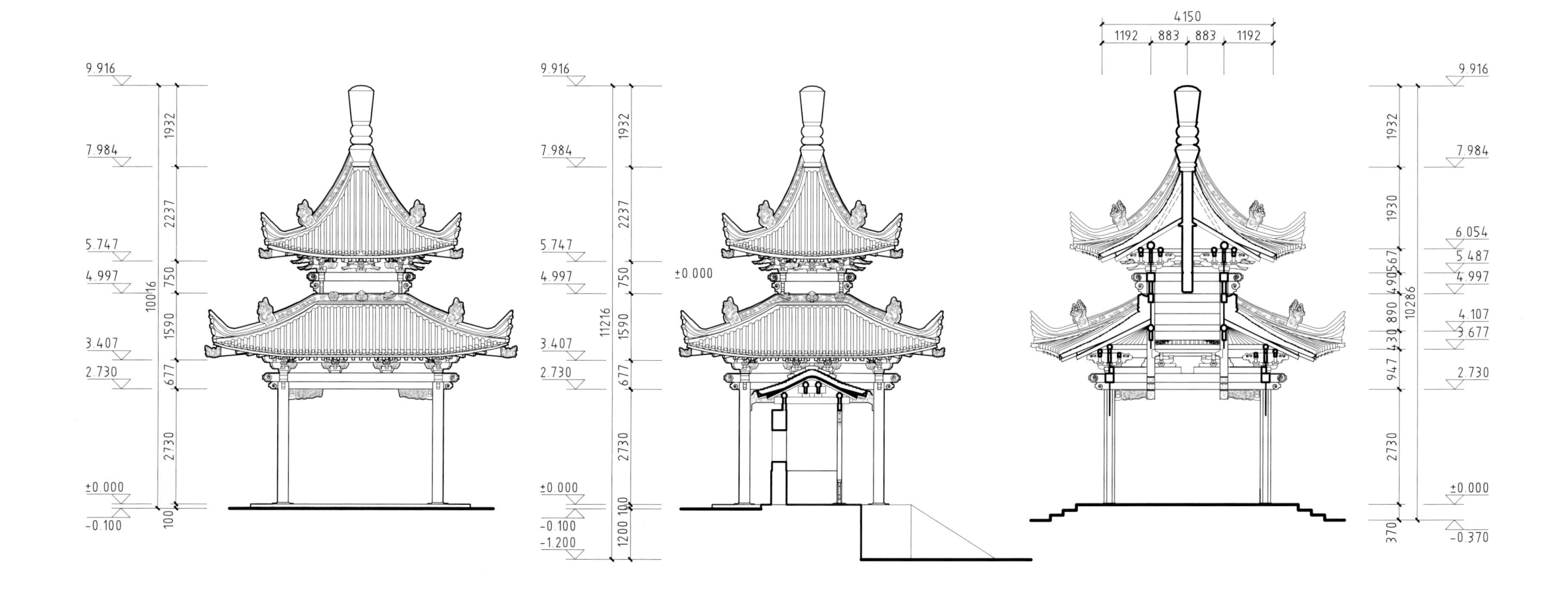

玉泉院通天亭南立面图
South elevation of Yuquan Courtyard's Tongtian Pavilion

玉泉院通天亭东立面图
East elevation of Yuquan Courtyard's Tongtian Pavilion

玉泉院通天亭横剖面图
Cross-section of Yuquan Courtyard's Tongtian Pavilion

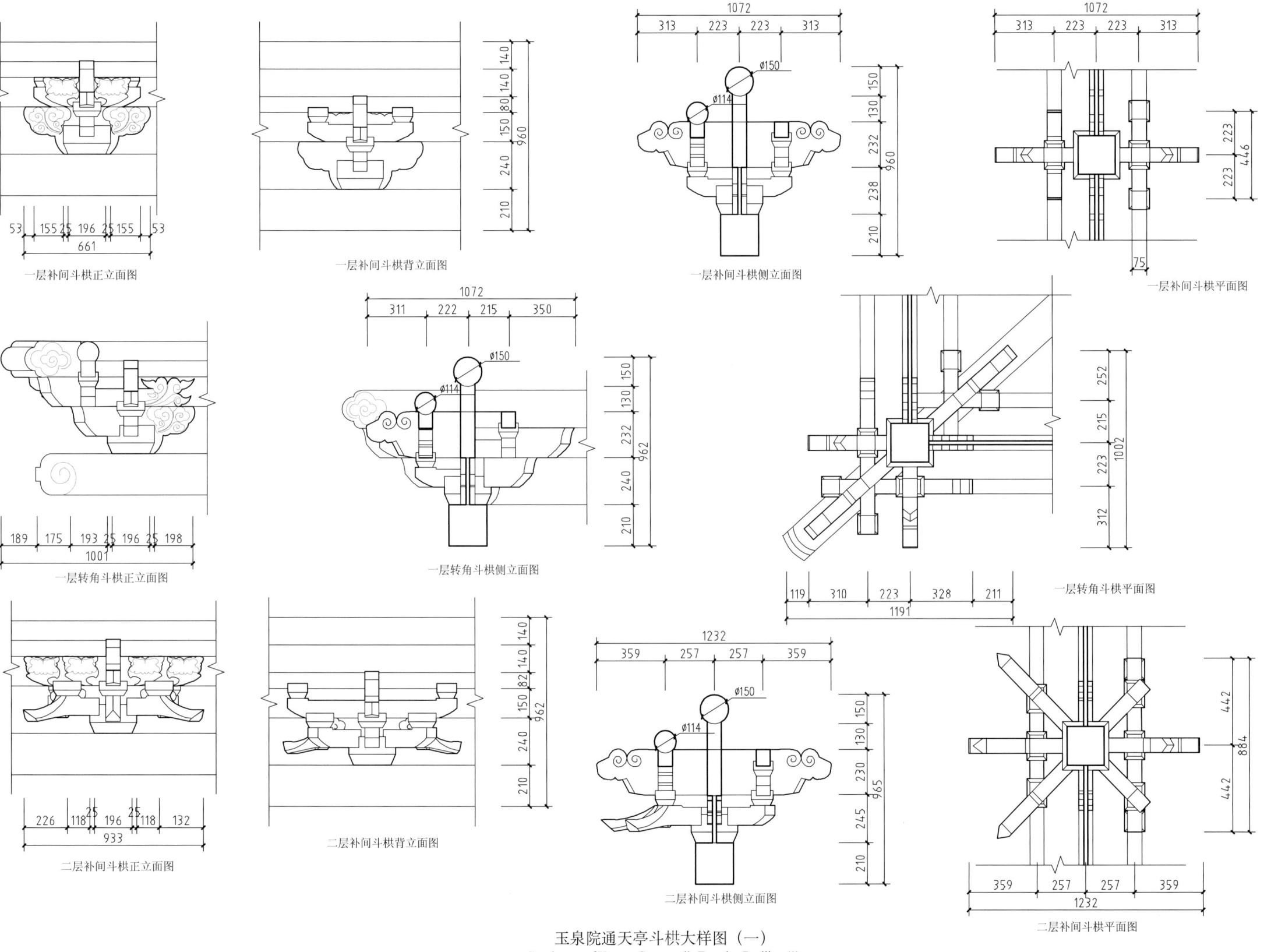

一层补间斗栱正立面图

一层补间斗栱背立面图

一层补间斗栱侧立面图

一层补间斗栱平面图

一层转角斗栱正立面图

一层转角斗栱侧立面图

一层转角斗栱平面图

二层补间斗栱正立面图

二层补间斗栱背立面图

二层补间斗栱侧立面图

二层补间斗栱平面图

玉泉院通天亭斗栱大样图（一）
Bracket set of Yuquan Courtyard's Tongtian Pavilion (1)

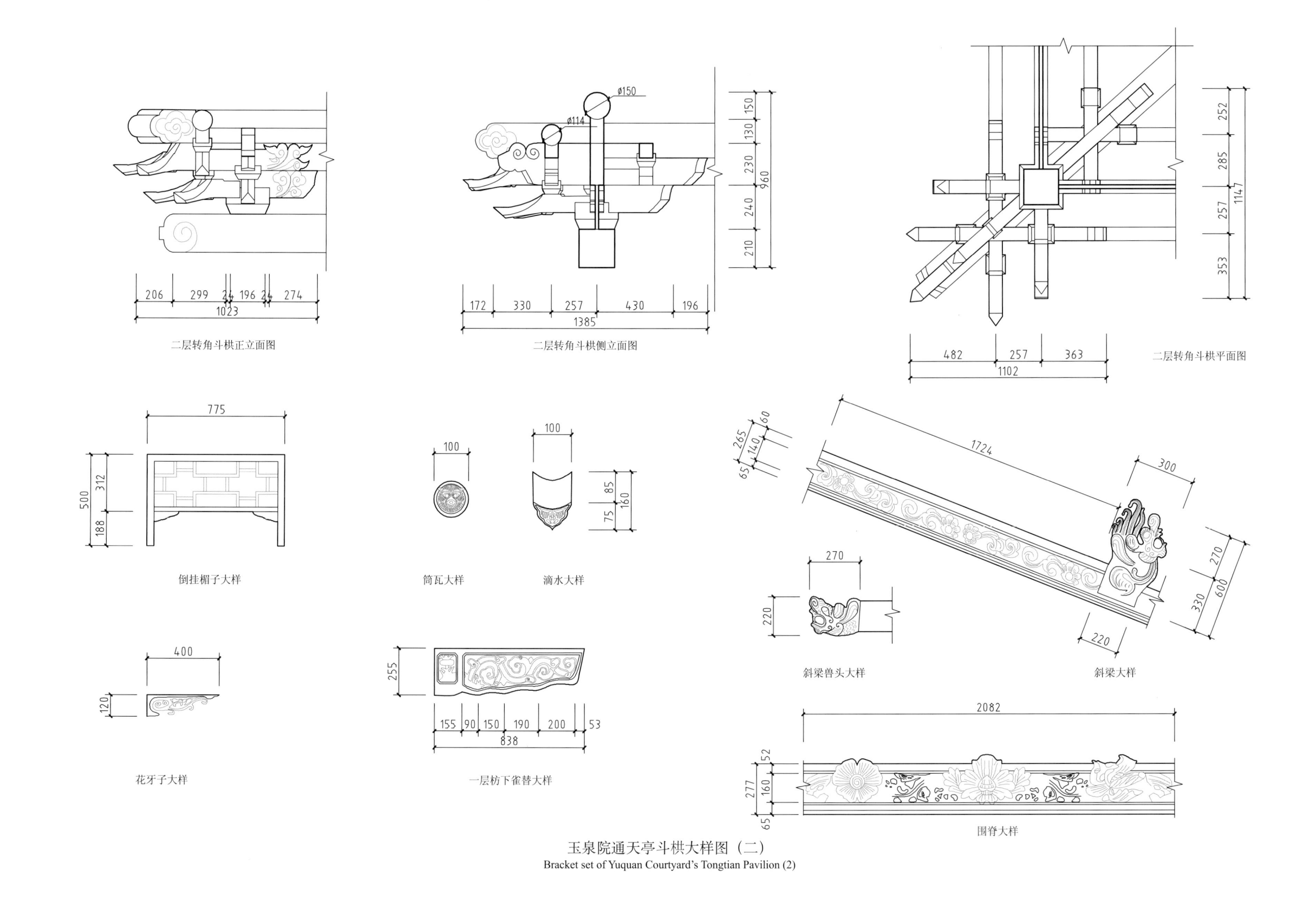

二层转角斗栱正立面图

二层转角斗栱侧立面图

二层转角斗栱平面图

倒挂楣子大样

筒瓦大样

滴水大样

斜梁兽头大样

斜梁大样

花牙子大样

一层枋下雀替大样

围脊大样

玉泉院通天亭斗栱大样图（二）
Bracket set of Yuquan Courtyard's Tongtian Pavilion (2)

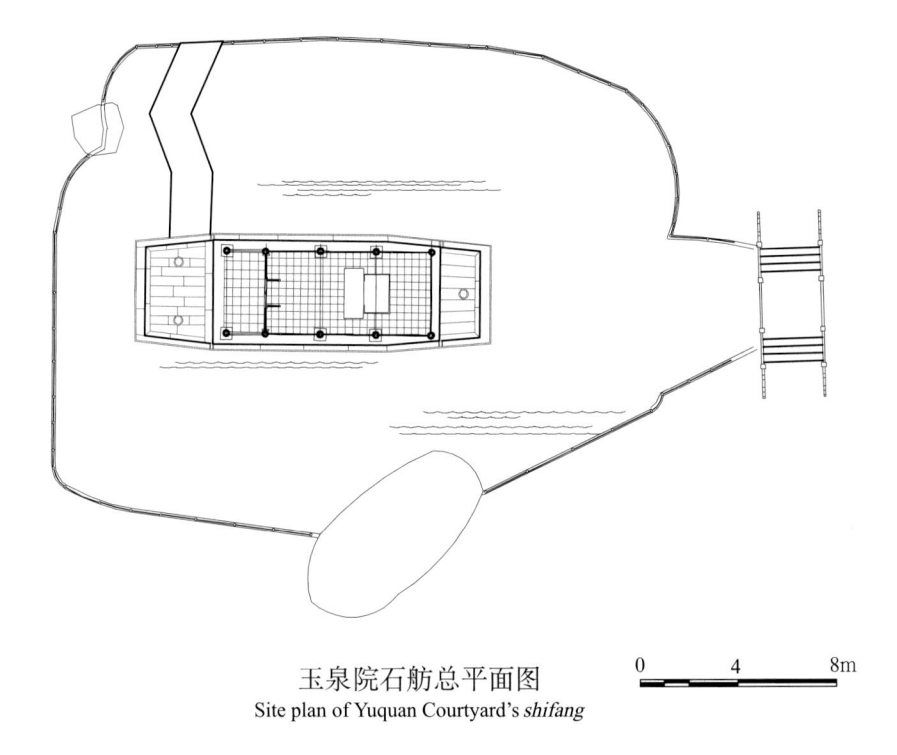

玉泉院石舫总平面图

0 4 8m

Site plan of Yuquan Courtyard's *shifang*

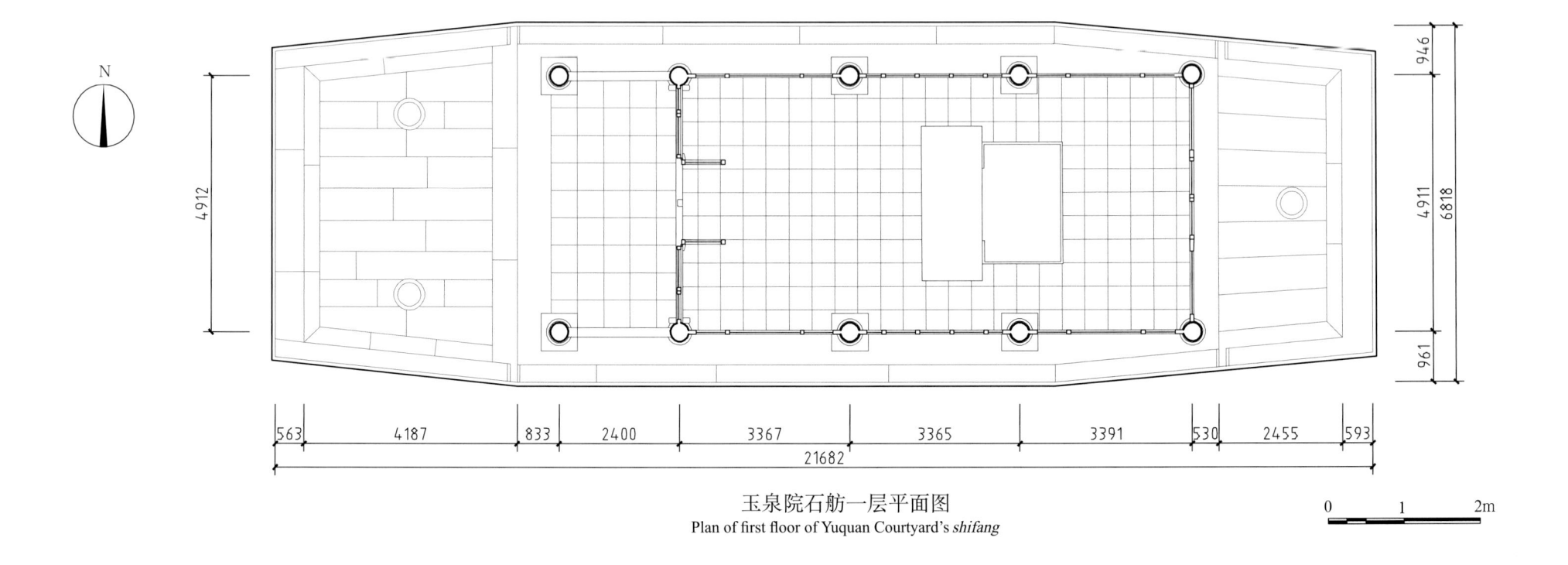

N

玉泉院石舫一层平面图

0 1 2m

Plan of first floor of Yuquan Courtyard's *shifang*

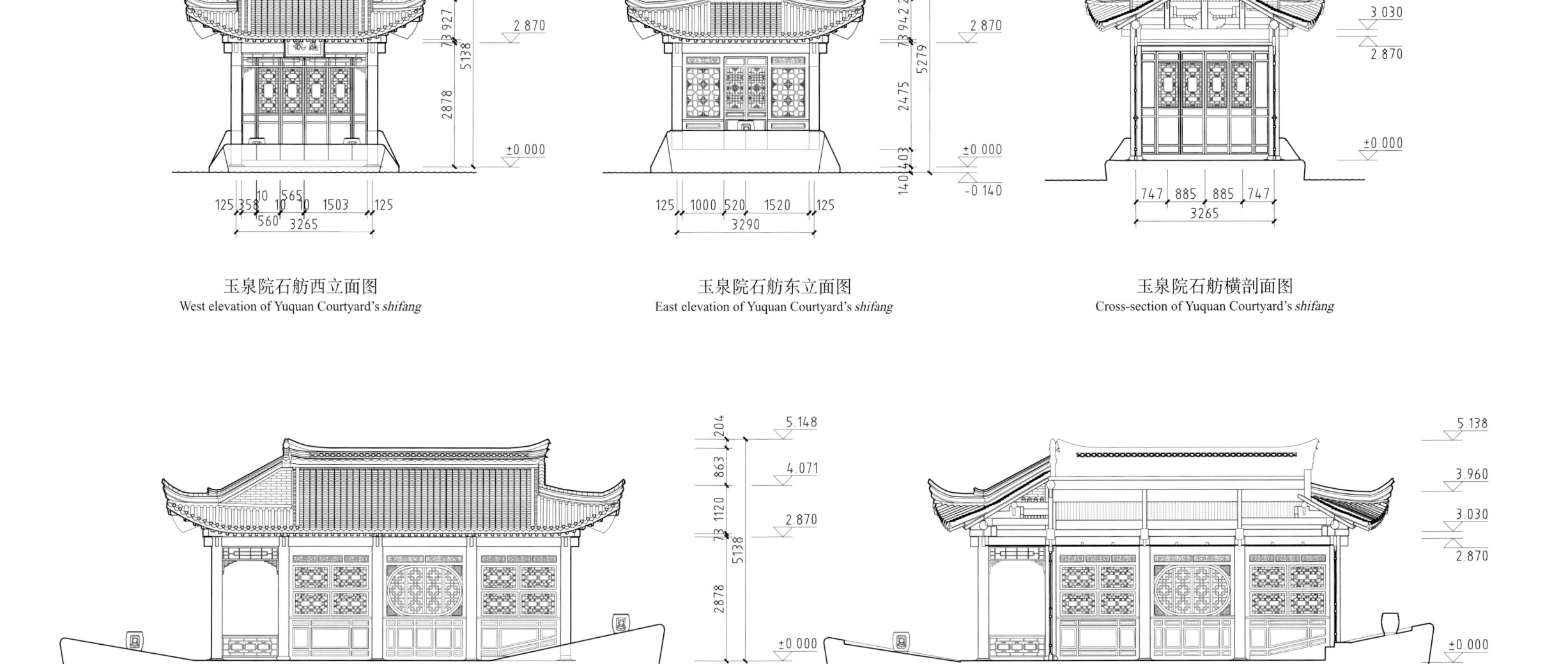

玉泉院石舫西立面图
West elevation of Yuquan Courtyard's *shifang*

玉泉院石舫东立面图
East elevation of Yuquan Courtyard's *shifang*

玉泉院石舫横剖面图
Cross-section of Yuquan Courtyard's *shifang*

玉泉院石舫南立面图
South elevation of Yuquan Courtyard's *shifang*

玉泉院石舫纵剖面图
Longitudinal section of Yuquan Courtyard's *shifang*

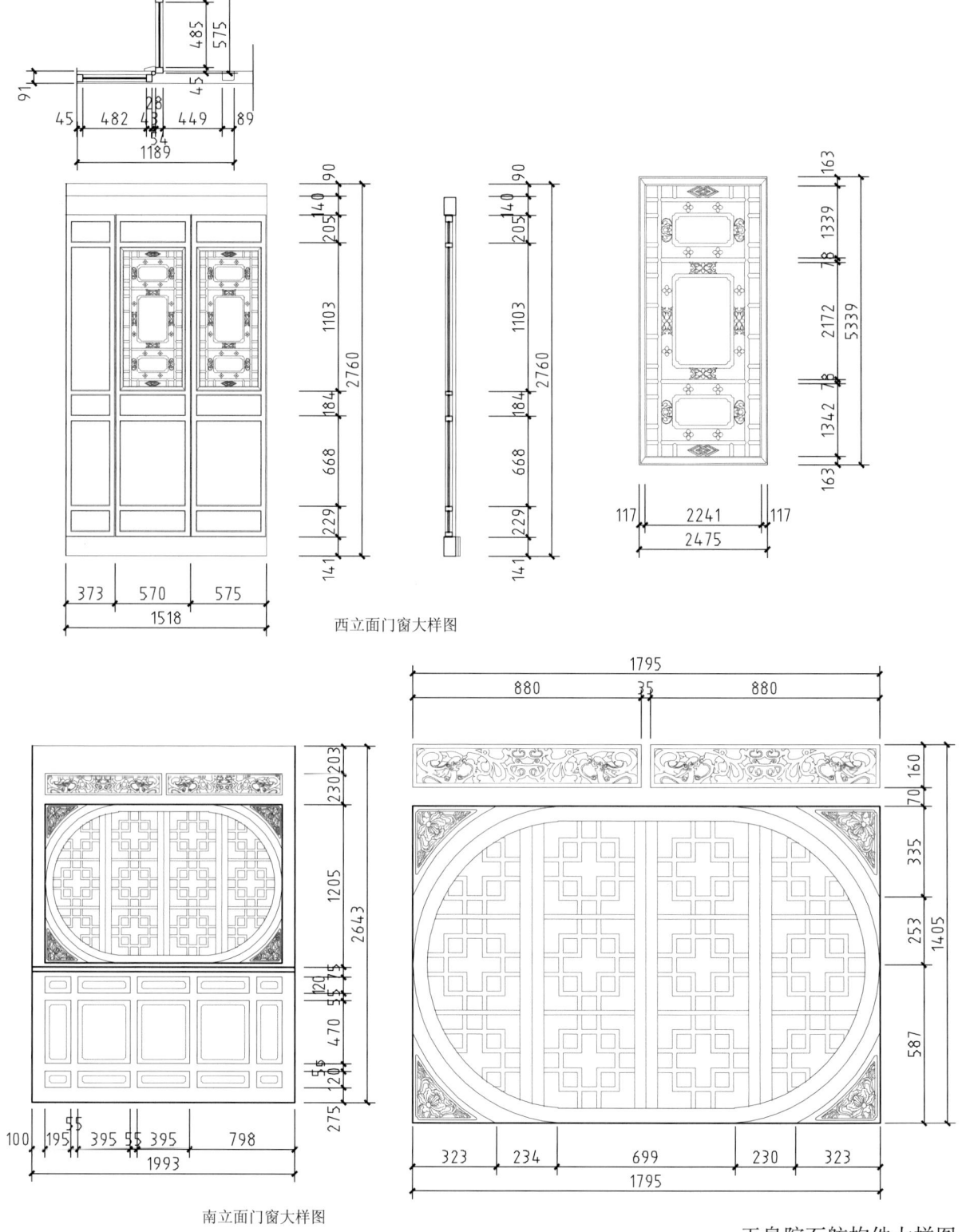

西立面门窗大样图

南立面门窗大样图

东立面门窗大样图

玉泉院石舫构件大样图（一）
Structural component of Yuquan Courtyard's *shifang* (1)

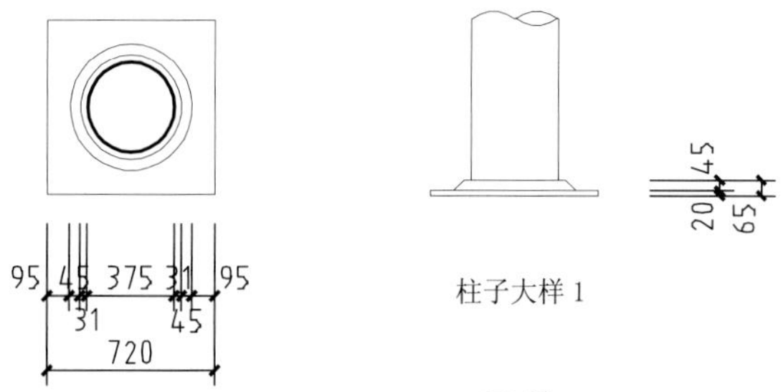

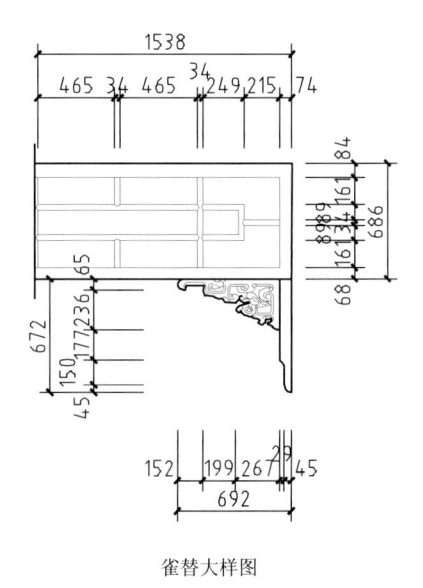

柱子大样 1

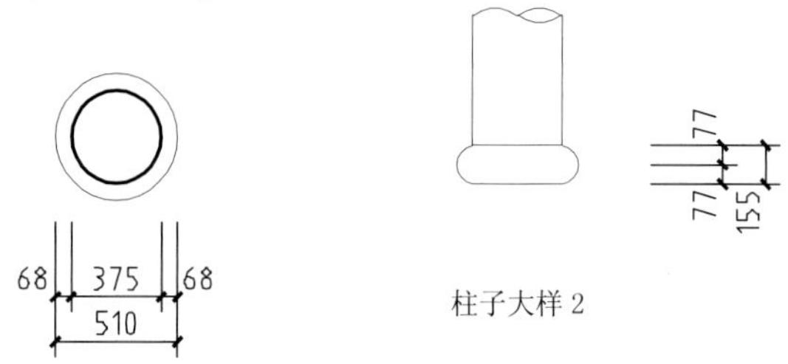

柱子大样 2

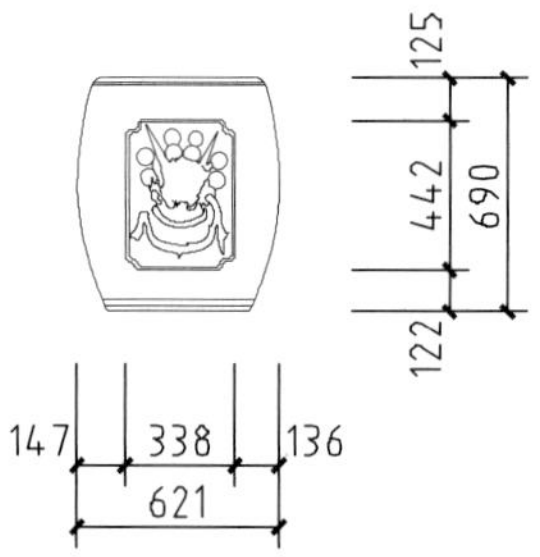

石鼓大样

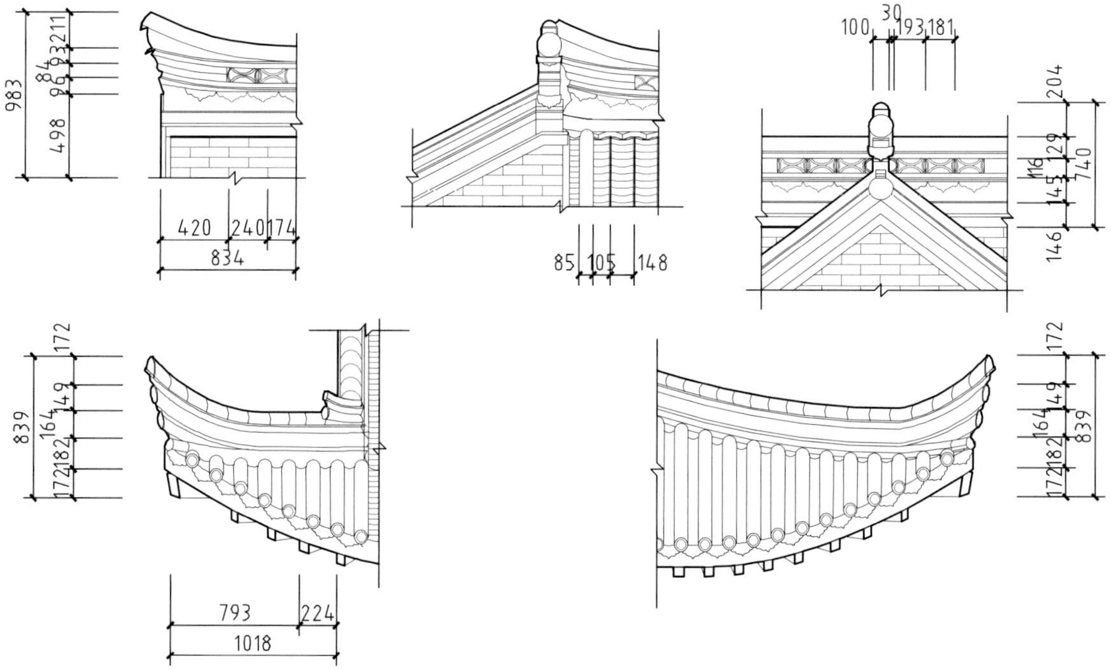

屋顶大样图

雀替大样图

玉泉院石舫构件大样图（二）
Structural component of Yuquan Courtyard's *shifang* (2)

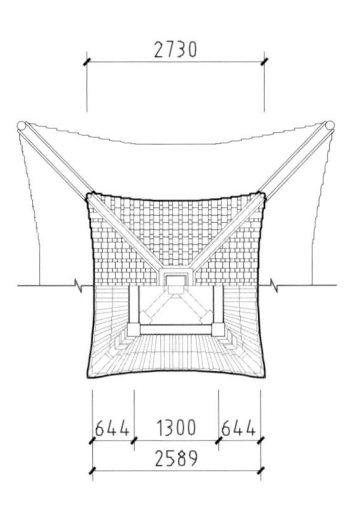

玉泉院纳凉亭下檐梁架仰视及屋顶平面图
Plan of lower eaves-framework as seen from below and roof plan
of Yuquan Courtyard's Naliang Pavilion

玉泉院纳凉亭上檐梁架仰视及屋顶平面图
Plan of upper eaves-framework as seen from below and roof plan
of Yuquan Courtyard's Naliang Pavilion

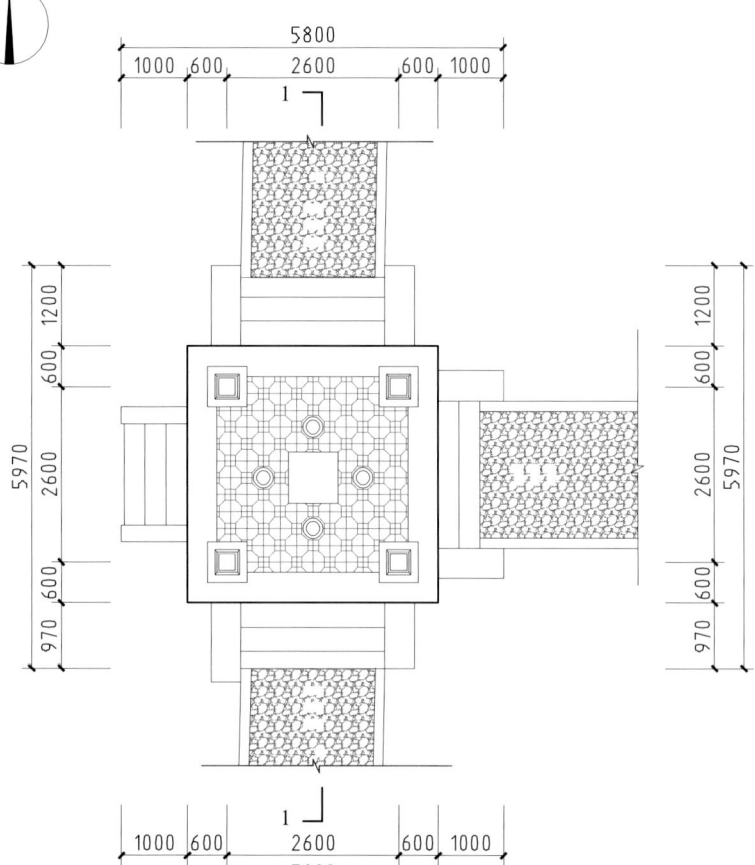

玉泉院纳凉亭平面图
Plan of Yuquan Courtyard's Naliang Pavilion

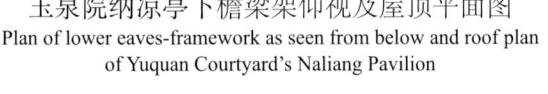
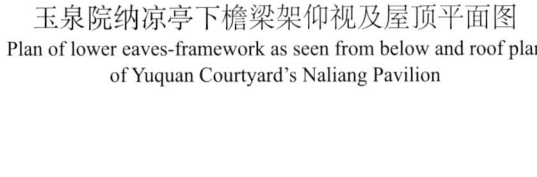
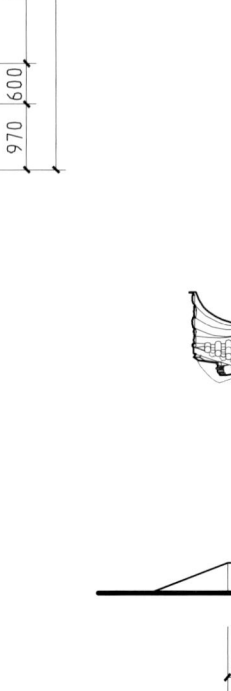

玉泉院纳凉亭立面图
Elevation of Yuquan Courtyard's Naliang Pavilion

玉泉院纳凉亭剖面图
Section of Yuquan Courtyard's Naliang Pavilion

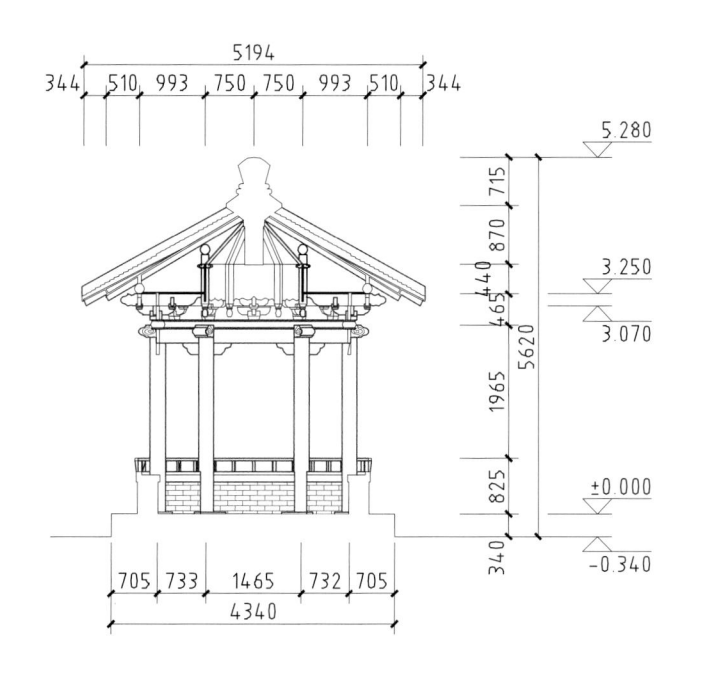

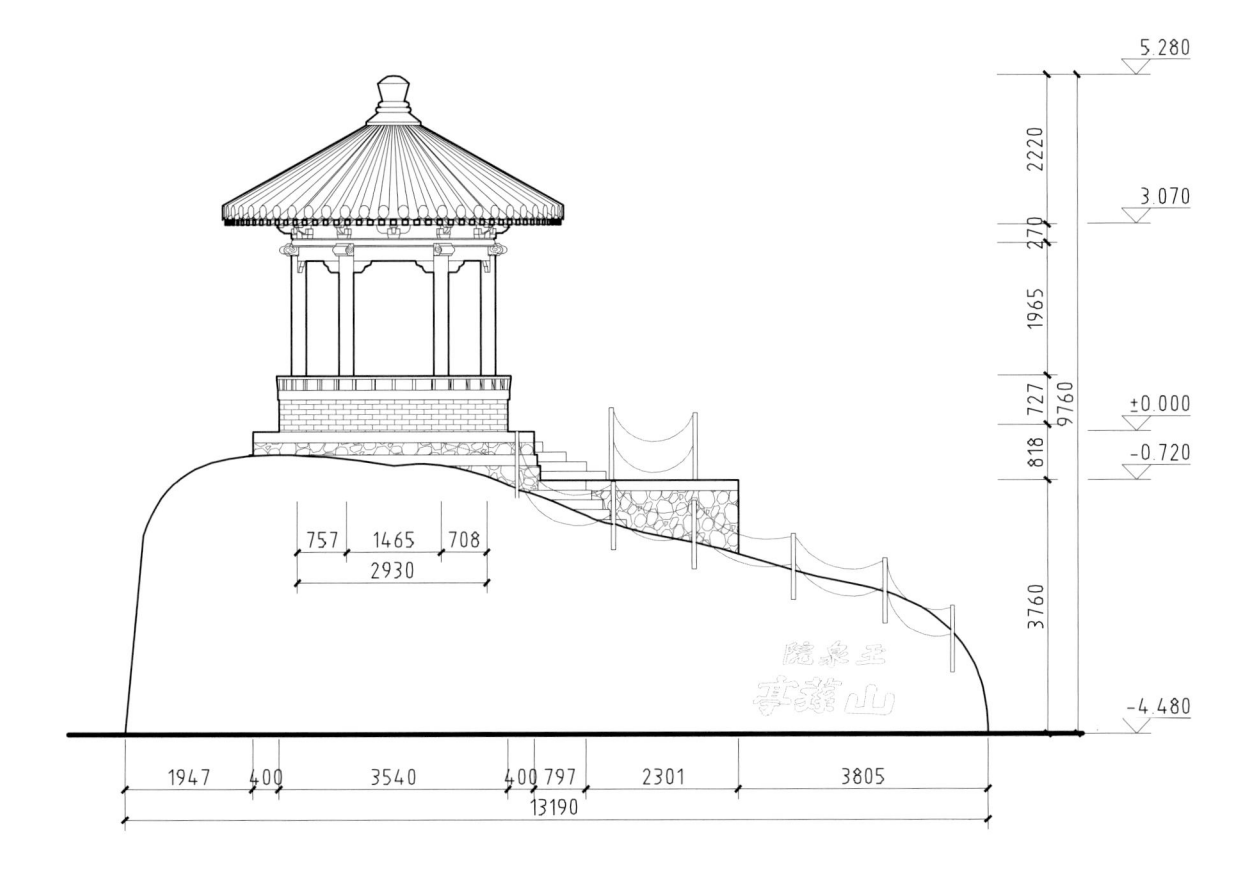

玉泉院山荪亭 A−A 剖面图
Section A-A of Yuquan Courtyard's Shansun Pavilion

玉泉院山荪亭正立面图
Front elevation of Yuquan Courtyard's Shansun Pavilion

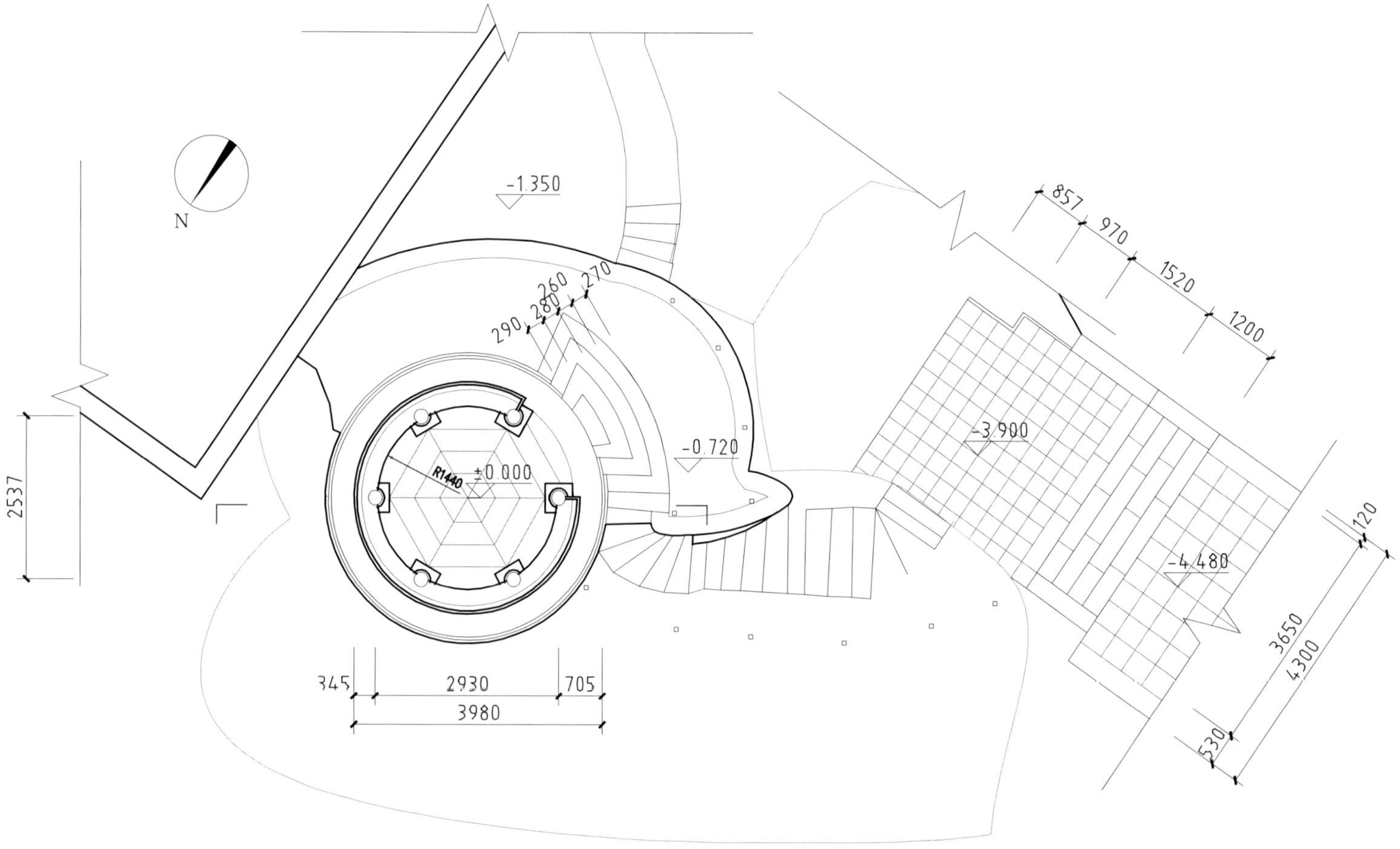

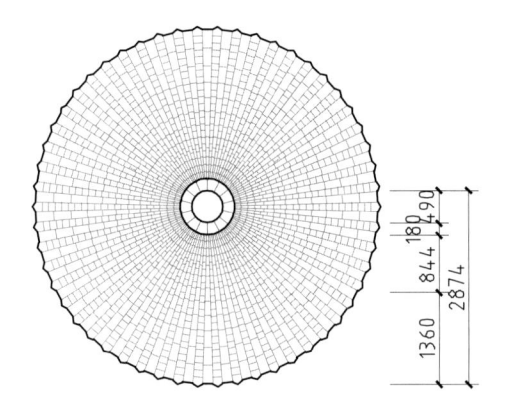

玉泉院山荪亭屋顶平面图
Roof plan of Yuquan Courtyard's Shansun Pavilion

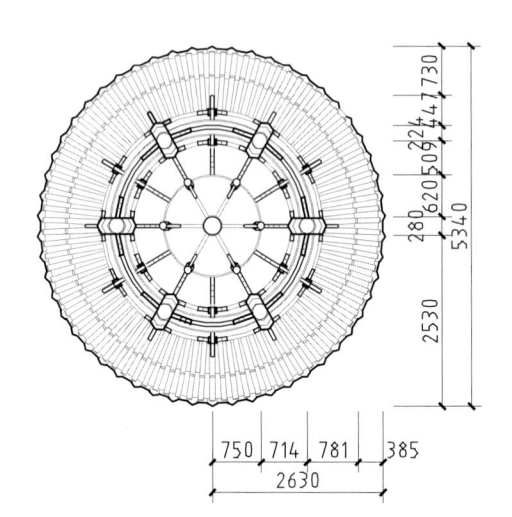

玉泉院山荪亭平面图
Plan of Yuquan Courtyard's Shansun Pavilion

玉泉院山荪亭梁架仰视平面图
Plan of framework of Yuquan Courtyard's Shansun Pavilion as seen from below

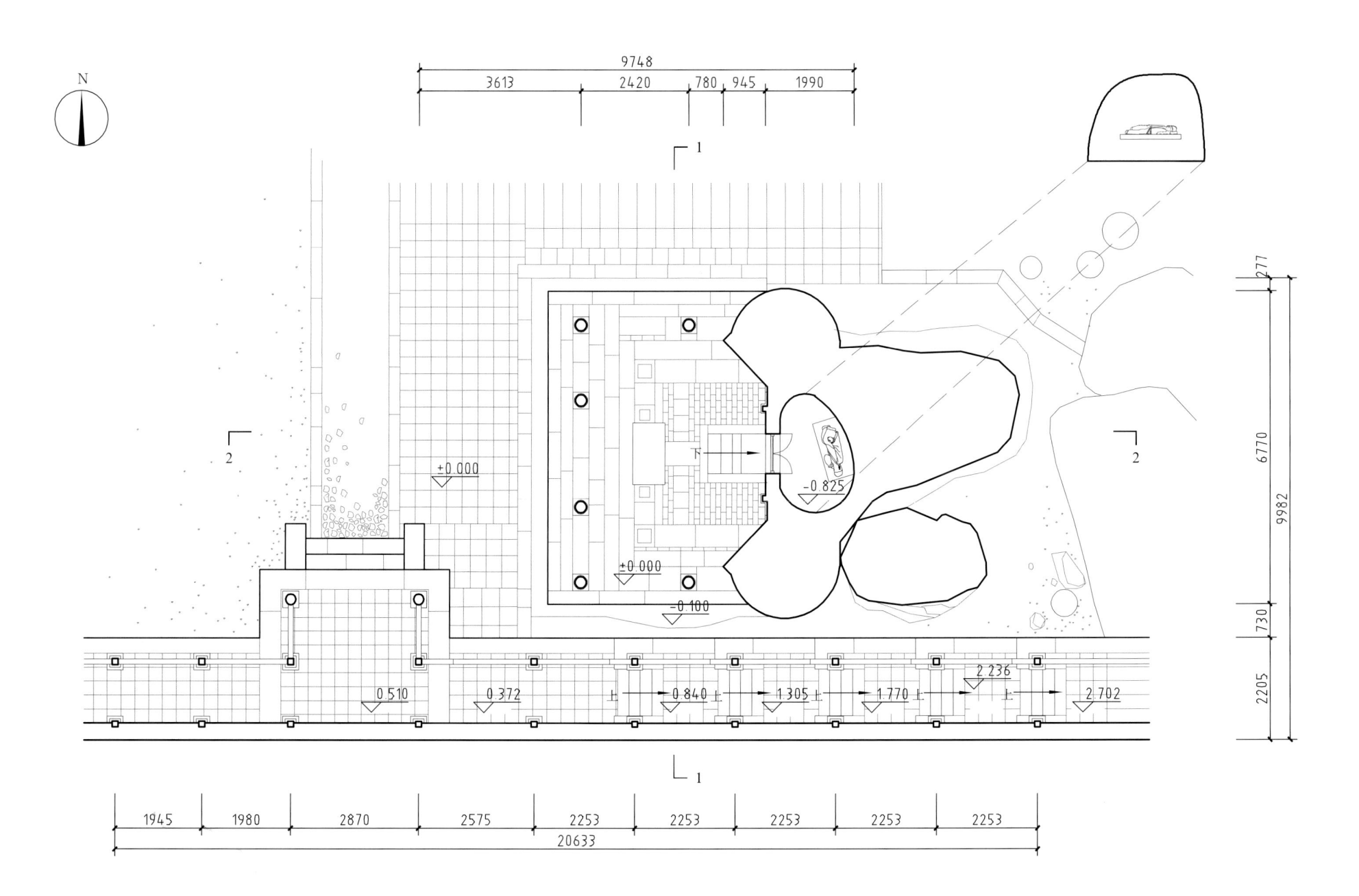

玉泉院陈祖洞平面图
Plan of Yuquan Courtyard's Chenzu Cave

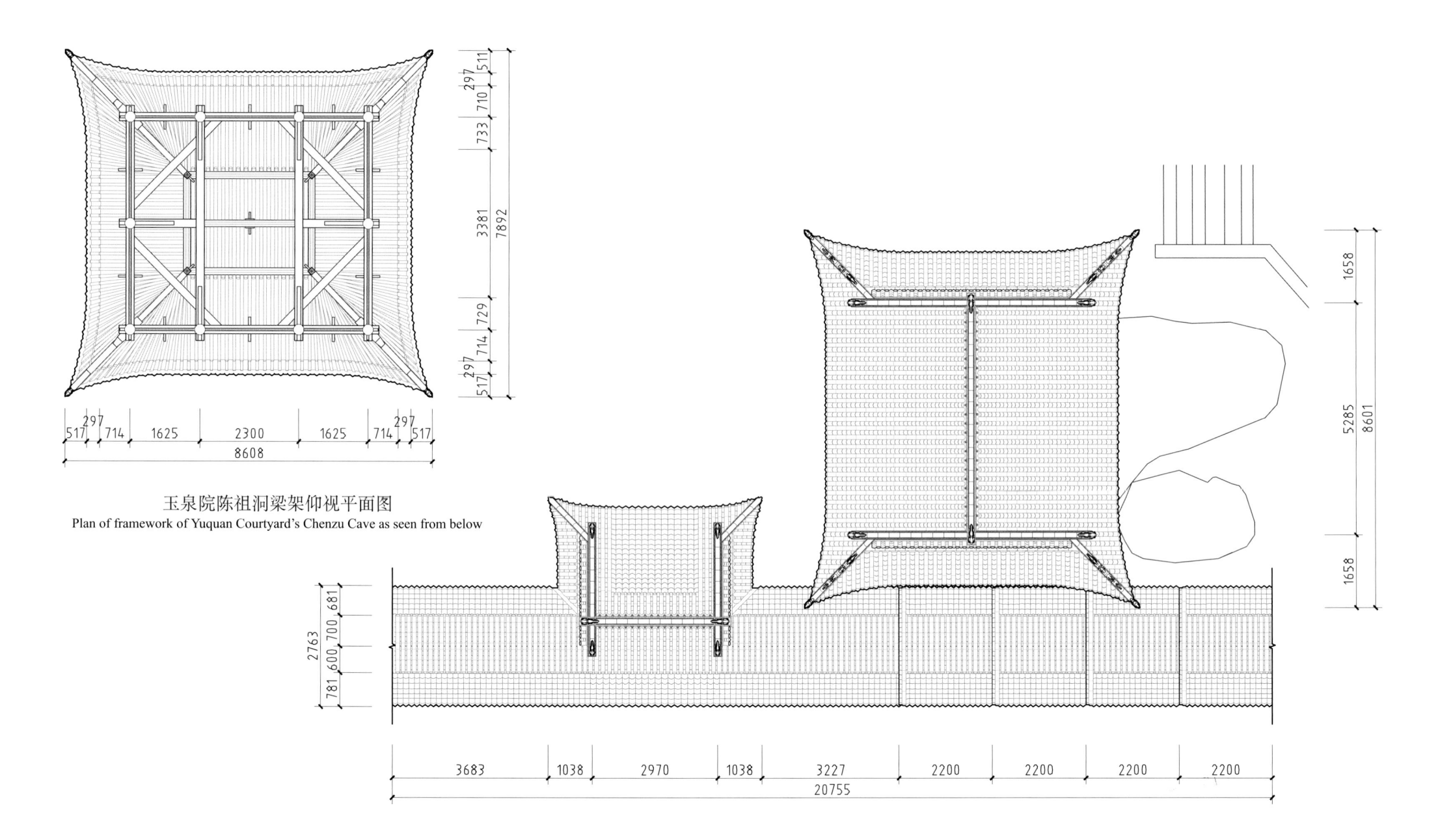

玉泉院陈祖洞梁架仰视平面图
Plan of framework of Yuquan Courtyard's Chenzu Cave as seen from below

玉泉院陈祖洞屋顶平面图
Roof plan of Yuquan Courtyard's Chenzu Cave

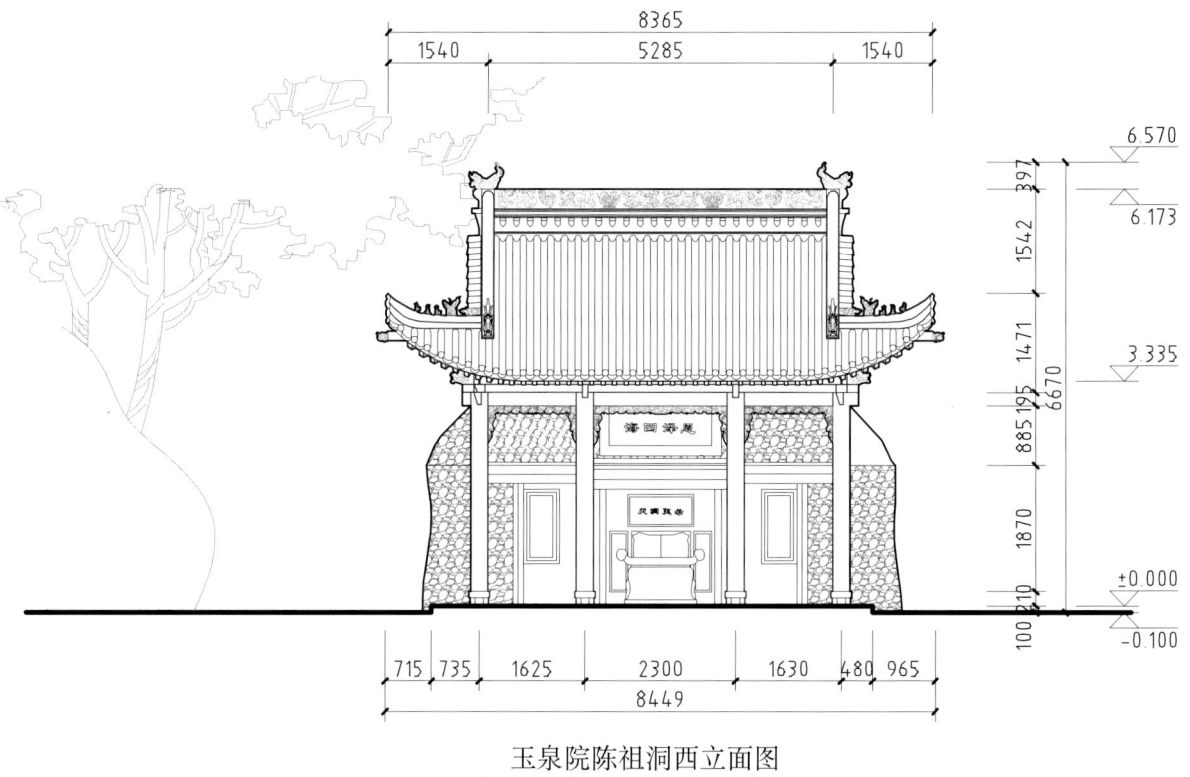

玉泉院陈祖洞西立面图

West elevation of Yuquan Courtyard's Chenzu Cave

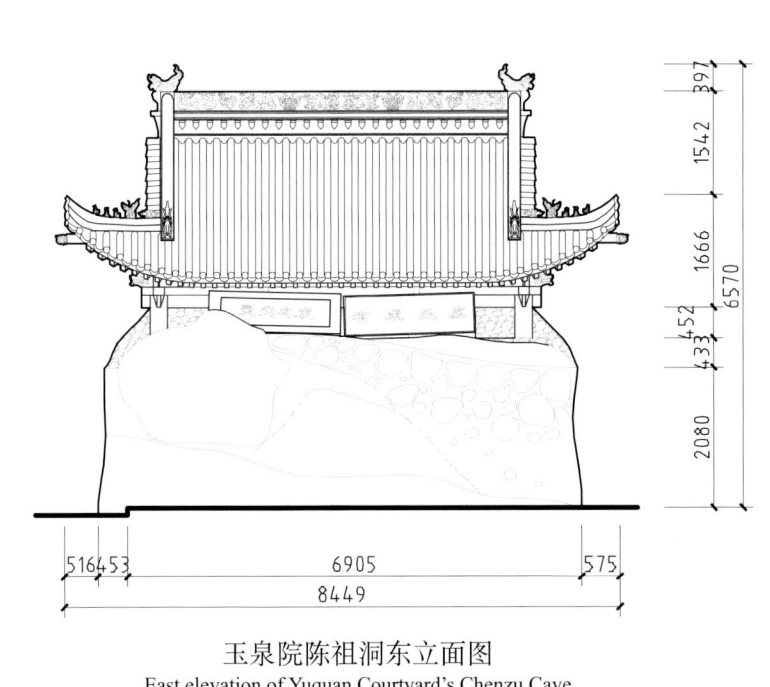

玉泉院陈祖洞东立面图

East elevation of Yuquan Courtyard's Chenzu Cave

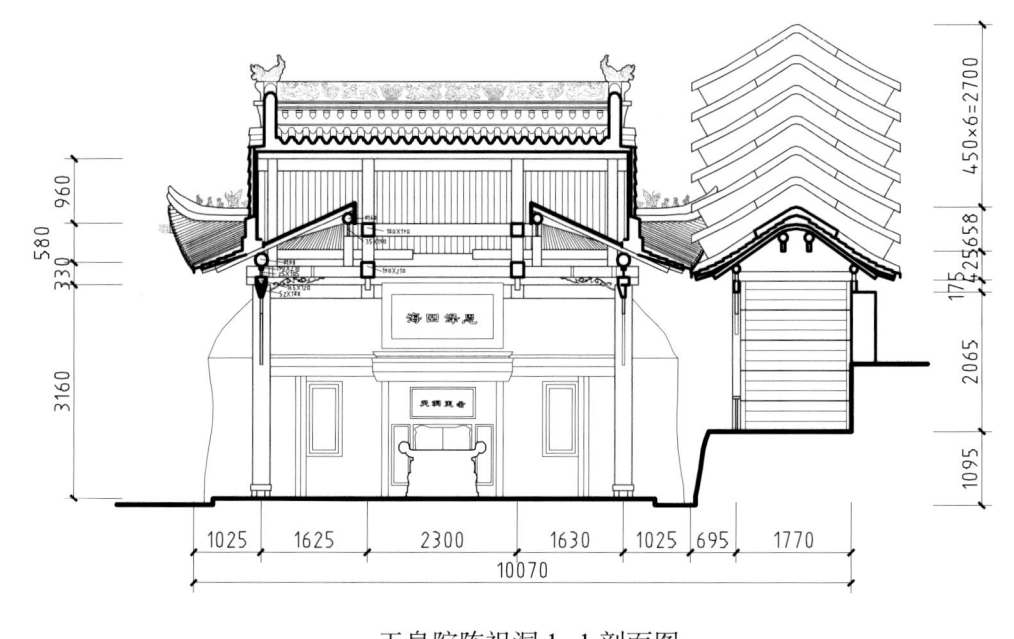

玉泉院陈祖洞 1—1 剖面图

Section 1-1 of Yuquan Courtyard's Chenzu Cave

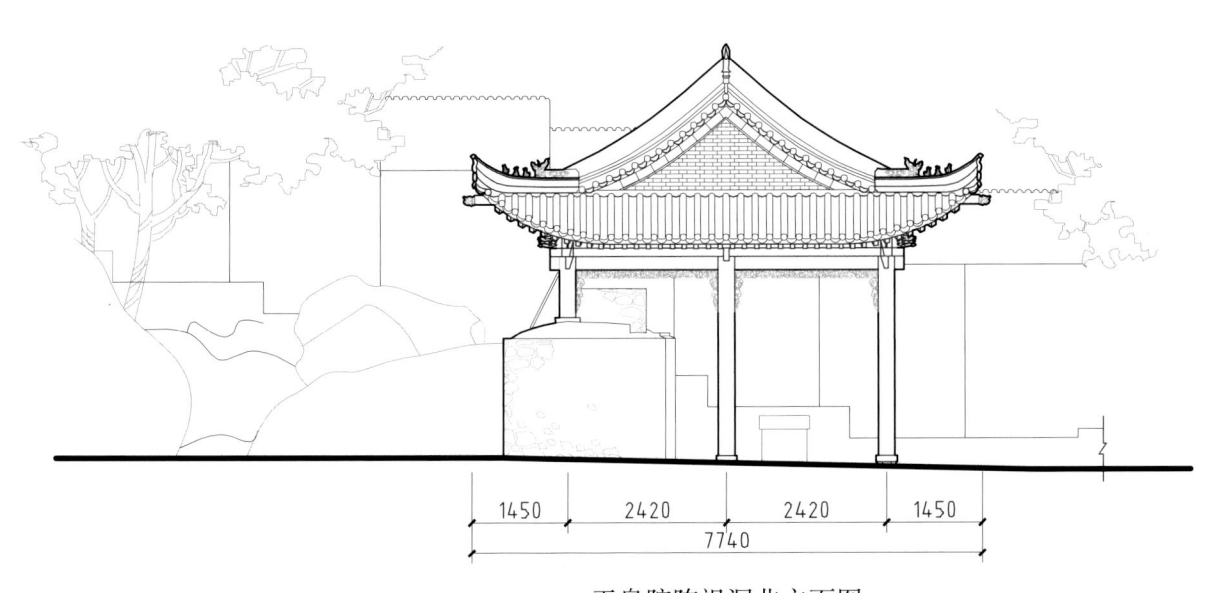

玉泉院陈祖洞北立面图

North elevation of Yuquan Courtyard's Chenzu Cave

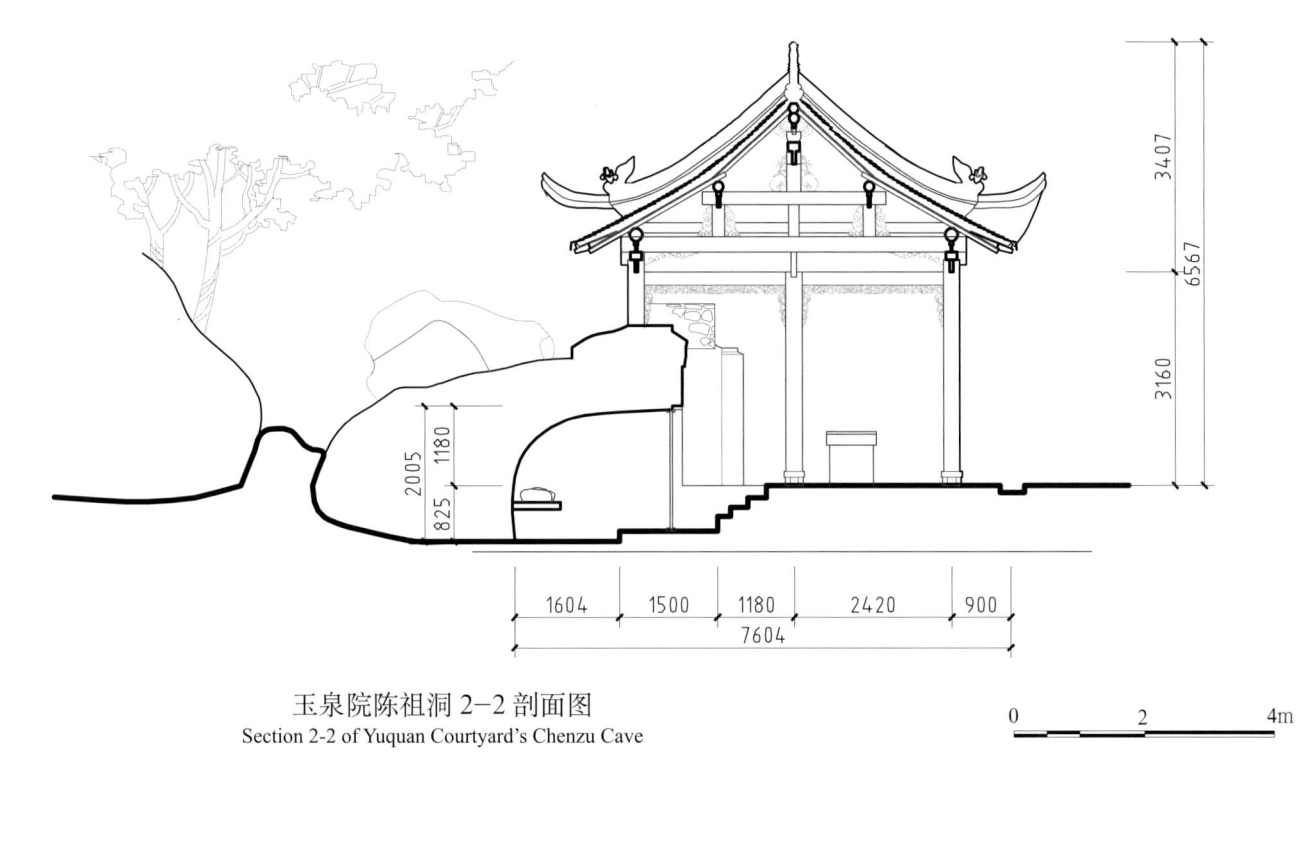

玉泉院陈祖洞 2-2 剖面图
Section 2-2 of Yuquan Courtyard's Chenzu Cave

0   2   4m

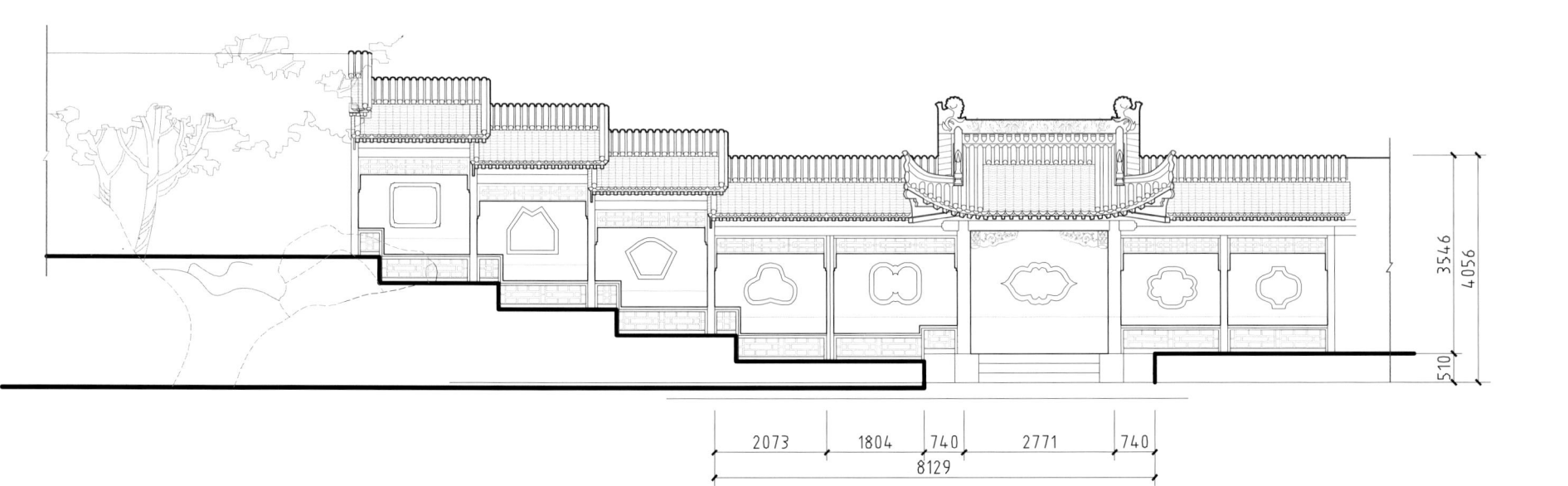

玉泉院陈祖洞长廊北立面图
North elevation of Yuquan Courtyard's Chenzu Cave's gallery

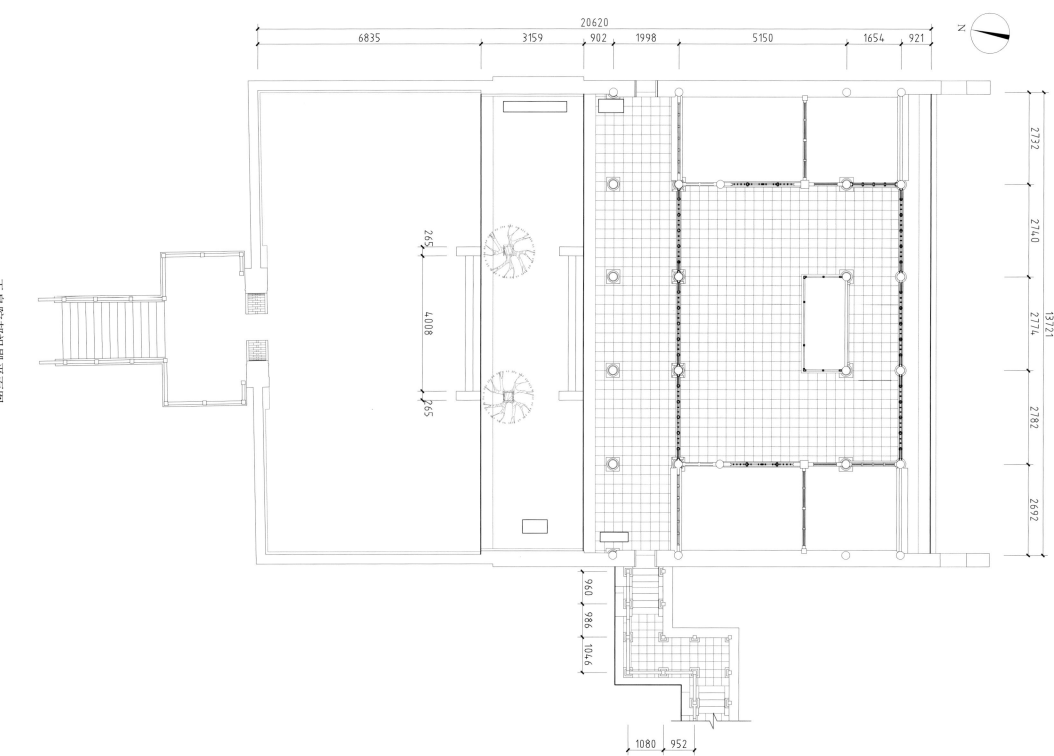

玉泉院郝祖殿平面图
Plan of Yuquan Courtyard's Haozu Cave

中国古建筑测绘大系·宗教建筑 —— 西岳庙与玉泉院

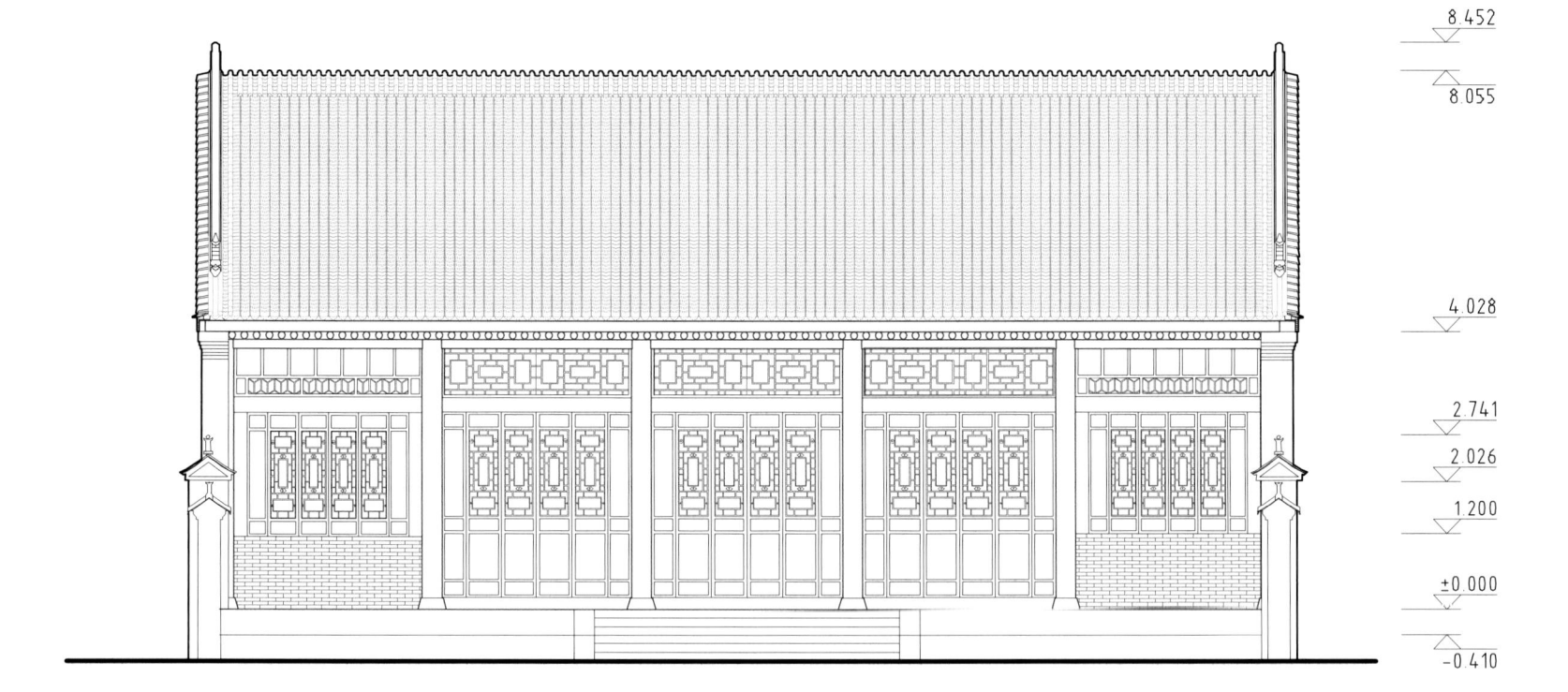

8 452

8 055

4 028

2 741

2 026

1 200

±0.000

-0 410

玉泉院郝祖殿正立面图
Front elevation of Yuquan Courtyard's Haozu Cave

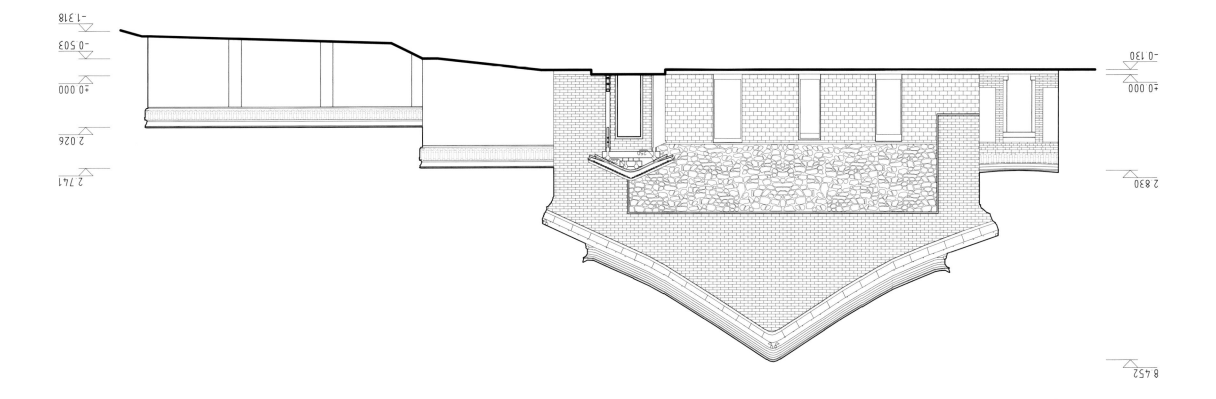

-1318
-0 503
±0 000
2 026
2 741

-0 130
±0 000
2 830
8 452

250

中国古建筑测绘大系·宗教建筑——华山道教建筑与雕塑

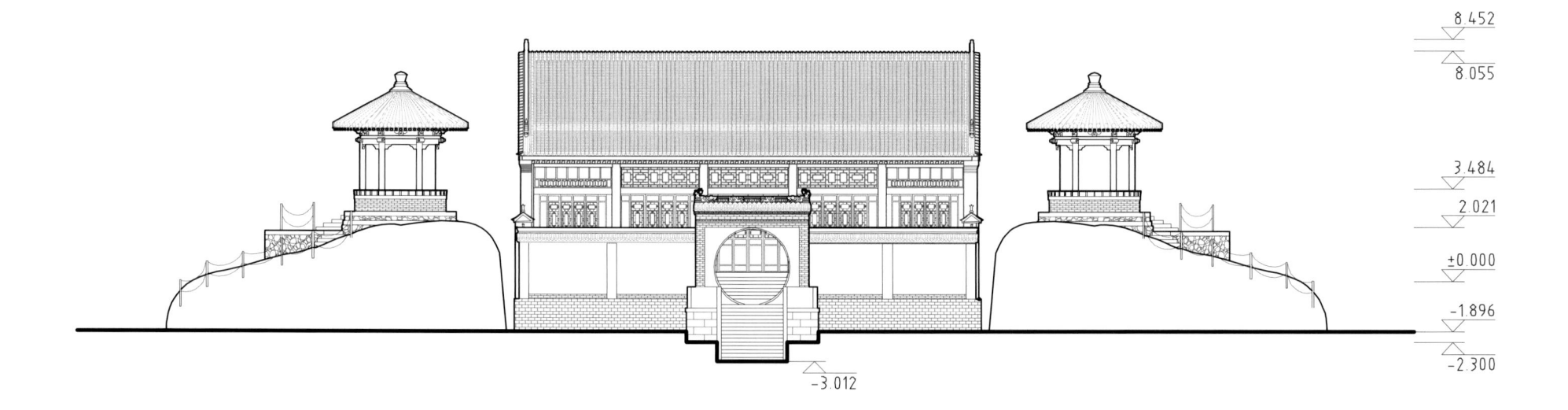

8.452

8.055

3.484

2.021

±0.000

-1.896

-2.300

-3.012

玉泉院郝祖殿总立面图
Site elevation of Yuquan Courtyard's Haozu Cave

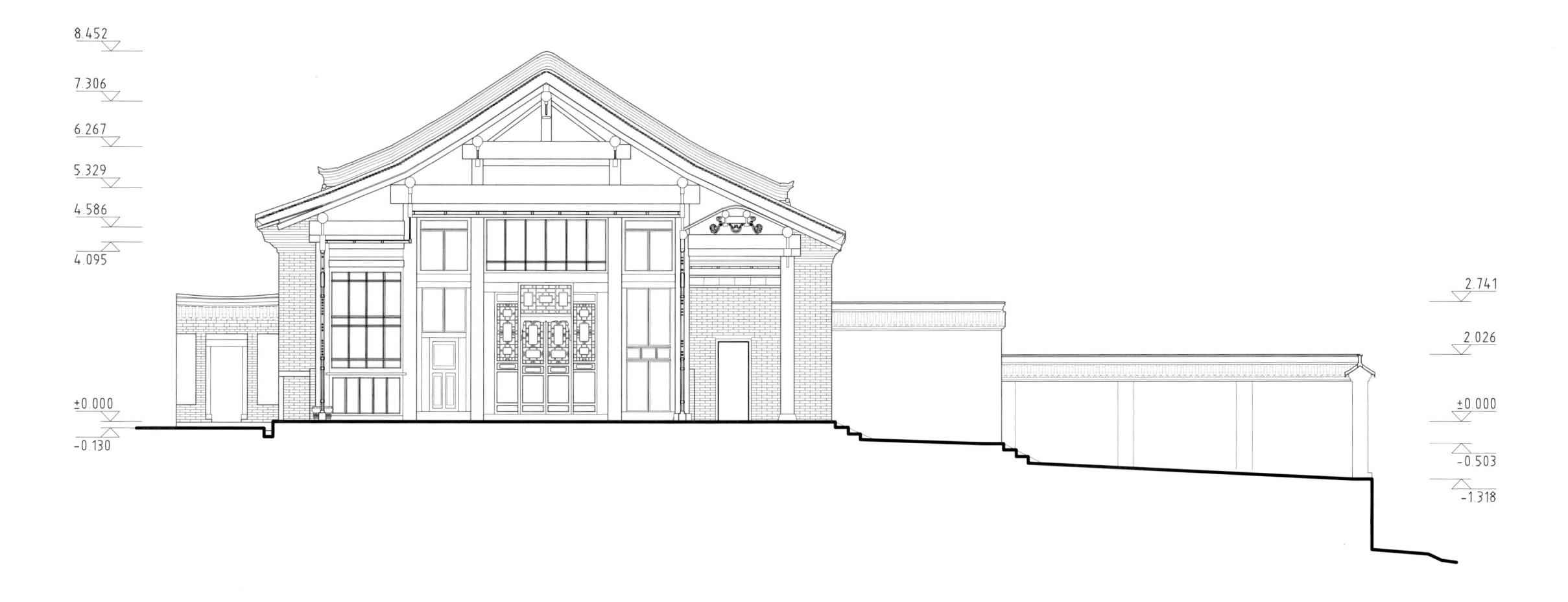

8.452

7.306

6.267

5.329

4.586

4.095

2.741

2.026

±0.000

-0.130

±0.000

-0.503

-1.318

玉泉院郝祖殿纵剖面图
Longitudinal section of Yuquan Courtyard's Haozu Cave

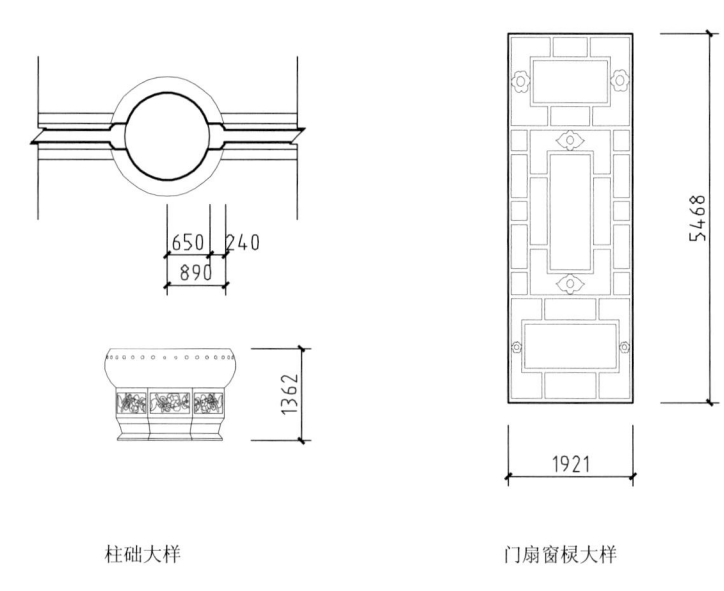

柱础大样

门扇窗棂大样

门扇窗棂大样

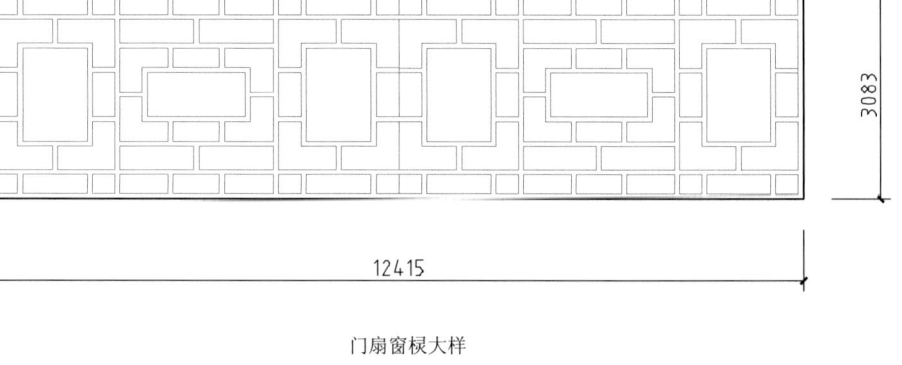

檐廊

门扇窗棂大样

门扇大样

玉泉院郝祖殿构件大样图
Structural component of Yuquan Courtyard's Haozu Cave

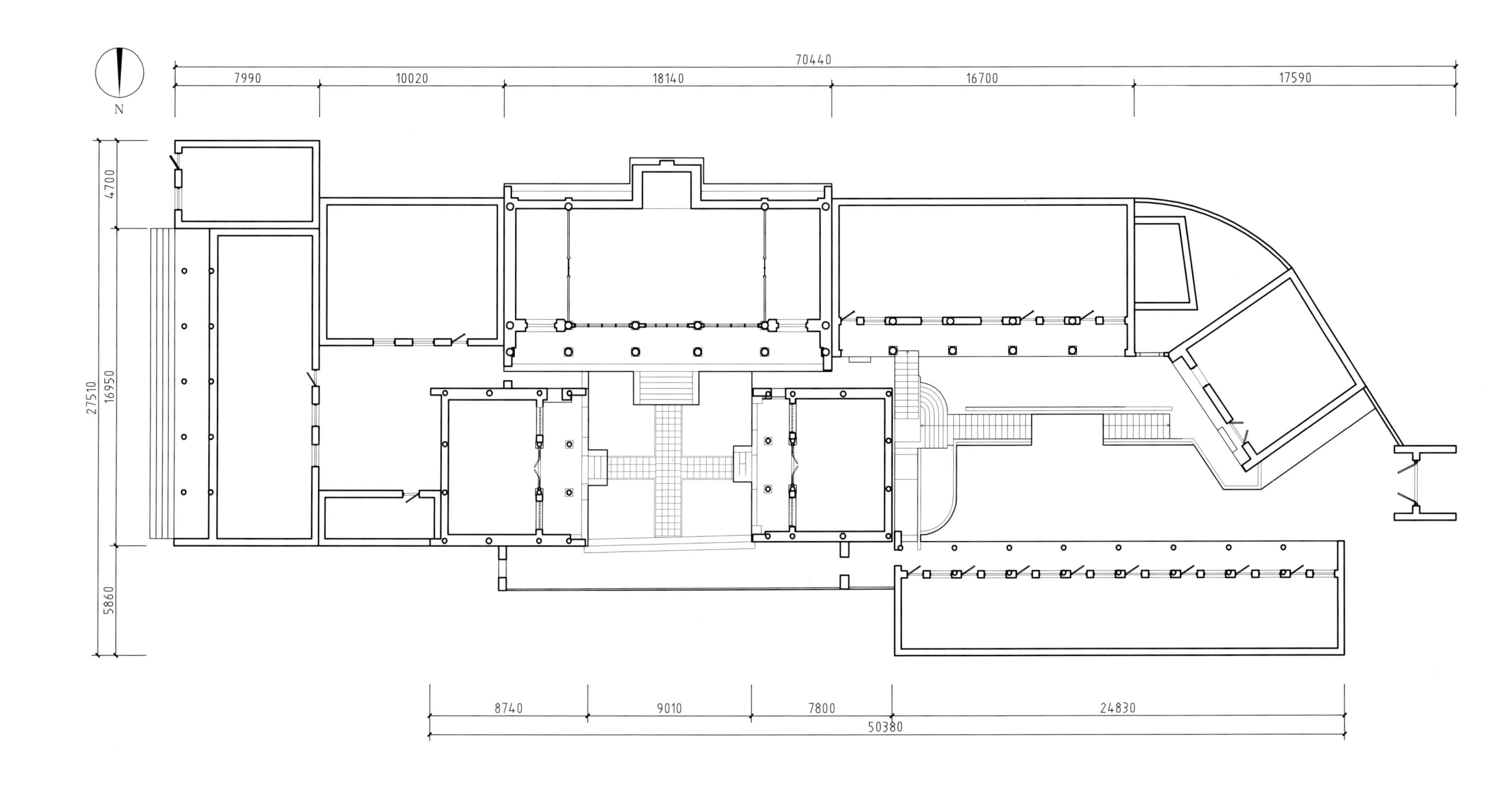

玉泉院陈祖殿院落平面图
Group plan of Yuquan Courtyard's Chenzu Hall

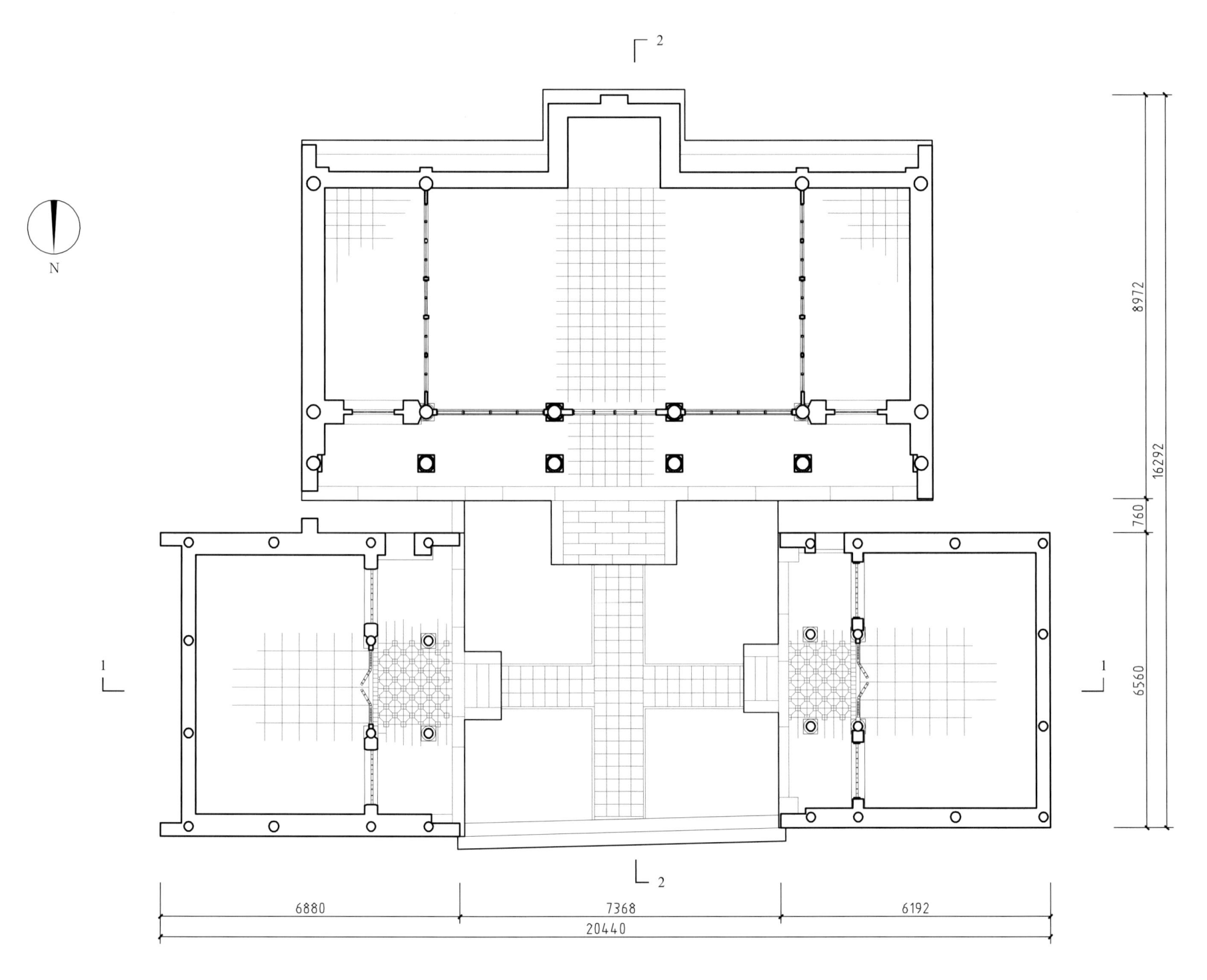

玉泉院陈祖殿院落平面图（局部）
(Partial) Group plan of Yuquan Courtyard's Chenzu Hall's courtyard

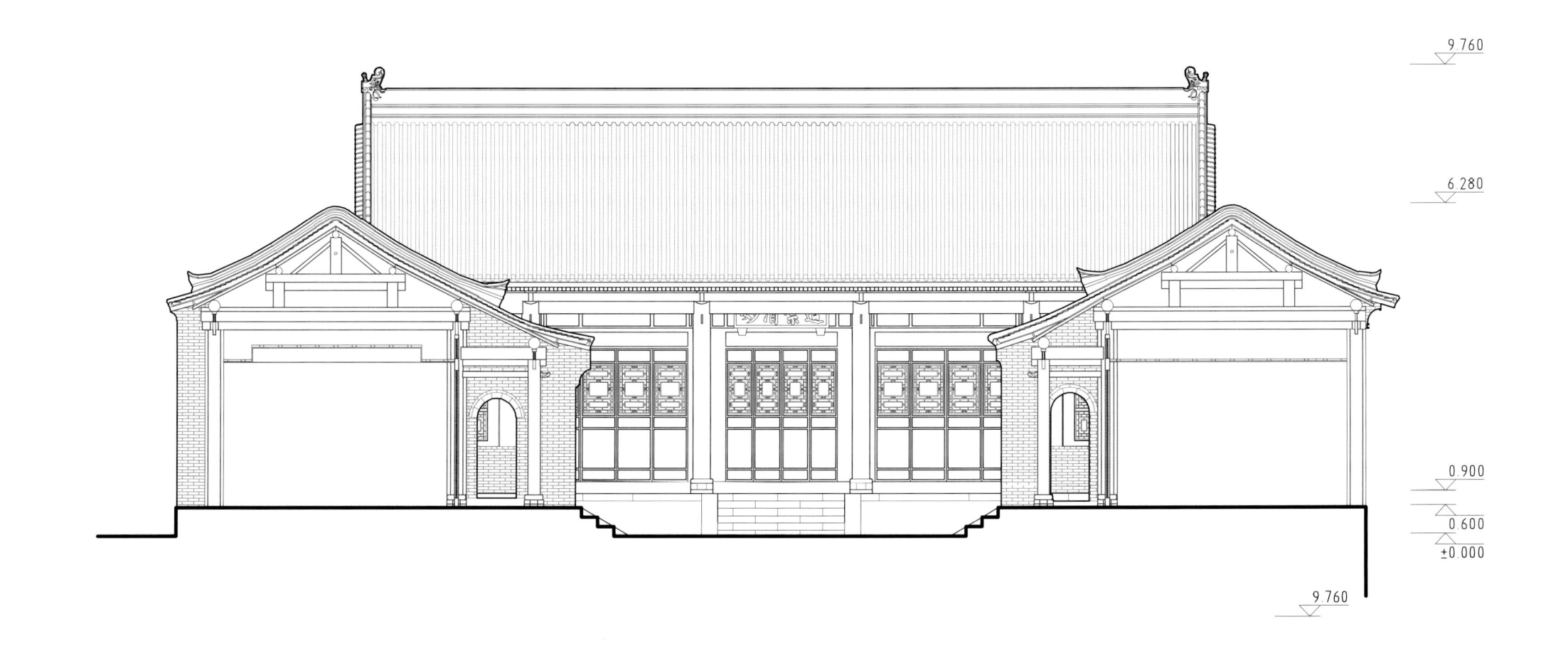

9.760

6.280

0.900

0.600

±0.000

9.760

玉泉院陈祖殿院落横剖面图
Cross-section of Yuquan Courtyard's Chenzu Hall's courtyard

6.280

0.900

±0.000

-0.600

0   2   4m

玉泉院陈祖殿院落纵剖面图
Longitudinal section of Yuquan Courtyard's Chenzu Hall's courtyard

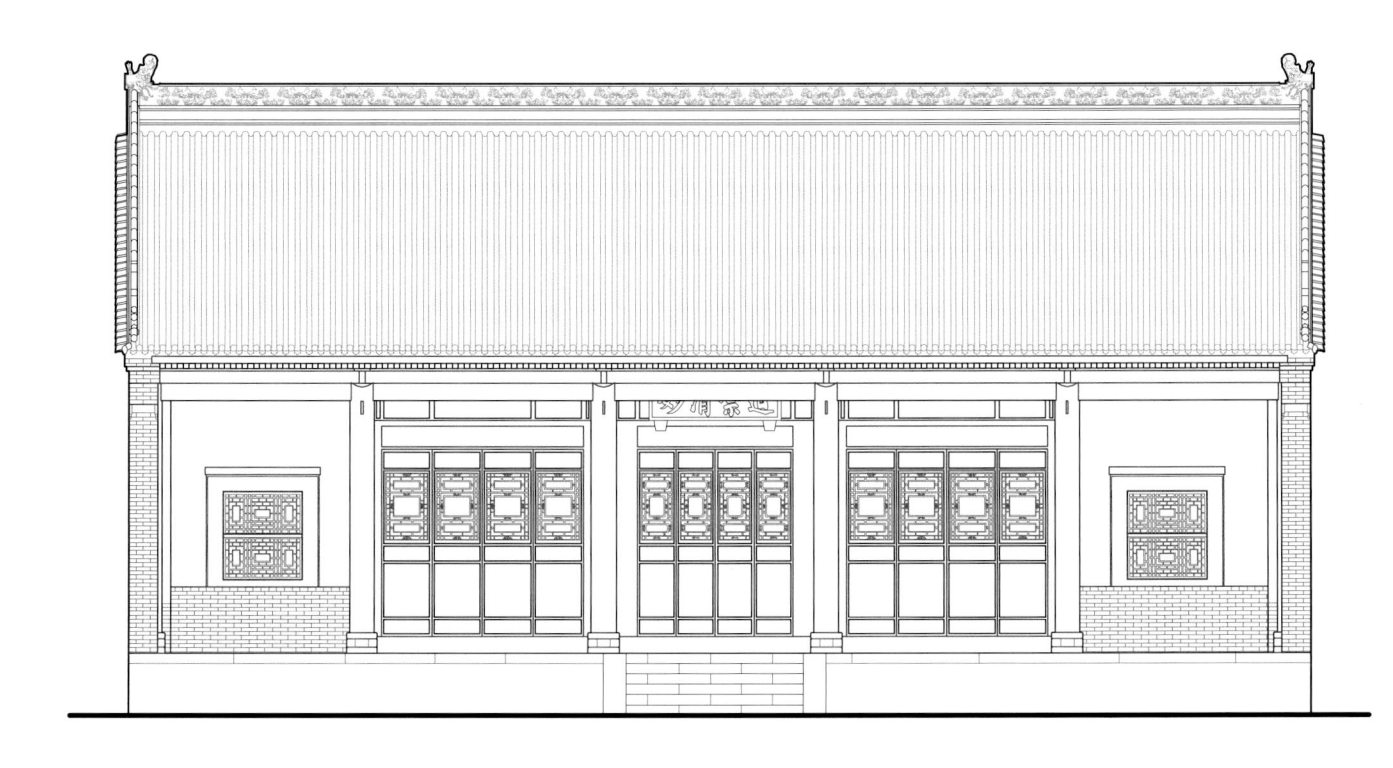

玉泉院陈祖殿立面图
Elevation of Yuquan Courtyard's Chenzu Hall

0　　2　　4m

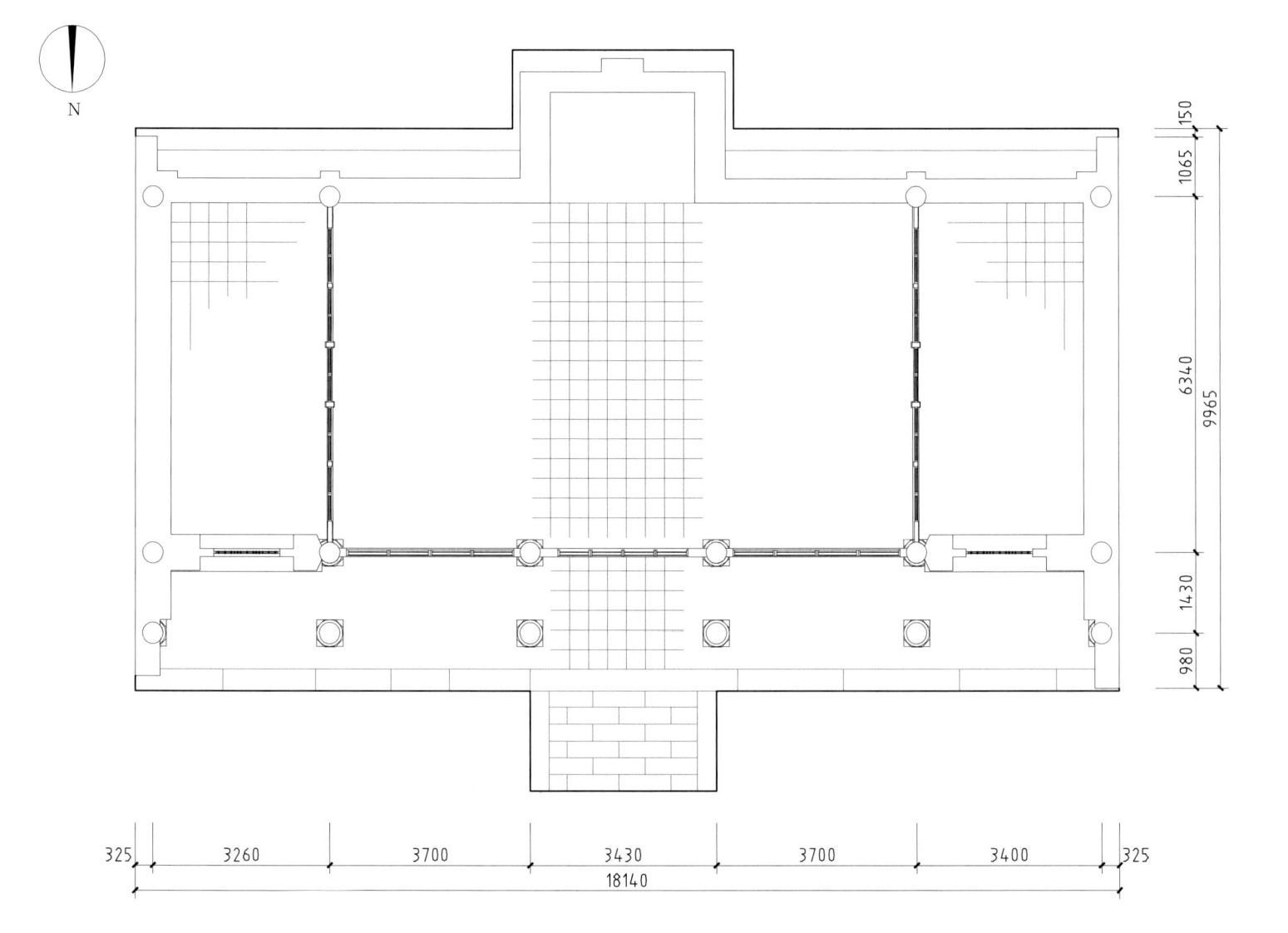

玉泉院陈祖殿平面图
Plan of Yuquan Courtyard's Chenzu Hall

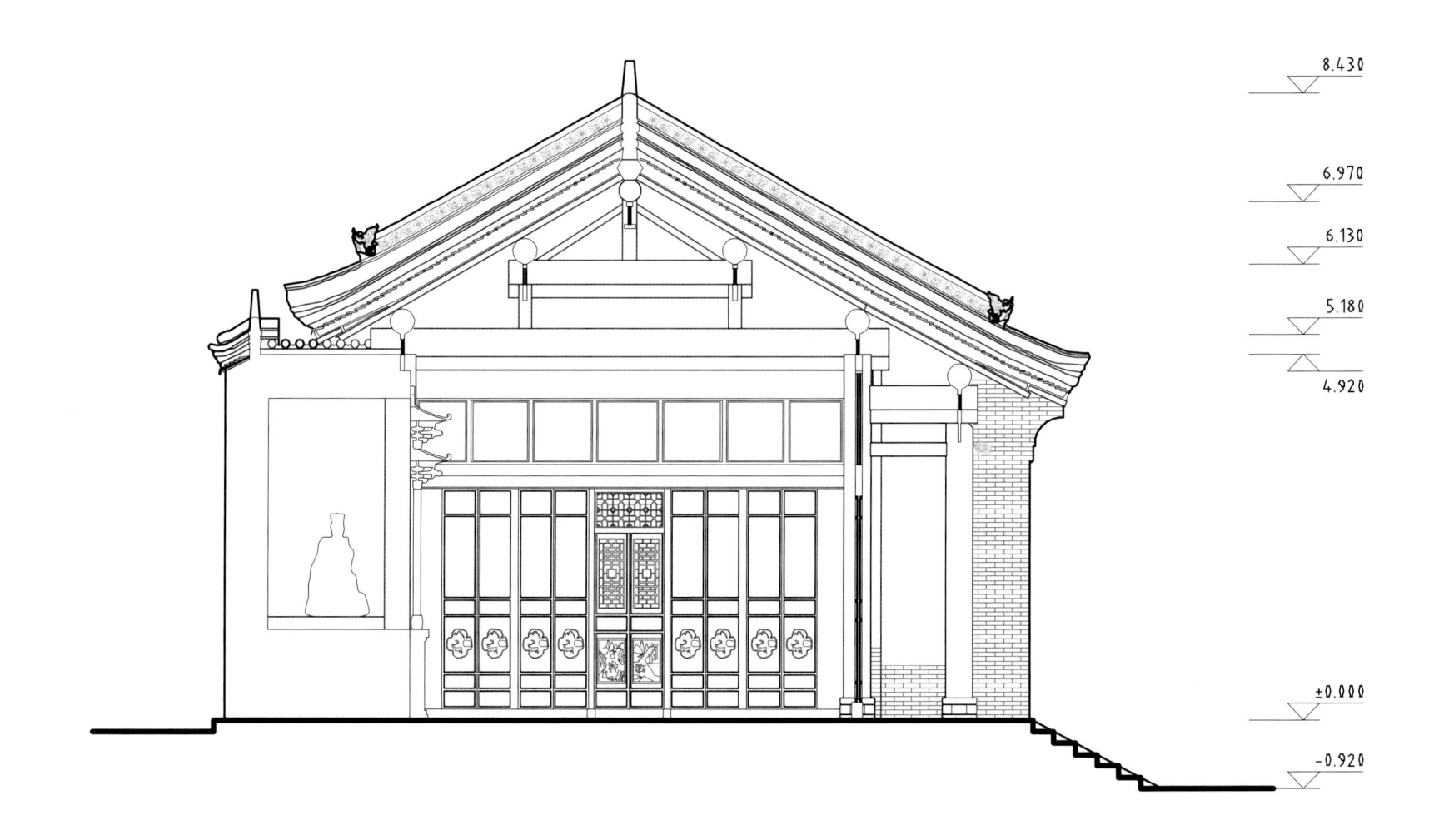

8.430

6.970

6.130

5.180

4.920

±0.000

-0.920

玉泉院陈祖殿横剖面图
Cross-section of Yuquan Courtyard's Chenzu Hall

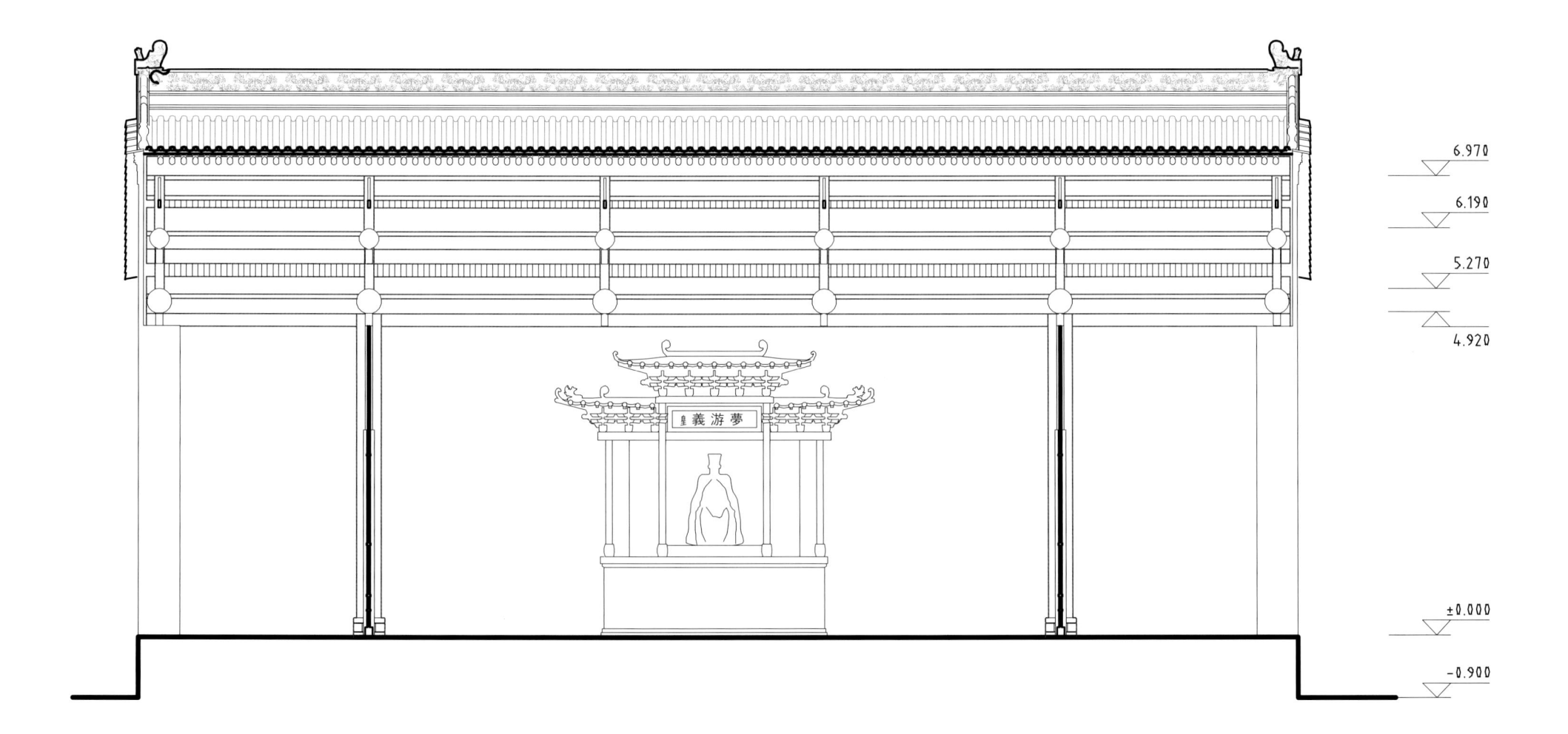

6.970

6.190

5.270

4.920

±0.000

−0.900

玉泉院陈祖殿纵剖面图
Longitudinal section of Yuquan Courtyard's Chenzu Hall

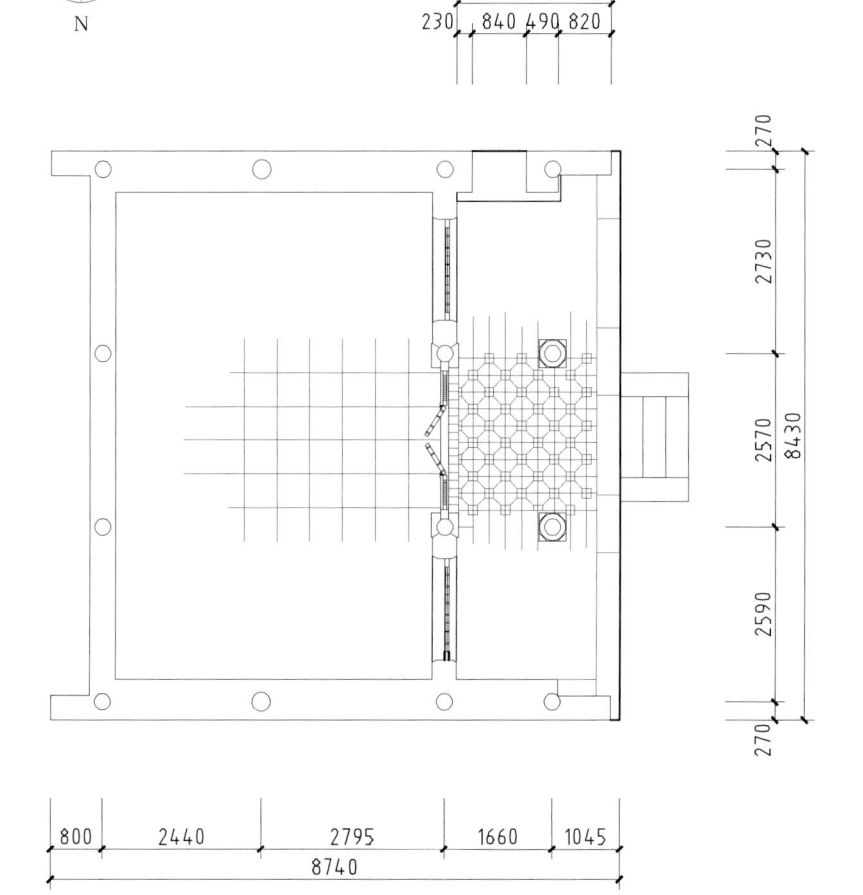

玉泉院陈祖殿东配殿平面图
Plan of Yuquan Courtyard's Chenzu Hall's east *peidian*

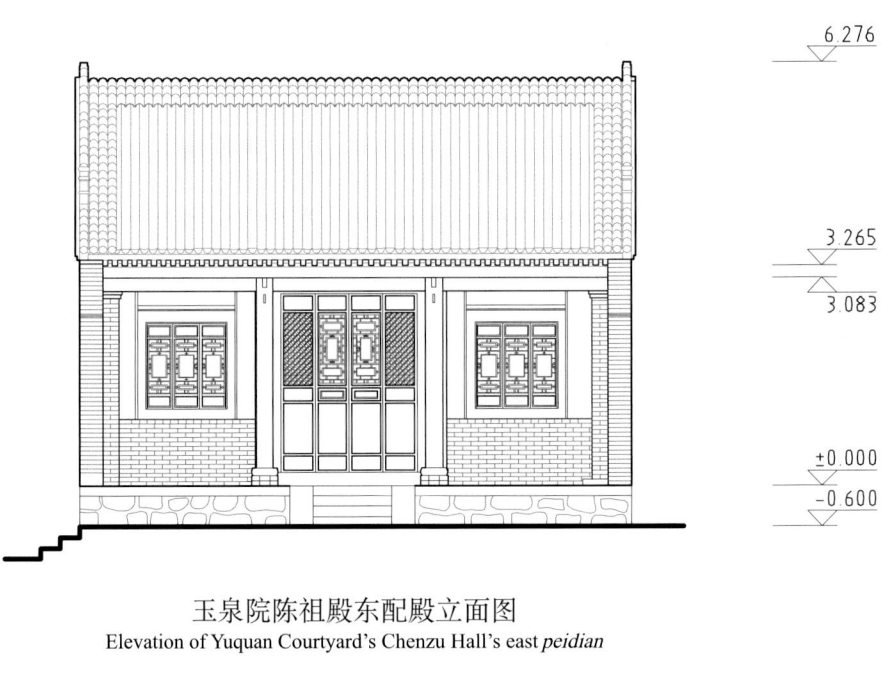

玉泉院陈祖殿东配殿立面图
Elevation of Yuquan Courtyard's Chenzu Hall's east *peidian*

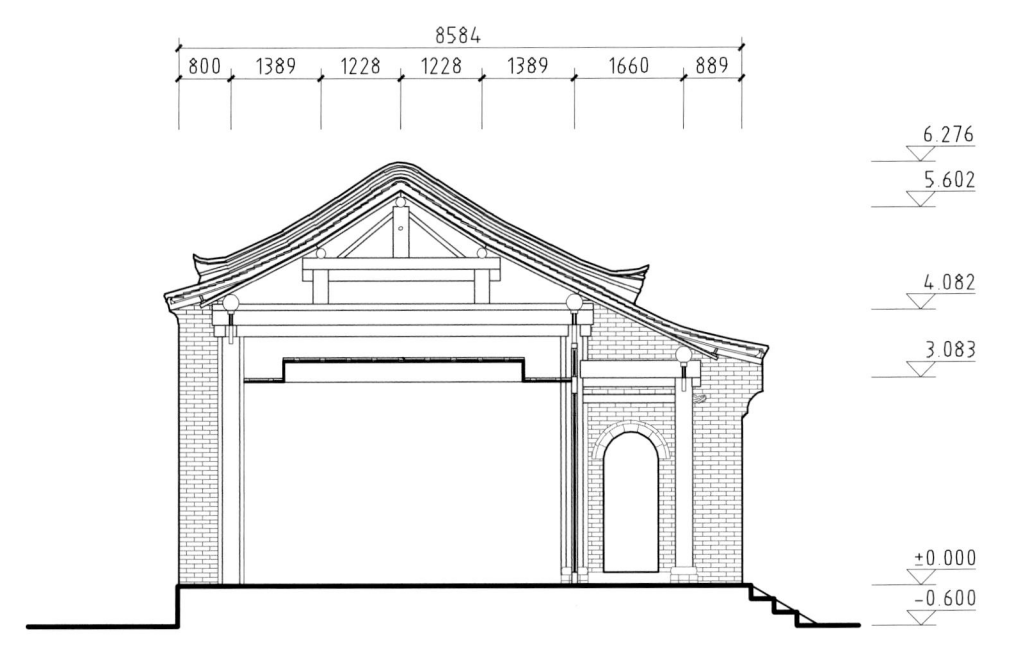

玉泉院陈祖殿东配殿横剖面图
Cross-section of Yuquan Courtyard's Chenzu Hall's east *peidian*

0   2   4m

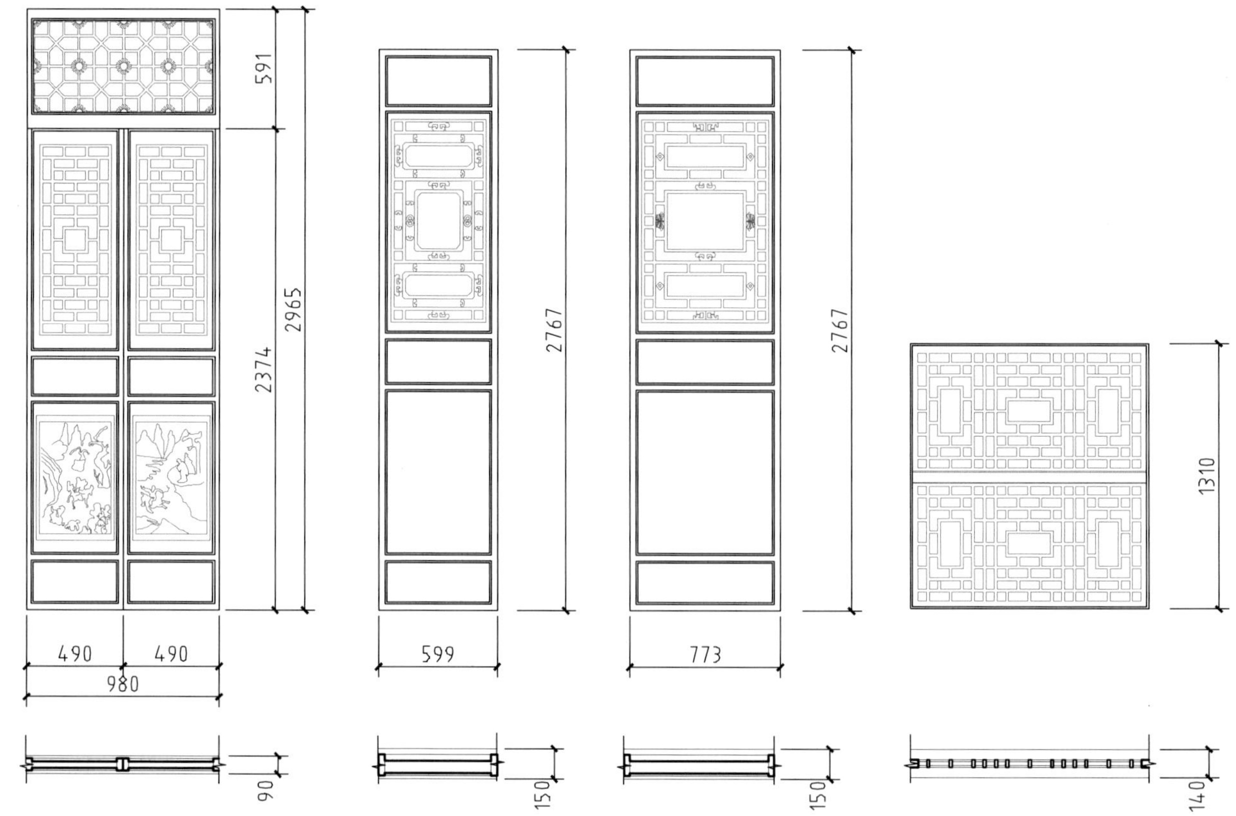

正殿门窗大样图

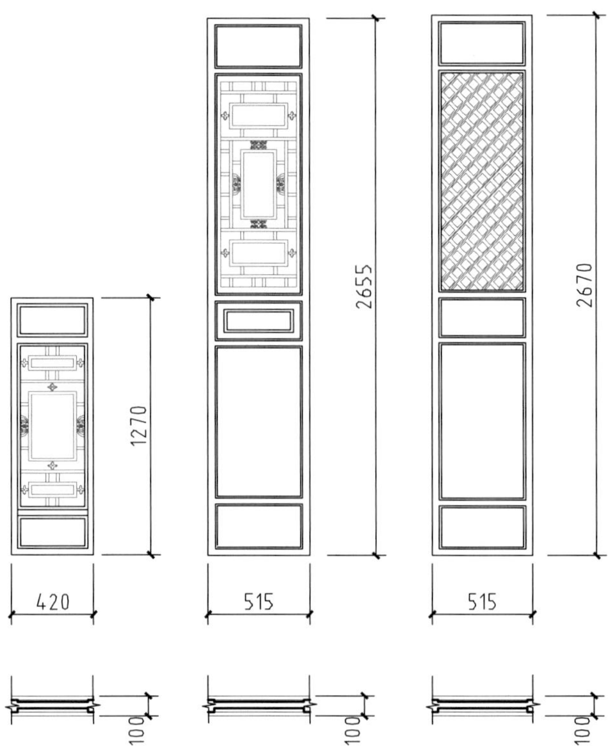

东配殿门窗大样图

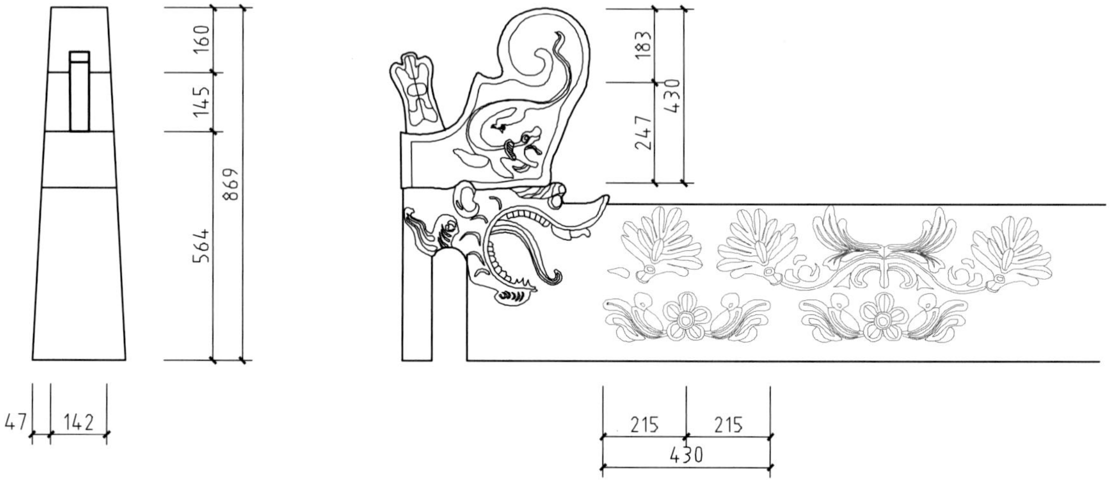

屋顶吻兽大样图

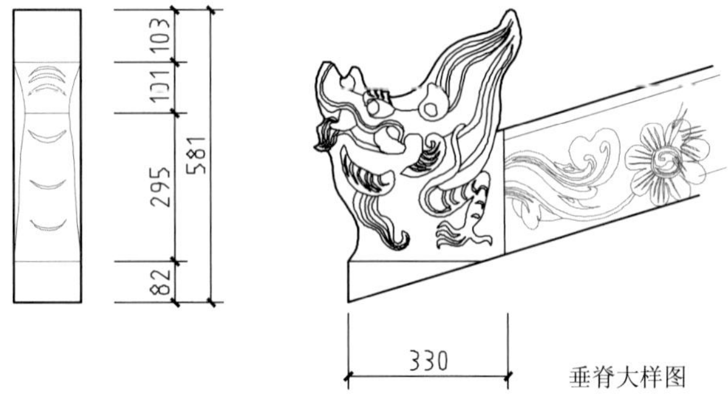

垂脊大样图

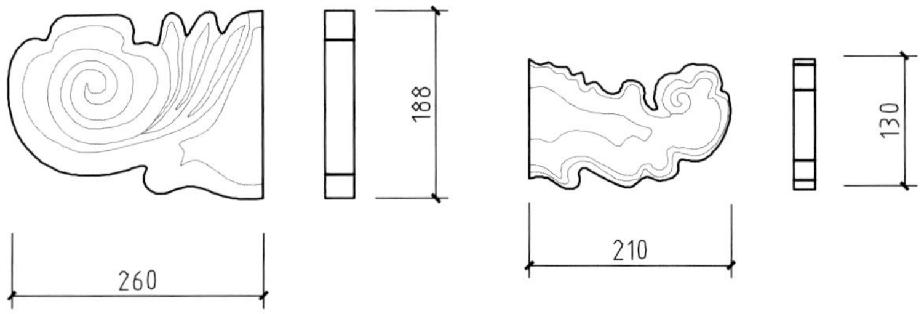

玉泉院陈祖殿构件大样图
Structural component of Yuquan Courtyard's Chenzu Hall

云头大样图

参与测绘及相关工作的人员名单

一、纯阳观测绘人员
指导教师：敖仕恒
测绘学生：李奇文　韩少扬　梁孙发　钟曼琳

二、翠云宫测绘人员
指导教师：李路珂　白昭薰
测绘学生：李倩怡　李益勋　王禹　谭舒丹　范冬阳　刘辉

三、大朝元洞测绘人员
指导教师：李路珂　白昭薰
测绘学生：王禹　刘辉　李益勋

四、金锁关无上洞测绘人员
指导教师：李路珂　白昭薰

五、玉女宫测绘人员
指导教师：李路珂　白昭薰
测绘学生：李倩怡　李益勋　王禹　谭舒丹　范冬阳　刘辉

六、群仙观测绘人员
指导教师：杨博
测绘学生：张艺凡　张晨阳　王宁

Name List of Participants Involved in Surveying and Related Works

1. Surveying and Mapping of Chunyang Taoist Temple

Supervising Instructor: AO Shiheng

Team Members: LI Qiwen, HAN Shaoyang, LIANG Sunfa, ZHONG Manlin

2. Surveying and Mapping of Cuiyun Palace

Supervising Instructor: LI Luke, BAI Zhaoxun

Team Members: LI Qianyi, LI Yixun, WANG Yu, TAN Shudan, FAN Dongyang, LIU Hui

3. Surveying and Mapping of First Dawn Cave

Supervising Instructor: LI Luke, BAI Zhaoxun

Team Members: WANG Yu, LIU Hui, LI Yixun,

4. Surveying and Mapping of Jinsuoguan Wushang Cave

Supervising Instructor: LI Luke, BAI Zhaoxun

5. Surveying and Mapping of Yu'nv Palace

Supervising Instructor: LI Luke, BAI Zhaoxun

Team Members: LI Qianyi, LI Yixun, WANG Yu, TAN Shudan, FAN Dongyang, LIU Hui

6. Surveying and Mapping of Qunxian Taoist Temple

Supervising Instructor: YANG Bo

Team Members:ZHANG Yifan, ZHANG Chenyang, WANG Ning

7. Surveying and Mapping of Xiyue Temple

Supervising Instructor: LIU Chang, JIA Jun, ZHENG Liang, DUAN Zhijun, YUAN Lin, ZHAO Xiaomei

Team Members: MA Mingfei, LIU Gang, HUA Shan, LI Zifei, QIAN Xiaoqing, QI Wenjun, SI Zhijie, WANG Meixuan, ZHANG Zhang, GUO Jizheng, ZHAO Qi, DING Yanan, LIANG Jianwei, YANG Dongying, RONG Xiao, WU Hao, TONG Lei, WANG Liying, YI Erwei, YANG Chenghan, ZHANG Ting, GONG Chenxi, PENG Fei, PENG Xuelin, ZHANG Qiao, WU Xijia, ZHANG Yuchen, LIN Jingyi, LI Guanglong

8. Surveying and Mapping of Yuquan (Jade Spring) Court

Supervising Instructor: LIAO Huinong, HE Congrong, Marianna Shevchenko, XIN Huiyuan, XU Tong

Team Members: LU Xiangyu, XIE Kuangzheng, JIA Zheng, YU Yang, WEI Xinghan, YANG Xi, ZHANG Xuan, WU Yifan, QI Junjie, XIAO Wenda, WANG Lu, XU Hui, CHEN Jianfei, CHEN Xiaolan, ZHU Yuan, TANG Renjie, TANG Xiaohu, XING Ke, LIN Wanting, LI Xin, AN Chen, WANG Yujiang, ZHANG Qi, CHEN Yunhe, SUN Zhe, Ahmed Tazrin

9. Editor of Drawings and Related Works

Drawings Arrangement: LI Jing

Drawings Editor: YANG Bo, TANG Henglu, SHAN Menglin, MAI Linlin, HU Jingfu

Translator in Chief: Alexandra Harrer

Translation Members: Alexandra Harrer, Michael Norton

图书在版编目（CIP）数据

华山岳庙与道观＝MOUNT HUA'S YUEMIAO AND
TAOIST TEMPLES/ 清华大学建筑学院编写；王贵祥等主
编 .—北京：中国建筑工业出版社，2019.12
（中国古建筑测绘大系 . 宗教建筑）
ISBN 978-7-112-24558-1

Ⅰ . ①华… Ⅱ . ①清… ②王… Ⅲ . ①华山—寺庙—
建筑艺术—图集 Ⅳ.①TU-885

中国版本图书馆CIP数据核字（2019）第286225号

丛书策划 / 王莉慧
责任编辑 / 李 鸽 陈海娇
英文审稿 / ［奥］荷雅丽（Alexandra Harrer）
书籍设计 / 付金红
责任校对 / 王 烨

中国古建筑测绘大系·宗教建筑

华山岳庙与道观

清华大学建筑学院 编写

王贵祥 刘 畅 廖慧农 李路珂 主编

Traditional Chinese Architecture Surveying and Mapping Series: Religious Architecture
MOUNT HUA'S YUEMIAO AND TAOIST TEMPLES
Compiled by School of Architecture, Tsinghua University
Edited by WANG Guixiang, LIU Chang, LIAO Huinong, LI Luke

*

中国建筑工业出版社出版、发行（北京海淀三里河路9号）
各地新华书店、建筑书店经销
北京方舟正佳图文设计有限公司制版
北京雅昌艺术印刷有限公司印刷
*
开本：787 毫米 ×1092 毫米 横 1/8 印张：28 字数：742 千字
2021 年 6 月第一版 2021 年 6 月第一次印刷
定价：**238.00 元**
ISBN 978-7-112-24558-1
（35115）